CRM Series in Mathematical Physics

The Centre de recherches mathématiques (CRM) was created in 1968 to promote research in pure and applied mathematics and related disciplines. Among its activities are thematic programs, workshops, summer schools, postdoctoral programs, and publishing. The CRM receives funding from the Natural Sciences and Engineering Research Council (Canada), the Ministère des Finances and FRQNT (Québec), the Simons Foundation (USA), the National Science Foundation (USA), and its partner universities (Université de Montréal, Concordia University, McGill University, Université Laval, Université du Québec à Montréal and Université de Sherbrooke). For more information visit www.crmath.ca

The CRM Series in Mathematical Physics includes monographs, lecture notes, and proceedings based on research pursued and events held at the Centre de recherches mathématiques.

Manu Paranjape · Daniel Carney · Bum-Hoon Lee ·
Taejin Lee
Editors

Proceedings of Gravity, Strings and Fields

A Conference in Honour of Gordon Semenoff

Springer

Editors
Manu Paranjape
Département de Physique, Complexe des Sciences
Université de Montréal
Montréal, QC, Canada

Bum-Hoon Lee
Department of Physics
Sogang University
Seoul, Korea (Republic of)

Daniel Carney
Lawrence Berkeley National Laboratory
Berkeley, CA, USA

Taejin Lee
Department of Physics
Kangwon National University
Gangneung-si, Korea (Republic of)

ISSN 2627-7654 ISSN 2627-7662 (electronic)
CRM Series in Mathematical Physics
ISBN 978-3-031-91265-8 ISBN 978-3-031-91266-5 (eBook)
https://doi.org/10.1007/978-3-031-91266-5

This Springer imprint is published by the registered company Springer Nature Switzerland AG
The registered company address is: Gewerbestrasse 11, 6330 Cham, Switzerland

Preface

For the past 40 years, Gordon Semenoff has been one of Canada's foremost theoretical physicists, making significant contributions across multiple domains, including quantum field theory, string theory, gravity, and condensed matter physics. His groundbreaking work on graphene—predating its very discovery—demonstrates remarkable foresight. His research on chiral anomalies was both profound and insightful, while his exhaustive studies on fermion number fractionalization have left a lasting mark.

Semenoff's influence extends beyond his research; his long and distinguished list of collaborators speaks to the depth of his impact on the field.

As he reaches the milestone of 70, we sought to honor his achievements with a conference bringing together his colleagues and collaborators—those whose work has been deeply intertwined with his insights. This gathering fostered a vibrant exchange of ideas, reflecting the ongoing influence of his contributions. Each invited participant had the opportunity to present their work, and this volume serves as a lasting testament to both the conference and Semenoff's enduring legacy in theoretical physics.

Montréal, Canada

Manu Paranjape

Contents

Late Time Acceleration of Our Universe Caused by the Absorption of Baby Universes

Jan Ambjorn and Yoshiyuki Watabiki

Abstract If our Universe is allowed to absorb baby universes, one obtains a modified Friedmann equation that can explain the late time acceleration of our Universe and there is no need for a cosmological constant. In addition the modified Friedmann equation favors the value of the Hubble constant obtained by local measurements.

Keywords Friedmann equation · Hubble constant · Baby universes

1 Introduction

Causal Dynamical Triangulations (CDT) is an attempt to define a non-perturbative theory of quantum gravity (see [1, 2] for reviews). It is formulated in a proper-time gauge and one can explicitly perform a rotation to Euclidean signature where the integrand in the path integral is then changed from e^{iS_L} to e^{-S_E}, the Euclidean action S_E being real. This implies that the four-dimensional CDT theory can be addressed by Monte Carlo simulations. One of the outcomes of these simulations is that when the topology of space is T^3 one observes the following effective action as a function of the three-volume $V(t)$ at proper-time t:

$$S = \frac{1}{\Gamma} \int dt \left(\frac{\dot{V}^2}{2V} + \Lambda V \right). \tag{1}$$

J. Ambjorn (✉)
The Niels Bohr Institute, Copenhagen University, Copenhagen, Denmark
e-mail: ambjorn@nbi.dk

IMAPP, Radbaud University Nijmegen, Nijmegen, The Netherlands

Y. Watabiki
Tokyo Institute of Technology, Department of Physics, High Energy Theory Group, Meguro-ku, Tokyo, Japan
e-mail: watabiki@th.phys.titech.ac.jp

M. Paranjape et al. (eds.), *Proceedings of Gravity, Strings and Fields*,
CRM Series in Mathematical Physics,
https://doi.org/10.1007/978-3-031-91266-5_1

This is remarkable, because it is essentially the Hartle-Hawking minisuperspace action (including the rotation of the conformal factor) [3]. The rotation of the conformal factor was proposed as a solution to the problem of the unboundedness from below of the Euclidean Einstein-Hilbert action. Also CDT solves the problem, but in CDT the effective action arises by integrating out (via the Monte Carlo simulations) all other degrees of freedom than $V(t)$, while Hartle and Hawking by hand restricted the geometry to only depend on $V(t)$.

Recall the standard minisuperspace approximation, where one uses the metric

$$ds^2 = -N^2(t)dt^2 + a^2(t)d\Omega_3, \qquad d\Omega_3 = \sum_{i=1}^{3} dx_i^2. \tag{2}$$

Introduce

$$v(t) = \frac{1}{\kappa}a^3(t), \quad \kappa = 8\pi G, \tag{3}$$

where G the gravitational constant. Then the minisuperspace Einstein-Hilbert action is

$$S = \int dt \left(-\frac{\dot{v}^2}{3Nv} - \lambda N v \right), \tag{4}$$

where λ is the cosmological constant. The Hubble parameter $H(t)$ is defined as

$$H(t) \equiv \frac{\dot{a}}{a} = \frac{1}{3}\frac{\dot{v}}{v}. \tag{5}$$

The Hamiltonian corresponding to (4) is

$$\mathcal{H}(v, p) = Nv\left(-\frac{3}{4}p^2 + \lambda \right), \tag{6}$$

where p denotes the momentum conjugate to v. Variation with respect to $N(t)$ ensures that the classical solutions are "on shell", i.e. $\mathcal{H}(v, p) = 0$ for a solution. In the following we will for simplicity gauge fix $N(t) = 1$, but only consider classical solutions corresponding to $\mathcal{H}(v, p)$ that satisfy $\mathcal{H}(v, p) = 0$. Such a classical solution is de Sitter spacetime where

$$v(t) = v(t_0)\,\mathrm{e}^{\sqrt{3\lambda}\,t}. \tag{7}$$

The analytic continuation of (4) by Hartle and Hawking, including the rotation of the conformal factor, will result in the following action:

$$S_{hh} = \int dt \left(\frac{\dot{v}^2}{3Nv} + \lambda N v \right), \tag{8}$$

Hartle and Hawking were mainly interested in using (8) in the path integral. Whatever result one would obtain, one would eventually have to rotate back to Lorentzian signature, if one is interested in cosmological applications, and if the quantum theory has a classical limit, this limit should be given by (4) and (6). Thus we will not expect the late time aspects of cosmology to be directly affected by the quantum aspects of gravity. Similarly, since CDT is intended to be a quantum gravity theory, the result (1) indicates CDT will not provide us with new insight about the late time cosmology. "Traditional" quantum gravity might affect the early time universe (cure Big Bang singularities etc.), not the late time universe. However, if we allow for some more "untraditional" quantum phenomena like the absorption and emission of so-called baby-universes, this situation can change[1]

2 A Universe that Absorb and Emit Baby Universes

By allowing our Universe to absorb and emit baby universes, we are really dealing with a multi-verse theory. In order to be able to perform some calculation in such a multi-verse theory, we will work in a minisuperspace approximation where the spatial universe at a give proper time t is characterized by its spatial volume $v(t)$. In a corresponding quantum theory we will denote a state with spatial volume v by $|v\rangle$. Let now $\hat{\mathcal{H}}$ denote the quantum mechanical Hamiltonian corresponding to the action S_{hh} given by (8):

$$\hat{\mathcal{H}}^{(0)} = v\Big(-\frac{3}{4}\frac{d^2}{dv^2} + \lambda\Big). \tag{9}$$

This can now be viewed as a single universe Hamiltonian and it was already encountered in the study of two-dimensional CDT [5, 6]. There is a well defined Hilbert space corresponding to $\hat{\mathcal{H}}^{(0)}$ and the multi-verse Hilbert space will now be the Fock space constructed from the single universe states $|v\rangle$. A natural many-universe Hamiltonian that allows for a universe to split in two or two universes to merge to a single universe is then

$$\hat{H} = \hat{H}^{(0)} - g\int dv_1 \int dv_2\; \Psi^\dagger(v_1)\Psi^\dagger(v_2)\; (v_1+v_2)\Psi(v_1+v_2) \tag{10}$$
$$-g\int dv_1 \int dv_2\; \Psi^\dagger(v_1+v_2)\; v_2\Psi(v_2)\; v_1\Psi(v_1) - \int \frac{dv}{v}\,\rho(v)\Psi^\dagger(v),$$

where

$$\hat{H}^{(0)} = \int_0^\infty \frac{dv}{v}\; \Psi^\dagger(v)\hat{\mathcal{H}}^{(0)}\, v\Psi(v), \qquad \rho(v) = \delta(v), \tag{11}$$

and where $\Psi^\dagger(v)$ and $\Psi(v)$ are creation and annihilation operators for single universes of spatial volume v. The existence of the last term in Eq. (10) implies that a

[1] Similar ideas have recently been advocated in [4].

universe can be created from the Fock vacuum $|0\rangle$ if the spatial volume is zero and therefore the formal Fock vacuum is not stable.

The Hamiltonian (10) was introduced in [7] and there exists a truncation, denoted generalized 2d CDT (GCDT) that can be solved analytically [8, 9]. It follows the evolution of our Universe in time and allows other universes (called baby universes) to merge with our universe during this time evolution, but does not allow our Universe to split in two. This is illustrated in Fig. 1. The effective Hamiltonian (the so-called inclusive Hamiltonian first introduced in [10]) can be found from (10) as follows: replace the quantum field $\Psi(v)$, representing the disappearance of a universe of spatial volume v by $\psi(v) + \Psi(v)$, where $\psi(v)$ is viewed as a classical field representing the distribution of the spatial volumes of the baby universes being absorbed. Let $F(p)$ denote the Laplace transform of $\phi(v) = v\psi(v)$:

$$\phi(v) = \phi_0 + \phi_1 v + \cdots, \quad F(p) = \int_0^\infty dv\, e^{-pv}\phi(v) = \frac{\Gamma(1)\phi_0}{p} + \frac{\Gamma(2)\phi_1}{p^2} + \cdots \tag{12}$$

After some algebra one obtains for the quadratic part of $\hat{H}$ that determines the propagation of the universe:

$$\hat{H}_{\text{eff}} = \hat{H}^{(0)} - 2g \int dv \Psi^\dagger(v)\, F\left(\frac{d}{dv}\right) v\Psi(v), \tag{13}$$

and we can write

$$\hat{\mathcal{H}}_{\text{eff}} = \hat{\mathcal{H}}_0 - 2gF\left(\frac{d}{dv}\right) v. \tag{14}$$

In GCDT $F(p)$ is determined by the self-consistency requirement that our Universe, modified by the impact $\phi(v)$ of baby universes, should be identical to the baby universes it absorbs. We refer to [8] (or the book [11]) for the details of this determination. We will allow for more general $F(p)$. To obtain the classical effective Hamiltonian relevant for cosmology we have to make the analytic rotation back from the Hartle-Hawking metric and replace $-id/dv$ by the classical momentum p conjugate to v. In this way we obtain

$$\mathcal{H}_{\text{eff}} = v\Big(-\frac{3}{4}p^2 + \lambda - 2gF(p)\Big) = -vf(v). \tag{15}$$

The on shell solution $v(t)$ satisfies

$$\dot{v} = \frac{\partial \mathcal{H}_{\text{eff}}}{\partial p} = -vf'(p), \quad f(p) = 0. \tag{16}$$

This implies that p is constant and one has (for suitable $f(v)$) an exponentially growing solution, just like the de Sitter solution (7). In particular we can have an exponentially growing solution even if the cosmological constant $\lambda = 0$. In such

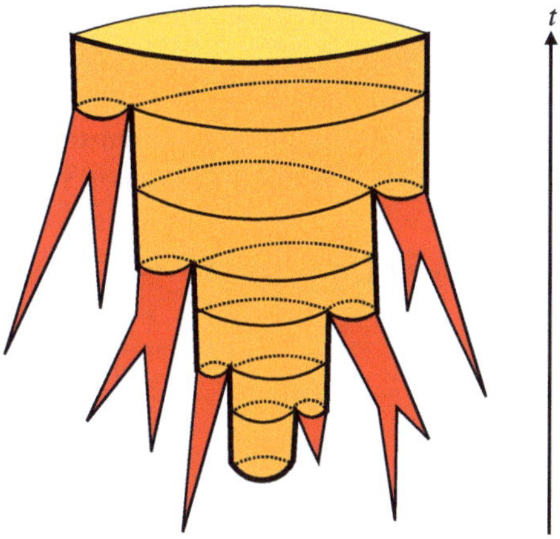

Fig. 1 Our Universe (orange), represented as one-dimensional circles, propagating in time, with baby universes (red) merging and increasing the spatial volume

a case the exponential growth is caused by the impact of other universes on our Universe, as illustrated in Fig. 1.

In the following we will consider the simplest choice of $F(p)$, namely the one corresponding the $\phi(v) = \phi_0$. This is the limit of any $\phi(v)$ where the impact volume of the baby universes is zero, so with this choice it is really appropriate to denote the incoming universes "baby" universes. In addition we will assume that the cosmological constant is zero. Choosing for convenience $\phi_0 = 3/4$ we can now write

$$\mathcal{H}_{\text{eff}} = -vf(v), \quad f(p) = \frac{3}{4}\Big(p^3 + \frac{2g}{p}\Big), \tag{17}$$

and we find the exponential growth

$$v(t) = v(t_0)\, \mathrm{e}^{\frac{3}{2}(2g)^{1/3}t}. \tag{18}$$

3 Including CDM Matter

We now include matter in our cosmological model. Since we are only interested in the late time cosmology, we include it as a CDM density $\rho_{\text{m}}(v)$ added to the Hamiltonian (17). We do not include the matter density in the baby universes, since they are precisely baby universes with infinitesimal volume. Thus the Hamiltonian has the form

$$\mathcal{H}[v,p] = v\,(-f(p) + \kappa\rho_{\rm m}(v)), \quad f(p) = \frac{3}{4}\Big(p^2 + \frac{2g}{p}\Big), \quad v\rho_{\rm m}(v) = v_0\rho_{\rm m}(v_0), \tag{19}$$

where v_0 and $\rho_{\rm m}(v_0)$ denote the values at the present time t_0. Let us discuss the solution of the eoms for arbitrary $f(p)$ in (19). The eoms simplify since $v\rho_{\rm m}(v)$ is constant.

$$\dot{v} = \frac{\partial \mathcal{H}}{\partial p} = -vf'(p), \quad \text{i.e.} \;\; 3\,\frac{\dot{a}}{a} = \frac{\dot{v}}{v} = -f'(p), \tag{20}$$

$$\dot{p} = -\frac{\partial \mathcal{H}}{\partial v} = f(p), \quad \text{i.e.} \;\; t = \int_{-\infty}^{p} \frac{dp}{f(p)}. \tag{21}$$

By construction any solution to (20)–(21) will satisfy $\mathcal{H} = \text{const}$, and we are interested in the "on-shell" solutions $\mathcal{H} = 0$, which by (19) implies that

$$f(p) = \kappa\rho_{\rm m}(v) = \kappa\rho_{\rm m}(v_0)\frac{v_0}{v} = f(p_0)\frac{v_0}{v} = f(p_0)(1+z)^3, \tag{22}$$

where p_0 denotes the value of p at present time t_0 and z denotes the redshift at time t, i.e.

$$z(t) + 1 = \frac{a(t_0)}{a(t)} = \Big(\frac{v(t_0)}{v(t)}\Big)^{1/3}. \tag{23}$$

Equation (22) *is the generalized Friedmann when Eq.* (20) *is used to express p in terms of* $\dot{a}/a$. In [12–14] we studied this Friedmann equation for $f(p)$ given by (19) and called it *the modified Friedmann equation*. We will restrict ourselves to that case in the following, and one can then integrate (21) analytically to find

$$t = \int_{-\infty}^{p} \frac{dp'}{f(p')} = \frac{4}{3\sqrt{3}\,a}\Big(\arctan\frac{2p-a}{\sqrt{3}\,a} + \frac{1}{2\sqrt{3}}\log\frac{(p-a)^2+ap}{(p+a)^2} + \frac{\pi}{2}\Big), \tag{24}$$

where $a = (2g)^{1/3}$. A more general study for other functions $f(v)$, including the $f(p)$ corresponding the GCDT, can be found in [15].

4 Comparison with Late Cosmological Data

Our model has two coupling constants coupling constant, κ and g. We consider κ fixed by local experiments and we will not discuss it any further. If our model should describe late cosmology well, the coupling constant g has to be quite small. The situation is the same as for the cosmological constant in the standard ΛCDM model, where the cosmological constant λ also must be small. In both cases they will play no role in the time development of the Universe for $t < t_{\rm LS}$, the time of last scattering. Thus we can calculate the time of last scatting $t_{\rm LS}$ using standard cosmology without any reference to g (or λ). Knowing $t_{\rm LS}$ we can obtain $p_{\rm LS}$ also without any reference to g (or λ) from

$$t_{\rm LS} = \int_{-\infty}^{p_{\rm LS}} \frac{dp}{\frac{3}{4}\,p^2 + \frac{1}{3}\,\kappa\rho_{\rm r}(v(p))}, \qquad \frac{3}{4}\,p^2 = \kappa\rho_{\rm m}(v) + \kappa\rho_{\rm r}(v). \tag{25}$$

We have included in the formula the radiation density, since that cannot be ignored all the way down to $t = 0$ in the region $t < t_{\rm LS}$. This formulas allows us to write $p_{\rm LS}(t_{\rm LS})$, still without any reference to g, and finally we can also write

$$f(p_{\rm LS}) = \frac{3}{4}p_{\rm LS}^2, \tag{26}$$

since $p_{\rm LS}^2 \gg g/|p_{\rm LS}|$ for $t \leq t_{\rm LS}$. For $t > t_{\rm LS}$ we expect that the CDM $\rho_{\rm m}(v)$ will be a good approximation to the matter density, and under this assumption g and t_0 are determined from the values the H_0 and $z_{\rm LS}$, the red shift at the time of last scattering. This is interesting since H_0 and $z_{\rm LS}$ can be obtained by measurements that are almost independent of cosmological models[2].

Given H_0 and $z_{\rm LS}$ we can first obtain p_0 (p at t_0, our present time) for a given value of g from (20)

$$f'(p_0) = -3H_0\,. \tag{27}$$

Next the Friedmann equation (22) applied at $p = p_{\rm LS}$ reads

$$f(p_0) = \frac{f(p_{\rm LS})}{(1+z_{\rm LS})^3} = \frac{3p_{\rm LS}^2}{4(1+z_{\rm LS})^3}\,. \tag{28}$$

Since we know $p_{\rm LS}$ from (25) and we know $f(p_0)$ as a function of g, this Friedmann equation will determine g. Finally $t_0 - t_{\rm LS}$ is the determined from (21) (or (24) for the the specific case of the modified Friedmann equation).

We have determined the coupling constant g and t_0 from H_0 and $z_{\rm LS}$. There are presently two values of H_0 that do not agree within 5σ (the so-called H_0 tension). One value is deduced from "local" measurements, using various "space candles" (SC) techniques [16]. We denote it $H_0^{\rm SC}$ ($= 73.04 \pm 1.04$ km/s/Mpc), and it is almost independent of cosmological models. The other value is deduced from the CMB data created at $t_{\rm LS}$. It is using cosmological models in a number of ways, among those to extrapolate to present time t_0. This value we denote $H_0^{\rm CMB}$ and it is model dependent. The value $H_0^{\rm CMB} = 67.4 \pm 0.5$ km/s/Mpc quoted in [17], and the one we will use here, refers the ΛCDM model.

A good check of our cosmological model is to compare its low z (late time) predictions with (almost model independent) low z measurements. This is done in Fig. 2. It is seen that our modified Friedmann equation agrees very well with the small z data if we use $H_0^{\rm SC}$ and not well if we use $H_0^{\rm CMB}$. By contrast the ΛCDM model agrees well with the data when we use $H_0^{\rm CMB}$ and not well when we use $H_0^{\rm SC}$.

---

[2] More precisely H_0 can be (and is) determined by local measurements. This is the value we denote $H_0^{\rm SC}$ below. Similarly, observations allow us to determine the temperature $T(t_0)$ of the CMB. $T(t_{\rm LS})$ can be calculated by atomic physics and is to a large extent independent of the cosmological model, as is also the statement that $T(t_{\rm LS})/T(t_0) = a(t_0)/a(t_{\rm LS}) = 1 + z_{\rm LS}$.

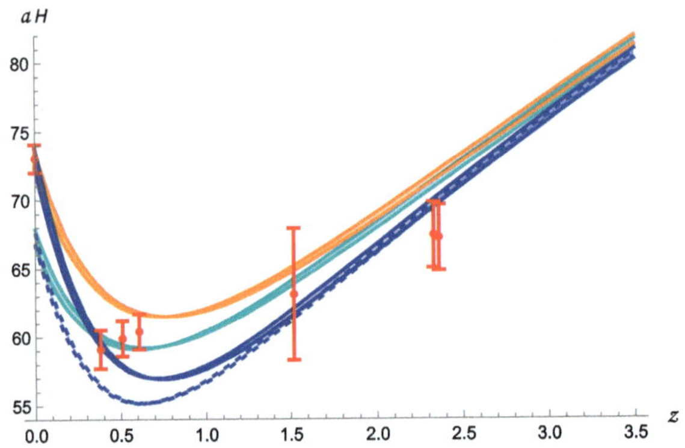

Fig. 2 The blue curves show $H(z)a(z)$, $a(z) = 1/(1+z)$, for the modified Friedmann equation starting from H_0^{SC} and H_0^{CMB} (the dotted blue curve). The green curve shows $H(z)a(z)$ based on the ΛCDM model with H_0^{CMB} and λ^{CMB} (the cosmological constant) taken from [17]. The orange curve shows $H(z)a(z)$ based on the ΛCDM model with λ chosen such that it starts with H_0^{SC}. Inserted in red are (apart from the SC data at $z \approx 0$) also independent data from other observations: first three points from the baryon acoustic oscillation data [18], the next from quasars [19] and the last two data points from Ly–α measurements [20, 21]

Table 1 The reduced χ^2 values for the two models, as described in the text

	$\chi^2_{red}(\Lambda)$	$\chi^2_{red}(\mathrm{MF})$
H_0^{SC}	3.5	1.8
H_0^{CMB}	1.2	5.6

Table 2 The present days times for the two models as described in the text

	$t_0(\Lambda)$	$t_0(\mathrm{MF})$
H_0^{SC}	13.3 Gyr	13.9 Gyr
H_0^{CMB}	13.8 Gyr	14.4 Gyr

In Table 1 we have listed the reduced χ^2 values, $\chi^2_{red}(\Lambda)$ for the ΛCDM model and $\chi^2_{red}(\mathrm{MF})$ for the modified Friedmann (MF) equation, obtained by comparing the calculated $H(z)$ with the observe $H(z)$. Finally, in Table 2 we list the calculated values of the present days time t_0 in the four cases.

We conclude that if H_0^{CMB} should be proven to be the correct value, our modified Friedmann equation is ruled out. Similarly, should H_0^{SC} turn out to be correct, the ΛCDM model seems to have a problem, while the modified Friedmann equation seems to do well.

The density fluctuations of matter is another "local" observable that will be available for $z < 2$. Presently the error bars on these data are too large to provide a serious testing of the modified Friedmann equation. However, this will most likely change dramatically with the new observations to be obtained by the Euclid satellite.

5 Discussion

We have proposed a modified Friedmann equation that results in a late time cosmology that is different from the late time cosmology provided by the ΛCDM model. It does not require a cosmological constant, the late time acceleration of our Universe being generated by the bombardment of the Universe by baby universes. The model fits the low z data very well provided that the $H_0^{\mathrm{SC}} = 73.04 \pm 1.04$ km/s/Mpc is the correct value of the Hubble constant.

The model has a formal equation of state parameter $w(z) < -1$. In standard cosmology this is a sign that some unphysical degrees of freedom have been added to the system. However, in the modified model this is not the case. In fact the term $(-3g/2p) \cdot v$ added to the effective Hamiltonian is more like a time dependent cosmological constant term, but without usual problem that a time dependent cosmological constant will break the invariance under time-reparametrization. An increasing time dependent cosmological will result in $w(z) < -1$ in an expanding universe. In our model $-3g/2p$ is an increasing function of t since $p(t)$ increases from large negative values towards the value $-(2g)^{1/3}$ for $t \to \infty$ where it will act as an ordinary cosmological term with value $3(2g)^{2/3}/4$. $w(z)$ changes monotonically from $-3/2$ at $z = \infty$ ($t = 0$) to -1 for $z = -1$ ($t = \infty$).

Before observations showed that the cosmological constant was positive (but very small), many favored that $\lambda = 0$, maybe caused by some not fully understood dynamics, like the one offered by Coleman's mechanism [22]. In some sense it might be easier to explain why $\lambda = 0$ than to explain the very small value observed today. What we have argued here is that even if one can show that $\lambda = 0$, we can still have an accelerating late universe, the acceleration due to baby universe absorption. In order to fit observations our coupling constant g also has to be very small. However, it appears in a larger multi-verse theory and one could hope that being able to solve this theory might also help us understanding the smallness of g. The program to understand this theory has been started [23–25], but it is still work in progress.

References

1. J. Ambjorn, A. Goerlich, J. Jurkiewicz, R. Loll, Nonperturbative quantum gravity. Phys. Rept. **519**, 127–210 (2012). [arXiv:1203.3591 [hep-th]]
2. R. Loll, Quantum gravity from causal dynamical triangulations: a review. Class. Quant. Grav. **37**(1), 013002 (2020). [arXiv:1905.08669 [hep-th]]
3. J.B. Hartle, S.W. Hawking, Wave function of the universe. Phys. Rev. D **28**, 2960–2975 (1983)
4. Y. Hamada, H. Kawai, K. Kawana, Baby universes in 2d and 4d theories of quantum gravity. JHEP **12**, 100 (2022). [arXiv:2210.05134 [hep-th]]
5. J. Ambjorn, R. Loll, Nonperturbative Lorentzian quantum gravity, causality and topology change. Nucl. Phys. B **536**, 407–434 (1998). [arXiv:hep-th/9805108 [hep-th]]
6. J. Ambjorn, R. Loll, J.L. Nielsen, J. Rolf, Euclidean and Lorentzian quantum gravity: lessons from two-dimensions. Chaos Solitons Fractals **10**, 177–195 (1999). [arXiv:hep-th/9806241 [hep-th]]

7. J. Ambjorn, R. Loll, Y. Watabiki, W. Westra, S. Zohren, A string field theory based on causal dynamical triangulations. JHEP **05**, 032 (2008). [arXiv:0802.0719 [hep-th]]
8. J. Ambjorn, R. Loll, W. Westra, S. Zohren, Putting a cap on causality violations in CDT. JHEP **12**, 017 (2007). [arXiv:0709.2784 [gr-qc]]
9. J. Ambjorn, R. Loll, Y. Watabiki, W. Westra, S. Zohren, A matrix model for 2D quantum gravity defined by causal dynamical triangulations. Phys. Lett. B **665**, 252–256 (2008). [arXiv:0804.0252 [hep-th]]
10. H. Kawai, N. Kawamoto, T. Mogami, Y. Watabiki, Transfer matrix formalism for two-dimensional quantum gravity and fractal structures of space-time. Phys. Lett. B **306**, 19–26 (1993). arXiv:hep-th/9302133 [hep-th]
11. J. Ambjorn, *Elementary Introduction to Quantum Geometry* (Taylor & Francis, 2022). ISBN 978-1-03-233555-1. [arXiv:2204.00859 [hep-th]]
12. J. Ambjorn, Y. Watabiki, A modified Friedmann equation. Mod. Phys. Lett. A **32**, 1750224 (2017). [arXiv:1709.06497 [gr-qc]]
13. J. Ambjorn, Y. Watabiki, Easing the Hubble constant tension. Mod. Phys. Lett. A **37**, 2250041 (2022). [arXiv:2111.05087 [gr-qc]]
14. J. Ambjorn, Y. Watabiki, The large scale structure of the Universe from a modified Friedmann equation. Mod. Phys. Lett. A **38**, 2350068 (2023). [arXiv:2208.02607 [gr-qc]]
15. J. Ambjorn, Y. Watabiki, Is the present acceleration of the Universe caused by merging with other universes? JCAP **12**, 011 (2023). [arXiv:2308.10924 [gr-qc]]
16. A.G. Riess, W. Yuan, L.M. Macri, et al., A comprehensive measurement of the local value of the Hubble constant with 1 km s^{-1} Mpc^{-1} uncertainty from the Hubble space telescope and the SH0ES Team. Astrophys. J. Lett. **934**(1), L7 (2022). [arXiv:2112.04510 [astro-ph.CO]]
17. N. Aghanim et al., (Planck), Planck 2018 results. VI. Cosmological parameters. Astron. Astrophys. **641**, A6 (2020). [arXiv:1807.06209 [astro-ph.CO]]
18. S. Alam et al., (BOSS), The clustering of galaxies in the completed SDSS-III Baryon Oscillation Spec- troscopic Survey: cosmological analysis of the DR12 galaxy sample. Mon. Not. Roy. Astron. Soc. **470**, 2617–2652 (2017). [arXiv:1607.03155 [astro-ph.CO]]
19. P. Zarrouk et al., The clustering of the SDSS-IV extended Baryon Oscillation Spectroscopic Survey DR14 quasar sample: measurement of the growth rate of structure from the anisotropic correlation function between redshift 0.8 and 2.2. Mon. Not. Roy. Astron. Soc. **477**, 1639–1663 (2018). [arXiv:1801.03062 [astro- ph.CO]]
20. J.E. Bautista, et al., Measurement of baryon acoustic oscillation correlations at $z = 2.3$ with SDSS DR12 Lyα-Forests. Astron. Astrophys. **603**, A12 (2017). [arXiv:1702.00176 [astro-ph.CO]]
21. A. Font-Ribera et al., [BOSS], Quasar-Lyman α forest cross-correlation from BOSS DR11: Baryon acoustic oscillations. JCAP **05**, 027 (2014). [arXiv:1311.1767 [astro-ph.CO]]
22. S.R. Coleman, Why there is nothing rather than something: a theory of the cosmological constant. Nucl. Phys. B **310**, 643 (1988)
23. J. Ambjorn, Y. Watabiki, A model for emergence of space and time. Phys. Lett. B **749**, 149 (2015). ([arXiv:1505.04353 [hep-th]])
24. J. Ambjorn, Y. Watabiki, Creating 3, 4, 6 and 10-dimensional spacetime from W3 symmetry. Phys. Lett. B **770**, 252 (2017). ([arXiv:1703.04402 [hep-th]])
25. J. Ambjorn, Y. Watabiki, Models of the Universe based on Jordan algebras. Nucl. Phys. B **955**, 115044 (2020). [arXiv:2003.13527 [hep-th]]

Comments on Graviton Detection

Daniel Carney

Abstract It is possible to make a detector which clicks after absorbing a single graviton. Similarly, it is possible to make a gravitational wave detector which can see the quantum noise induced by certain highly squeezed states of the graviton. However, contrary to some recent arguments in the literature, observation of either or both of these signals would not constitute proof or even evidence that the gravitational field is quantized. This is a simple technical statement: a classical gravitational wave can produce the same output data in a detector. Here we explain this result, presented earlier in Carney et al. (Phys Rev D 109(4):044009, 2024, [1]), which is a straightforward extension of an ancient argument in quantum optics.

1 Introduction

By simple analogy with the other known fields of nature, it seems highly likely that the gravitational field is quantized, but experimental evidence for this idea is currently lacking. Theoretically, it is straightforward to perturbatively quantize general relativity around a background and treat the resulting graviton fluctuations as an effective quantum field theory [2]. This model makes very definite predictions, and these predictions can be measured in real experiments, as discussed below. However, the right question is whether we can form a reasonable alternative to the graviton picture and perform an experiment that can distinguish the two cases, i.e., do a measurement that would provide us with more Bayesian evidence that the gravitational field is quantized. As we will see, this is a substantially harder goal than simply confirming some prediction of graviton physics.

What kind of experiment could one do to test the quantization of gravity, i.e., the existence of the graviton? A seemingly obvious method would be to just "detect a

Dedicated to Gordon Semenoff on the occasion of his 70th birthday, and based on a talk at his festschrift held at the Université de Montréal, July 2023.

D. Carney (✉)
Physics Division, Lawrence Berkeley National Laboratory, Berkeley, CA, United States
e-mail: carney@lbl.gov

M. Paranjape et al. (eds.), *Proceedings of Gravity, Strings and Fields*,
CRM Series in Mathematical Physics,
https://doi.org/10.1007/978-3-031-91266-5_2

graviton". It may sound like it is impossibly difficult to detect single gravitons, but in the modern understanding of quantum measurement, it is actually pretty straightforward. Detecting a graviton would mean constructing a detector which clicks when it absorbs a graviton, much as a photodiode clicks when it absorbs a photon. It is easy to think of ways to do this. Examples include absorption through a quadrupole transition in an atom [3], using single-photon counters after converting the gravitons into photons through the Gertsenshtein effect [4, 5], or using single-phonon counters after converting the gravitons into phonons in a Weber bar [6, 7].

As remarkable as it would be to see such an event, we now have to ask the right question: would such a detection provide evidence for the existence of the graviton? The answer, unfortunately, is not really [1]. Consider the following pair of models for a detector coupled to an incident gravitational signal:

$$\hat{H}_{\rm q} = \hat{H}_{\rm det} - \int d^3\mathbf{x} \hat{h}_{\mu\nu} \hat{T}^{\mu\nu}_{\rm det} \tag{1.1}$$

$$\hat{H}_{\rm cl} = \hat{H}_{\rm det} - \int d^3\mathbf{x} h_{\mu\nu} \hat{T}^{\mu\nu}_{\rm det}. \tag{1.2}$$

In both models, the detector itself is treated as a quantum system. The difference is whether or not the gravitational field perturbation $h_{\mu\nu}$ is treated as a quantum field (denoted by a hat) or as a classical field (without a hat). In general, we might imagine that the classical field $h_{\mu\nu}$ is drawn from a classical random distribution $P_{\rm cl}(h_{\mu\nu})$. We are not specifying the internal dynamics of h, just treating it as some external signal incident on the detector. Here is the claim: any realistic "graviton" signal picked up by such a detector can equally well be explained by (1.1) or (1.2).[1]

In what follows, we will make this statement precise. We begin by discussing the analogous statement regarding the electromagnetic field and photon detection, which is an almost identical argument [11–13]. First we cover the case of photon detection and discuss what kinds of events can and cannot be explained by a classical model of electromagnetic radiation. We then extend these arguments to the gravitational case. In both the electromagnetic and gravitational cases, we emphasize that similar arguments apply to both "counting" detectors (like photodiodes or atomic absorbers) and to "linear" or "amplitude" detectors (like homodyne detectors and LIGO, which measures strain $h_{\mu\nu}$ directly). The latter case will clarify why measuring things like "quantum noise" in squeezed states with a detector like LIGO [14, 15] can similarly be explained by a classical radiation model except in very extreme, and physically unrealistic, scenarios. We emphasize this with an explicit calculation in LIGO in Sect. 5, which is the only new material appearing here that was not essentially in Ref. [1].

[1] Some attempts to make self-consistent dynamical models in the classical case include [8–10]. Taken literally, Eqs. (1.1) and (1.2) suggest violations of energy conservation, since there is no loss of energy in the field after detection, as emphasized in [6, 7]. However, this is irrelevant to the point of this paper, because the state of the field after it interacts with the detector is not measured. Thus this violation of energy conservation cannot be seen in the experiments described here.

2 Photoelectric Effect

Consider the electromagnetic analogues of the two models (1.1) or (1.2):

$$\hat{H}_{\text{q}} = \hat{H}_{\text{det}} - \int d^3\mathbf{x} \hat{A}_\mu \hat{J}^\mu_{\text{det}} \tag{2.1}$$

$$\hat{H}_{\text{cl}} = \hat{H}_{\text{det}} - \int d^3\mathbf{x} A_\mu \hat{J}^\mu_{\text{det}}. \tag{2.2}$$

Again the detector is a quantum, dynamic system; the electromagnetic field is being treated as non-dynamical, external perturbation; and the difference between the models is whether the electromagnetic field is quantized $\hat{A}$ or a classical variable A drawn in general from some probability distribution $P_{\text{cl}}(A)$. The question is what kind of state $|\psi\rangle$ of the quantized electromagnetic field and detector observable could be used to distinguish between the two models (2.1) and (2.2).

Let's begin with a famous *non*-example [11–13]: detection of individual clicks in a photodiode based on the photoelectric effect. We model the photodiode $\hat{H}_{\text{det}}$ as a bunch of electrons, each with a spectrum

$$\hat{H}_{\text{det}} = -\Delta \left|g\right\rangle \left\langle g\right| + \sum_{E=0}^{\infty} E \left|E\right\rangle \left\langle E\right|. \tag{2.3}$$

Here $\Delta > 0$ is the energy gap between the ground state, in which the electron is bound to the atomic lattice, and the conduction band states $|E\rangle$ representing unbound electrons (often then referred to as "photoelectrons"). We are treating this as a discrete spectrum with no degeneracy for simplicity. Properly we should have a sum over $N \gg 1$ individual electrons in $\hat{H}_{\text{det}}$ but for notational simplicity we will suppress this (Fig. 1).

The detector "clicks" when an electron transitions from $|g\rangle \to |E\rangle$ for any $|E\rangle$. In the quantized radiation model (2.1), this is explained by the electron absorbing

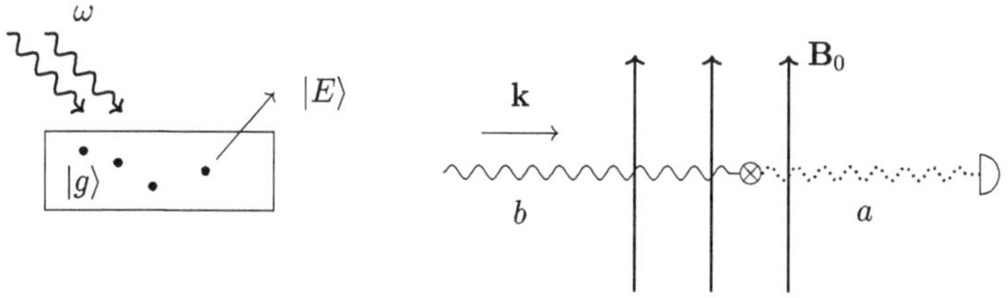

Fig. 1 Left: a single-photon detector, based on the photoelectric effect. Incident electromagnetic radiation excites bound electrons $|g\rangle \to |E\rangle$, which are then sent through an amplifier and detected. Right: a single-graviton detector, based on the Gertsenshtein effect. Incident gravitational radiation is converted to X-ray frequency light through the interaction $S = \int d^4x T^{EM}_{\mu\nu} h^{\mu\nu}$ in the presence of a large external field $\mathbf{B}_0$, and the subsequent X-ray photons are then detected

a photon. Historically, this explanation and the observation of individual photoelectron excitations was viewed as evidence for the quantization of the electromagnetic field [16]. However, these clicks can equally well be explained by both (2.1) and (2.2). They both lead to precisely the same formula for the rate of photoelectron excitation, as we now show explicitly.

We can start with the classical case. Consider for simplicity a monochromatic incident field $\mathbf{A}(\mathbf{x},t) = \mathbf{N}_0 \left[b e^{ik\cdot x} + \text{h.c.} \right]$, where $\mathbf{N}_0$ is some fixed vector, b is the (complex) Fourier amplitude for the mode, $\mathbf{k}$ is the wavevector and $\omega = |\mathbf{k}|$ is the frequency, and we are working in $A^0 = 0$ gauge. The electron has $\hat{\mathbf{J}}(\mathbf{x}) = (e\hat{\mathbf{p}}/m)\delta^3(\mathbf{x} - \hat{\mathbf{x}})$, where $\hat{\mathbf{x}}$ is the position operator of the electron. Assuming the electron is localized near the origin to better precision than the wavelength of the incident field $\lambda = 1/\omega$, the $\mathbf{J} \cdot \mathbf{A}$ coupling takes the simple form

$$\hat{V}_{\text{cl}}(t) = -\frac{e}{m}\hat{\mathbf{p}} \cdot \mathbf{N}_0 \left[b e^{-i\omega t} + \text{h.c.} \right]. \tag{2.4}$$

The amplitude for an electron to excite into a given conduction state $|E\rangle$ in some small time interval δt is given by the usual formula from first-order perturbation theory

$$\begin{aligned} \mathcal{M}(g \to E) &= -i \int_{-\delta t/2}^{\delta t/2} dt \, \langle E | V_{\text{cl},I}(t) | g \rangle \\ &= i\frac{e}{m} \int_{-\delta t/2}^{\delta t/2} dt \, \mathbf{N}_0 \cdot \langle E | \hat{\mathbf{p}} | g \rangle \left[b e^{-i(E+\Delta-\omega)t} + b^* e^{-i(E+\Delta+\omega)t} \right] \\ &\approx i\frac{2e}{m} \mathbf{N}_0 \cdot \langle E | \hat{\mathbf{p}} | g \rangle \, b \frac{\sin\left[(E+\Delta-\omega)\,\delta t/2 \right]}{E+\Delta-\omega}. \end{aligned} \tag{2.5}$$

In the first line, the subscript I means interaction picture. To get to the last line, we assumed that $\delta t \gtrsim 1/\omega$, in which case the exponential in the first term integrates to the sinc function, while the exponential in the second term averages to zero. This is an excellent approximation in an optical photodiode for example, where $\omega \sim 0.1$ eV ~ 100 THz and even a fast photodiode would only bin the events at something like $\delta t \sim 1$ ns rates. The probability that the detector clicks in this time window is given by squaring this amplitude and summing over all possible conduction band states, which gives

$$\frac{d\mathcal{P}(g \to \text{any } E)}{dt} = \eta I_b \Theta(\omega - \Delta), \tag{2.6}$$

where we took the limit $\omega\delta t \gg 1$ to write the sinc^2 as $\delta(E + \Delta - \omega)\delta t$, and defined the incident intensity I_b (given the Fourier amplitude b) and detector efficiency η as

$$\begin{aligned} I_b &= |b|^2 \\ \eta &= \frac{e^2}{m^2}\rho(\omega - \Delta)|\mathbf{N}_0 \cdot \langle \omega - \Delta|\hat{\mathbf{p}}|g\rangle|^2. \end{aligned} \tag{2.7}$$

We took the continuum limit of the conduction states $|E\rangle$, assigning them a density of states $\rho(E)$. The dimensionless quantity η gives the fraction of the power of an incident beam polarized along $\mathbf{N}_0$ which is absorbed by the detector. Modern photodiodes can achieve $\eta \gtrsim 0.9$, meaning they absorb $\gtrsim 90\%$ of photons in a beam aimed at them.

The quantum calculation is very similar. Consider the same electromagnetic mode $(\mathbf{k}, \omega)$, which we now expand into creation and annihilation operators

$$\hat{\mathbf{A}}(\mathbf{x}, t) = \mathbf{N}_0 e^{ik\cdot x}\hat{b} + \text{h.c.}, \tag{2.8}$$

where $\hat{b}$ is the annihilation operator for the mode and again $\mathbf{N}_0$ is a fixed overall vector. Now consider an incident coherent state $\hat{b}\,|\beta\rangle = \beta\,|\beta\rangle$ of this mode; we will generalize to an arbitrary state shortly. The amplitude of interest now has to allow for the field state to change. Let us assume that the state of the electromagnetic field is not measured after the interaction with the detector. Then the probability that a photoelectron is excited into a specific state $|E\rangle$ comes from averaging over the final states of the field,

$$\begin{aligned} \mathcal{P}(g \to E) &= \int d\alpha\, |\mathcal{M}(g, \beta \to E, \alpha)|^2 \\ &\approx \frac{e^2}{m^2}\left|\mathbf{N}_0 \cdot \langle E|\hat{\mathbf{p}}|g\rangle\right|^2 |\beta|^2 \frac{\sin^2\left[(E + \Delta - \omega)\,\delta t/2\right]}{(E + \Delta - \omega)^2}. \end{aligned} \tag{2.9}$$

Here $|\alpha\rangle$ is another coherent state. The calculation of $\mathcal{M}$ is identical to that in (2.5); we used the same arguments to drop the $e^{i(E+\Delta+\omega)t}$ term, as well as the completeness relation of the coherent states $\int d\alpha\,|\alpha\rangle\,\langle\alpha| = \pi$. Finally, we can integrate over the possible conduction states to get the total rate of probability for a click,

$$\frac{d\mathcal{P}(g \to \text{any } E)}{dt} = \eta I_\beta \Theta(\omega - \Delta), \tag{2.10}$$

where now $I_\beta = |\beta|^2$ is the intensity of the light in the state $|\beta\rangle$ and the efficiency η is the same as before.

The conclusion is straightforward. The prediction for the rate of photoelectron clicks in the detector is the same in either the model with quantized radiation or simple classical radiation, once one makes the identification $b \to \beta$. Notice all the features that are commonly taught as being due to the quantum nature of light persist: the click rate is proportional to the intensity of the light I, the clicks only occur when "energy is conserved" $\omega \geq \Delta$ (which in the classical case is a simple resonance condition between the light and the states above the electron gap), and the clicks themselves are of course discrete.

It should be emphasized that we are not saying that a purely classical model of the universe can reproduce the photoelectric effect. Rather, we are saying that a classical model of the *radiation field*, coupled to a quantum model of the detector, can reproduce the photoelectric effect. In the context of electromagnetism this is purely academic since we have many other tests of the quantization of the radiation field. However, this is *precisely* the situation in a gravitational wave detector at the present time: we know that the detectors are made of normal quantum matter, but we are looking for evidence that the gravitational radiation field itself is quantized.

Here we presented these rates with either a fixed-amplitude classical field (2.6) or quantum field (2.10). This can be easily generalized to an arbitrary state in either case. For the classical case, we want to allow for a randomly drawn field amplitude, say from a distribution $P_{\rm cl}(b)$. In the quantum case, we can instead use the Glauber-Sudarshan representation: for any state ρ of a harmonic oscillator, one can write

$$\rho = \int d\beta P(\beta) \left|\beta\right\rangle \left\langle\beta\right| , \tag{2.11}$$

where the weight function $P(\beta)$ is real, formally defined as the integral [11, 17]

$$P(\beta) = \frac{e^{|\beta|^2}}{\pi^2} \int d\alpha \left\langle -\alpha|\rho|\alpha\right\rangle e^{|\alpha|^2 - \beta^* \alpha + \beta\alpha^*} \tag{2.12}$$

and, when it exists, satisfies $\int d\beta P(\beta) = 1$. It is easy to track through the calculations of (2.6) and (2.10) to see that in an arbitrary state, either classical or quantum, the click rate is

$$\frac{d\mathcal{P}}{dt} = \eta \left\langle I\right\rangle \Theta(\omega - \Delta), \tag{2.13}$$

with the only difference being whether the bracket means a classical expectation $\langle I\rangle = \int db P_{\rm cl}(b) I_b$ or a quantum expectation $\langle I\rangle = \int d\beta P(\beta) I_\beta$.

3 Non-classicality Tests by Counting Photons

What Eq. (2.13) shows is that the expected rate $d\mathcal{P}/dt$ of discrete photoelectron detector clicks can always be explained equally well with a classical or quantum model of the electromagnetic radiation. In this sense, observation of the photoelectron effect does not demonstrate that the photon exists. The extension to gravitational radiation is immediate. The question, then, is what kind of observation *would* distinguish the classical and quantum models?

Generally speaking, what has been understood in quantum optics is that one can make sharp tests based on correlation statistics beyond averages, and given certain quantum states of the radiation field. This is similar in spirit to Bell inequality violations: it is possible to find correlation functions which are bound by some quantity in any classical model, but for which certain quantum states violate the bound. In the next subsections, we give explicit examples based on click counting statistics (e.g., variance in the click rate) and noise statistics (e.g., variance in the observed electric field strength).

First, we give a characterization of the types of "non-classical" quantum states of radiation we will be interested in. In a given quantum state ρ, the Glauber function $P(\beta)$ may or may not be positive for all β. If $P(\beta) \geq$ everywhere, then $P(\beta)$ is a probability distribution, so the state (2.11) can be interpreted as a classical ensemble of quantum coherent states or classical states of definite Fourier amplitude. However, if $P(\beta) < 0$ for some region of $\beta \in \mathbb{C}$, then $P(\beta)$ is not a probability distribution, and it turns out that in this case one can find observables that can only be explained by a quantum mechanical model of the radiation field. Thus people sometimes refer to states whose Glauber-Sudarshan representation is somewhere negative as "non-classical". This terminology is obviously a bit loaded. A more conservative convention would be to say that these states have Glauber negativity. Similar terminology is often used in the Wigner function representation.

Before moving on to gravity, let's discuss a concrete example of a "photon counting" measurement that can distinguish between (2.1) and (2.2). Suppose that we shine light on our photodetector for some time T and measure the total number of photoelectron clicks, N. If we do this many times, then from the measured values $N_1, N_2, \ldots, N_{n_{\text{runs}}}$ we can calculate the average $\overline{N} = \sum_i N_i / N_{n_{\text{runs}}}$ and variance $\Delta N^2 = \sum_i (N_i - \overline{N})^2 / N_{n_{\text{runs}}}$. Here is the statement: an arbitrary classical state with classical random distribution $P_{\text{cl}}(b)$ will always produce Poisson or super-Poisson statistics, $\Delta N^2 \geq \overline{N}$. However, certain "non-classical" states, for example squeezed states, can produce sub-Poisson statistics where $\Delta N^2 < \overline{N}$. Thus, observation of sub-Poisson click rates would necessarily imply that the electromagnetic field has to be quantized.

The proof is by direct calculation. First consider the classical case, where the Fourier amplitude in general is drawn from some distribution $P_{\text{cl}}(b)$. In each run of the experiment, we get some realization of this distribution, and it produces a Poisson-distributed set of clicks, with expected value $\overline{N}_b = \eta I_b T$ and variance $\Delta N_b^2 = \overline{N}_b$. Here the statistics come from the quantum nature of the *detector electrons* being measured. Now, performing many runs, each draws from $P_{\text{cl}}(b)$, which leads to even more randomness since we are averaging over multiple Poisson distributions. One finds [1, 13]

$$\Delta N^2 = \overline{N} + \eta^2 T^2 \int db P_{\text{cl}}(b) \left[I_b - \langle I \rangle \right]^2 . \tag{3.1}$$

Here, the first term $\overline{N} = \int da P_{\text{cl}}(b) \eta I_b T$ is the average number expected over all the runs. The second term is just the variance of the incident electromagnetic field intensity $\langle \Delta I^2 \rangle$. This is positive, so this equation shows that the observed click rates are always either Poisson or super-Poisson $\Delta N^2 \geq \overline{N}$. The noisier the input field (larger ΔI^2), the noisier the data.

Now consider the quantum model of radiation. A directly analogous calculation can be performed in an arbitrary state with Glauber-Sudarshan representation $P(\beta)$. In a given coherent state, as discussed above, the click statistics will be Poisson distributed with $\overline{N}_\beta = \eta I_\beta T$ and $\Delta N_\beta^2 = \overline{N}_\beta$. Now, with many runs, one might try to repeat the same argument as the classical case, averaging over $P(\beta)$. But in general $P(\beta)$ is not a true probability distribution, so we have to be slightly more careful, and in particular operator ordering changes the answer slightly. One finds [1, 13]

$$\Delta N^2 = \overline{N} + \eta^2 T^2 \int d\beta P(\beta) \left[I_\beta - \langle I \rangle \right]^2. \tag{3.2}$$

The first term is just $\overline{N} = \int d\beta P(\beta) I_\beta T$ as in the classical case. However, the intensity $I = b^\dagger b$ as an operator, so the quantum expectation $\langle I^2 \rangle = \langle b^\dagger b b^\dagger b \rangle \neq \int d\beta P(\beta) I_\beta^2 = \int d\beta P(\beta) |\beta|^4$, because one has a non-trivial contribution from the $[b, b^\dagger] = 1$ commutator. Less prosaically, the second term in Eq. (3.2) is not the variance of an operator, and in particular, if $P(\beta) < 0$ for some β, it is possible that this integral is *negative*. This is the case for example in for phase-squeezed states of the radiation field [18]. Thus quantum radiation fields allow for sub-Poisson statistics $\Delta N^2 < \overline{N}$.

Sub-Poisson statistics of the electromagnetic field were first experimentally observed in the 1980's [19]. They now form a routine method of reducing shot noise in practical experiments [20]. This observation provides very definitive evidence that the electromagnetic field is quantized, since this observation cannot be reproduced by any classical distribution of radiation fields.

4 Non-classicality Tests with Gravitational Radiation

Extending the arguments presented above to the case of gravitational wave or "graviton" detectors is mostly straightforward. The question is what kind of detection could distinguish a model where the gravitational radiation field is classical or quantized, in the sense of the models in Eqs. (1.1) and (1.2). The only subtlety is that most gravitational wave detectors currently do not measure the gravitational wave intensity—i.e., they do not actually count gravitons, but rather measure the strain waveform as a function of time. I will discuss the difference below.

First, however, we can quickly cover the case of an actual graviton counting detector, such as those suggested in [3, 4, 6]. Here we have a photodiode-like device which "clicks" on absorption of a single graviton. The question is whether a classical gravitational model can reproduce the same clicks. The answer is the same as in the discussion of the photoelectric effect above. The simple counting of such clicks proves nothing, i.e., cannot distinguish between the models (1.1) and (1.2). On the other hand, one might try to do something like observe sub-Poisson graviton counting statistics, as in Eq. (3.2). This would first of all require a source of squeezed gravitational radiation, which would be exotic. Worse, though, it would be unobservable, for a simple reason. Notice in Eq. (3.2), the deviation from Poisson statistics is proportional to the efficiency:

$$\Delta N^2 - \overline{N} \propto \eta^2 \sim \left(\frac{P_{\text{absorbed}}}{P_{\text{incident}}}\right)^2. \tag{4.1}$$

In a state-of-the-art photodiode, $\eta \sim O(1)$ is possible. But in a typical gravitational wave detector, $\eta \ll 1$ is astronomically small. In the CAST experiment, for example, $\eta \sim 10^{-33}$ [1]. What this means is that any "graviton detector" is going to see a Poisson distribution of clicks, and these can equally well be explained by classical gravitational radiation. The intuition is that gravity is weakly coupled to the detector, so the measurement is very weak, and we are trying to determine the underlying sub-Poisson statistics of the signal using an extremely minute sample set, which is impossible. The statistics are swamped instead by the detector's own noise.

What about strain detectors like LIGO? Here the observable is not the incident intensity $I \sim b^\dagger b$ but rather the incident strain field $h \sim b + b^\dagger$. These are "linear" detectors rather than "square law" detectors of the gravitational field. Non-classicality tests in linear observables are also available. The rough idea is that the two variables h and its conjugate $\pi_h \sim -i(b - b^\dagger)$ come with uncertainties $\Delta h = \Delta\pi_h = 1/2$ in the vacuum or a coherent state. By similar logic as above, a detector registering this level of noise could just as well be explained with an incoming classical wave. However, certain quantum states like squeezed states can achieve sub-vacuum levels of noise, say $\Delta h < 1/2$ (while $\Delta\pi_h > 1/2$ by Heisenberg uncertainty). It turns out that there are observables analogous to sub-Poisson statistics in such states that cannot be explained classically. However, much like sub-Poisson statistics, detecting these would require an impossibly efficient gravitational wave detector.

Let's give a simple, precise version of this idea as described in [1], and then connect it to more standard language involving noise power spectra of gravitational waves. See Fig. 2. Consider two modes, call them a and b, which interact through the linear exchange interaction

$$V = -ig(ab^\dagger - a^\dagger b). \tag{4.2}$$

Here a will be the mode we ultimately detect, while b is the mode with the signal of interest — either the gravitational wave or some electromagnetic mode. This is the effective interaction generated by, for example, the coupling $V = \int d^3\mathbf{x} h_{\mu\nu} T^{\mu\nu}$ once linearized around a large mass detector [21]. The interaction (4.2) generates a beamsplitting effect, in the sense that the Heisenberg equations $\dot{a} = i[V, a]$ and $\dot{b} = i[V, b]$ have the solutions

$$a(t) = a\cos(gt) - b\sin(gt), \quad b(t) = b\cos(gt) + a\sin(gt). \tag{4.3}$$

For $gt = \pi/4$, this is a 50/50 beamsplitter. For $gt \ll 1$, this is a mostly transmissive mirror. Suppose that we send in vacuum $|0\rangle$ to the detector mode a, and a squeezed state $S(r,\theta)\,|0\rangle$ in the signal mode b, where the squeezing operator is

$$S(r,\theta) := \exp\left\{\frac{1}{2}\left(re^{-i\theta}b^2 - re^{i\theta}b^{\dagger 2}\right)\right\}. \tag{4.4}$$

Now, we are going to detect the mode a' after the initial beamsplitter, say using a homodyne detector to measure $Y' = -i(a' - a'^{\dagger})/\sqrt{2}$ (the "phase quadrature", i.e., the phase of the light exiting the beamsplitter). See Fig. 2. We want to know what the variance of the output will be, as a function of the signal squeezing r.

To calculate this, we can work out how the squeezing and beamsplitter operations transform the various modes. Let's take $\theta = \pi$, i.e., the signal is squeezed in the signal's phase quadrature $h = -i(b - b^{\dagger})/\sqrt{2}$, which will connect to our discussion of LIGO below. We have

$$S^{\dagger} b S = b\cosh r + b^{\dagger}\sinh r, \quad S^{\dagger} b^{\dagger} S = b^{\dagger}\cosh r + b\sinh r. \tag{4.5}$$

Thus the detector mode, after mixing with this squeezed b mode, has phase

$$Y' = -i\frac{a - a^{\dagger}}{\sqrt{2}}\cos(gt) - i\frac{b - b^{\dagger}}{\sqrt{2}}e^{-r}\sin(gt), \tag{4.6}$$

where we used Eq. (4.3). The variance of the measurements is then given by

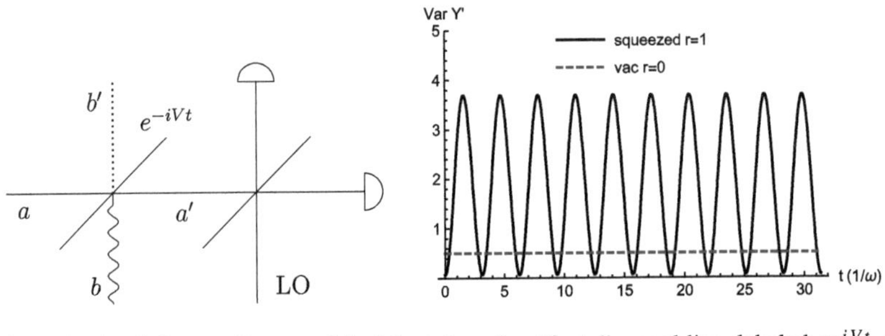

Fig. 2 Left: simple beamsplitter model of the interaction [first diagonal line, labeled e^{-iVt}, as in Eqs. (4.2) and (4.9)], followed by a homodyne measurement to read out $Y' = -i(a' - a'^{\dagger})/\sqrt{2}$. The quantum state of the signal mode after the interaction (b') is lost to the bath. Right: Variance in the detector quadrature output $\Delta Y'^2(t)$ as a function of time, with an incoming squeezed signal in a mode of frequency ω. We see that the noise oscillates between values above and below the vacuum noise value $1/2$. The large noise above the vacuum can be equally well explained with a *classical* signal, as discussed in the text. While the squeezed state does produce a large amount of noise (compared to vacuum), only the small sub-vacuum components require a quantum-mechanical signal to explain them

$$\Delta Y'^2 = \langle Y^2 \rangle \cos^2(gt) + \langle \pi_h^2 \rangle e^{-2r} \sin^2(gt) = \frac{1}{2}\left[\cos^2(gt) + e^{-2r}\sin^2(gt)\right], \quad (4.7)$$

where the expectation values are taken in vacuum. For zero squeezing $r = 0$, this is just $\Delta Y'^2 = 1/2$, the vacuum value. This is the case where both a, b are in vacuum (or coherent) states, mix at the beamsplitter, and therefore produce the same amount of output noise. For non-zero squeezing $r > 0$, however, the observed variance $\Delta Y'^2$ will oscillate in time between sub-vacuum $\Delta Y'^2 < 1/2$ and super-vacuum $\Delta Y'^2 > 1/2$ levels of noise. For comparison to the case of LIGO below, it is helpful to consider the $gt \ll 1$ limit, where the signal is very weakly coupled to the detector. In this case,

$$\Delta Y'^2 = \frac{1}{2}\left[1 - (1 - e^{-2r})g^2t^2 + \cdots\right] \leq \frac{1}{2}, \quad (4.8)$$

with equality only for $r = 0$. What is happening is that the sub-vacuum signal (the squeezed noise in b) is getting transduced onto the detector mode a. Note that by unitarity, this means that extra noise is now pushed to the signal mode b, which is not measured.

What if the mode b is treated purely classically? In this case, the interaction (4.2) becomes a simple unitary on the detector mode a, namely

$$V_{\mathrm{cl}} = -ig(a\beta^* - a^\dagger\beta), \quad (4.9)$$

where β is a c-number, possibly drawn from a classical distribution $P_{\mathrm{cl}}(\beta)$ as above. The detector mode after this interaction is

$$Y' = -i\frac{a - a^\dagger}{\sqrt{2}} - \frac{\beta + \beta^*}{\sqrt{2}}gt, \quad (4.10)$$

which is just the original operator X operator shifted by a c-number. In particular, its variance is

$$\Delta Y'^2 = \frac{1}{2} + \Delta h_{\mathrm{cl}}^2, \quad \Delta h_{\mathrm{cl}}^2 = \int d\beta P_{\mathrm{cl}}(\beta)\frac{(\beta + \beta^*)^2}{2}g^2t^2. \quad (4.11)$$

We see that the classical signal can produce vacuum or super-vacuum output noise $\Delta Y'^2 \geq 1/2$, but not sub-vacuum noise $\Delta Y'^2 < 1/2$. The analogy with Poisson and sub-Poisson statistics should be clear.

As a final detail, we note that these calculations were performed as if the two modes a, b have no time dependence other than the interaction, i.e., we are in the interaction picture. In quantum optics this is referred to as a rotating frame. In the lab frame, we also have sinusoidal time dependence $a \sim ae^{i\omega t}$ and so forth, where ω is the frequency of the relevant mode, which causes the squeezed quadrature to oscillate between X and Y (or h and π_h). In the right panel of Fig. 2, we show how this can lead to an oscillating variance in the detector output. The variance oscillates between sub-vacuum and super-vacuum values. The conclusion is the same: the sub-vacuum parts require a quantum signal mode to explain, but the super-vacuum parts can be explained by a purely classical signal.

5 LIGO and Squeezed Gravitons

Finally, let's explicitly connect our simple models above to the suggestion in [14, 15] that a highly squeezed gravitational signal could produce a large amount of "quantum noise". It is true that a squeezed state produces a lot of extra noise (i.e., the $\Delta\pi_h > 1/2$ above, which oscillates into the detector signal as in Fig. 2). However, the majority of this can be explained with a classical random signal, as we just discussed. It is only the sub-vacuum part that requires a quantum explanation. But this part is not observable in a real detector. We now make this precise by working out some numbers for LIGO. Despite the non-observability, the calculation is interesting, because it shows how the sub-vacuum graviton noise can be in principle transduced into a macroscopic detector.

In continuous linear detectors, the noise is characterized by the noise power spectral density (PSD),

$$S_{hh}(\nu) = \int dt\, e^{-i\nu t} \langle h(t)h(0)\rangle\,. \tag{5.1}$$

The actual observed noise power has a component both from the signal (i.e., the gravitational field) as well as noise due to the detector itself, in particular noise from the quantum readout as in both the photodiode and amplitude detectors discussed above. We can express this in terms of the optical mode's variables $X = (a + a^\dagger)/\sqrt{2}$ and $Y = -i(a - a^\dagger)/\sqrt{2}$. In the usual discussion of LIGO-like detectors [21], one works in a coordinate frame where a suspended mirror with center-of-mass variables x, p couples to this optical mode through an optomechanical coupling $V = GxX$. A gravitational wave in this frame acts as an effective force $F^{\text{sig}} \sim mL\ddot{h}$, where L is the baseline of the interferometer. Gravitational waves are detected by the signal driving the mirror motion, which in turn drives the optical field, which is then measured, say in the Y basis. The noise power spectral density of this readout is given by an expression like

$$S_{YY}^{\text{out}} = |\chi_{YY}|^2 S_{YY}^{\text{in}} + |\chi_{YX}|^2 S_{XX}^{\text{in}} + \cdots\,, \tag{5.2}$$

where the right-hand side contains a bunch of "input" noise PSDs S_{ij} and response functions χ_{ij}. These come from the input laser fluctuations, thermal forces on the mirror, and also include the gravitational noise of interest here. The question is: can a squeezed (or otherwise quantum) gravitational signal *reduce* the detector output $S_{YY}^{\text{out,quantum}}$ below the minimum possible noise $S_{YY}^{\text{out,classical}}$ which is consistent with a classical random gravitational signal?

To analyze this, we need to compute the noise PSD S_{YY}^{out}. This can be done using a standard method called the input-output formalism [21, 22], which essentially amounts to solving the linear equations of motion of the detector and assuming the level of noise injected by the various noise sources. In the input-output formalism, the phase and amplitude of the light coming out of the cavity is related to that coming in through the input-output relations

$$X_{\text{out}} = X_{\text{in}} + \sqrt{\kappa}X, \quad Y_{\text{out}} = Y_{\text{in}} + \sqrt{\kappa}Y. \tag{5.3}$$

To use these, we need to solve the equations of motion for the cavity light, i.e., for X and Y. The Heisenberg equations for the mirror motion take the form

$$\begin{aligned} \dot{x} &= m\omega^2 x \\ \dot{p} &= -\gamma_m p - \gamma_h p + F^{\text{in}} + F^{\text{sig}} + GX, \end{aligned} \tag{5.4}$$

The last term gives the coupling of the mirror motion to the optical readout. The mirror motion is harmonic with mass $m \approx 40$ kg and frequency $\omega \approx 1$ Hz. The coefficients γ_m and γ_h represent damping of the mirror respectively from coupling to any ambient thermalizing bath like gas (which exerts the random thermal force F^{in}), and damping of the mirror from coupling to the gravitational field (through F^{sig}). Both of these terms are usually neglected in LIGO; the former because it is subdominant to the noise coming from the coupling to X, and the latter because it is astronomically small. However, loss of information via γ_h will play a key role in what follows: it is the analogue to the unmonitored port in the beamsplitter model of Sect. 4.

In addition to the equations of motion for the mirror, we have the equations of motion for the optical field in the cavity,

$$\begin{aligned} \dot{X} &= -\frac{\kappa}{2}X + \sqrt{\kappa}X_{\text{in}} \\ \dot{Y} &= -\frac{\kappa}{2}Y + \sqrt{\kappa}Y_{\text{in}} + Gx. \end{aligned} \tag{5.5}$$

Here κ is the rate at which photons leak into and out of the cavity, i.e., the coupling to the optical modes outside of the cavity, in particular the laser field. We are in the frame co-rotating with the cavity and assume zero detuning between the laser and cavity mode.

The equations of motion can be solved easily by moving to the frequency domain with $Y(t) = \int d\nu e^{i\nu t} Y(\nu)$ and so forth. Inserting the solutions into the input-output relations, we obtain the following expression for the phase of the light coming out:

$$Y_{\text{out}} = \chi_{YY} Y_{\text{in}} + \chi_{YX} X_{\text{in}} + \chi_{YF} F_{\text{in}} + \chi_{Yh} h_{\text{in}}, \tag{5.6}$$

where the response functions are

$$\chi_{YY} = \frac{-i\nu - \kappa/2}{-i\nu + \kappa/2} = e^{i\phi}, \quad \chi_{YX} = \kappa G^2 \chi_c^2 \chi_m$$
$$\chi_{YF} = \sqrt{\kappa} G \chi_c \chi_m, \quad \chi_{Yh} = \sqrt{\kappa} G \chi_c \chi_m m^2 L^2 \nu^4, \tag{5.7}$$

in terms of the basic cavity and mechanical response functions

$$\chi_c(\nu) = \frac{1}{-i\nu + \kappa/2}, \quad \chi_m(\nu) = \frac{1}{m[(\omega^2 - \nu^2) - i\gamma\nu]}. \tag{5.8}$$

The total mechanical damping is the sum of the usual thermal damping term as well as the damping from loss into the gravitational field itself

$$\gamma = \gamma_m + \gamma_h. \tag{5.9}$$

In practice, in LIGO, the thermal noise effects (represented by $F_{\rm in}$ and γ_m) are subdominant to the quantum noise (coming from the $Y_{\rm in}$ and $X_{\rm in}$ terms). All of these are of course hugely dominant over the gravitational noise in usual signals, but here we are interested in an extreme case where the gravitational noise is important.

Finally, we can use these solutions to compute the detector's noise PSD. This can be easily obtained from Eq. (5.6) by what is called the Weiner-Khinchin theorem, which essentially says that $S_{xx}(\nu) \propto \langle |x(\nu)|^2 \rangle$ for any observable x (see [21] for a detailed treatment). Here this gives Eq. (5.2), i.e.,

$$S_{YY}^{\rm out} = |\chi_{YY}|^2 S_{YY}^{\rm in} + |\chi_{YX}|^2 S_{XX}^{\rm in} + |\chi_{YF}|^2 S_{FF}^{\rm in} + |\chi_{Yh}|^2 S_{hh}^{\rm in}. \tag{5.10}$$

The input noise from the laser can be modeled as simple vacuum fluctuations, where

$$S_{YY}^{\rm in} = S_{XX}^{\rm in} = 1/2. \tag{5.11}$$

This $1/2$ is exactly the vacuum-state variance of the harmonic oscillator variables used in the simple examples in the main text. In LIGO currently the situation is slightly more sophisticated since squeezed vacuum is inserted [20], but this is not going to change anything we are about to say, so for simplicity we will stick with standard vacuum input. The thermal noise is modeled as white noise

$$S_{FF}^{\rm in} = 2(n_T(\omega) + 1) m\omega\gamma_m, \tag{5.12}$$

where $n_T(\omega) = 1/(1 - e^{\omega/T})$ is the Bose-Einstein factor at the mechanical frequency and the $+1$ term represents vacuum fluctuations of the bath.

Now we want to compare the situation in which the gravitational field is purely classical or quantized. Following the main text, the idea is that for the quantized case, the detector output noise $S_{YY}^{\rm out}$ can be decreased below what is possible in the classical case, for example by transducing sub-vacuum squeezed noise from h into

Y. There are two main differences between the quantum and classical radiation field treatments. One is that in the classical case, the input noise $S_{hh}^{\text{in}} = 0$ is possible, whereas in the quantized case there is always vacuum noise. The other is that in the classical case, there can be zero damping $\gamma_h = 0$, for the same reason: unless h is quantized, there is no amplitude for the mirror to spontaneously dump energy into the field. These effects are related, as encoded in the fluctuation-dissipation theorem evaluated at zero temperature, which here reads [22, 23]

$$\gamma_h = x_0^2 m^2 L^2 \omega^4 \left[S_{hh}^{\text{in}}(\omega) - S_{hh}^{\text{in}}(-\omega) \right], \tag{5.13}$$

where $x_0 = 1/\sqrt{2m\omega}$ is the zero-point uncertainty of the mirror position. For classical h, the right-hand side is strictly equal to zero (the PSD is symmetric), while in the quantized case it equals 1, from vacuum fluctuations (i.e., from the term where you commute $[b, b^\dagger] = 1$). Using these two observations, we can compute the ratio of the detector's output noise in the classical and quantum cases:

$$\frac{S_{YY}^{\text{out,quantum}}}{S_{YY}^{\text{out,classical}}} = \frac{1 + |\chi_{YX}|^2 + 2|\chi_{YF}|^2 S_{FF}^{\text{in}} + 2|\chi_{Yh}|^2 e^{-2r} S_{hh}^{\text{vac}}}{1 + |\chi_{YX}|^2_{\gamma_h=0} + 2|\chi_{YF}|^2_{\gamma_h=0} S_{FF}^{\text{in}}}. \tag{5.14}$$

Here we used the vacuum laser values from Eq. (5.11), assumed a general squeezed graviton state with squeezing r, and we are specializing to the case of noiseless classical fluctuations in the denominator. The physics is directly analogous to our discussion of a quantum or classical local oscillator beamsplitter around Eqs. (4.2) and (4.9), where here the mechanical element interacting with the gravitational wave is playing the role of the beamsplitter.

We can get a very transparent approximate answer out of Eq. (5.14). Let's focus on LIGO, which observes at frequencies around $\nu_* \approx 1$ kHz. This scale is much larger than the mechanical resonance and cavity loss rates $\nu_* \gg \omega, \kappa$ which considerably simplifies the response functions in Eq. (5.7). Also, the photon-mirror coupling $G \propto \sqrt{P_L}$ depends on the input laser power and is therefore tuneable; LIGO chooses to tune it to $G^2 = [2\kappa|\chi_c(\nu_*)|^2|\chi_m(\nu^*)|]^{-1}$. This has the effect of minimizing the total quantum noise at ν_* (achieving the "Standard Quantum Limit" at ν_*), and sets $|\chi_{YX}|^2 = 1$.[2] Put together this simplifies the output noise to

$$S_{YY}^{\text{out}} = 1 + |\chi_m| S_{FF}^{\text{in}} + |\chi_m| m^2 L^2 \nu^4 S_{hh}^{\text{in}}, \tag{5.15}$$

in either the classical or quantum case, where here we are focusing on the noise PSD evaluated at the signal band $\nu_* \approx 1$ kHz. Now the key is that $|\chi_m|$ depends on the damping parameter γ_h, since it involves the response of the mirror to external forces [see Eq. (5.7)]. A simple Taylor expansion shows that it depends linearly on $\gamma_h \ll \gamma_m$,

[2] For the connoisseurs: doing this relies on knowing the value of γ. Here we are assuming that we know the full $\gamma = \gamma_m + \gamma_h$, for example through some calibration; if we instead used $\gamma = \gamma_m$ this would give a further $O(M_{\text{pl}}^{-2})$ correction.

$$|\chi_m| = |\chi_m|_0 \left(1 - \gamma_h \gamma_m m^2 \nu^2 |\chi_m|_0^2 + \cdots\right). \tag{5.16}$$

Using this and the scaling from the fluctuation-dissipation theorem [Eq. (5.13)], we can expand the ratio in Eq. (5.14) to lowest order in $1/M_{\rm pl}^2$. This gives

$$\frac{S_{YY}^{\rm out,quantum}}{S_{YY}^{\rm out,classical}} \approx 1 - \gamma_h \gamma_m m^2 \nu^2 |\chi_m|_0^3 S_{FF}^{\rm in} + |\chi_m|_0 m^2 L^2 \nu^4 e^{-2r} S_{hh}^{\rm vac}. \tag{5.17}$$

The basic structure of this answer is exactly the same as the simple beamsplitter case discussed in the main text in Eq. (4.8). The final term is an additional noise coming from the input graviton state. However, the second term shows a *loss* of noise at the same order; this is the low-noise graviton state being transduced into the mechanical state. Just like the beamsplitter example, however, the positive contribution can be made arbitrarily small with sufficient squeezing $r \to \infty$, while the reductive term is independent of r. Thus, a sufficiently squeezed graviton state can actually act to reduce the observed detector output below what is possible with a purely classical gravitational wave.

Unfortunately, it would be impossible to detect this signature of quantum gravity. To see it would require that the two correction terms here add up to some total which is negative and order one (which physically means that they subtract a meaningful fraction of the intrinsic detector quantum noise, represented by the 1 term). In the extreme case of a completely squeezed graviton state $r \to \infty$ we can entirely drop the positive term, and estimate the size of the noise reduction. For LIGO, this works out to be of order

$$\frac{S_{YY}^{\rm out,quantum}}{S_{YY}^{\rm out,classical}} \approx 1 - \gamma_h \gamma_m m^2 \nu^2 |\chi_m|_0^3 S_{FF}^{\rm in} + \cdots \approx 1 - 10^{-43}, \tag{5.18}$$

where we used $\nu = 1\,{\rm kHz}, m = 40\,{\rm kg}, L = 4\,{\rm km}, \gamma_m = 10^{-6}\,{\rm Hz}$ (a very conservative estimate; 10^{-8} is probably more accurate), the vacuum graviton PSD $S_{hh}^{\rm vac}(\nu) \approx \nu/M_{\rm pl}^2$ in the fluctuation-dissipation calculation of γ_h [see Eq. (5.13)], and set the mechanical bath to $T = 300$ K. The conclusion is clear: there is no chance of seeing this quantum-only signature in a real detector, even in the extreme case studied here where the gravitational wave is infinitely squeezed.

6 Comments

Given that all the fields of nature other than gravity are known to be quantized, perhaps the most natural guess is that gravity will be quantized similarly. However, given the difficulties encountered when trying to extrapolate this quantization to very high energies, it is worth seeking definitive experimental evidence that the field is quantized in the first place. A natural place to start looking is in the gravitational radiation field, now detectable by a variety of devices. Unfortunately, the considerations above suggest that any measurement of gravitational radiation with a realistic detector will be unable to distinguish classical gravitational waves from quantized ones. The remaining options appear to be tabletop experiments [24], which probe the quantization of the Newton potential (i.e., virtual graviton exchange), or measurements of the CMB [25].

Acknowledgements The scientific content of this paper is largely based on joint work with Valerie Domcke and Nick Rodd [1], and I am grateful to them for the wonderful collaboration we have had on this topic. This writeup is a slightly expanded version of a talk prepared for Gordon Semenoff's 70th birthday festschrift held at the Université de Montreal, July 2023. Gordon and I have been thinking about quantization effects in low-energy gravity since my very first paper on the topic [26]. I would like to congratulate him on his illustrious career, and thank him heartily for both our scientific collaboration and perhaps more importantly his mentorship over the years since we overlapped at UBC. My work at Berkeley Lab is supported by the U.S. DOE, Office of High Energy Physics, under Contract No. DEAC02-05CH11231, by DOE Quantum Information Science Enabled Discovery (QuantISED) for High Energy Physics grant KA2401032, and by the Heising-Simons Foundation.

References

1. D. Carney, V. Domcke, N.L. Rodd, Graviton detection and the quantization of gravity. Phys. Rev. D **109**(4), 044009 (2024). https://doi.org/10.1103/PhysRevD.109.044009, arXiv:2308.12988 [hep-th]
2. J.F. Donoghue, *Quantum General Relativity and Effective Field Theory* (2023). arXiv:2211.09902 [hep-th]
3. S. Boughn, T. Rothman, Aspects of graviton detection: graviton emission and absorption by atomic hydrogen. Class. Quant. Grav. **23** 5839–5852 (2006). https://doi.org/10.1088/0264-9381/23/20/006, arXiv:gr-qc/0605052
4. F. Dyson, Is a graviton detectable?. Int. J. Mod. Phys. A **28**, 1330041 (2013). https://doi.org/10.1142/S0217751X1330041X
5. A. Palessandro, Graviton-Photon Oscillations as a Probe of Quantum Gravity. arXiv:2405.01407 [gr-qc]
6. G. Tobar, S.K. Manikandan, T. Beitel, I. Pikovski, Detecting single gravitons with quantum sensing. arXiv:2308.15440 [quant-ph]
7. V. Shenderov, M. Suppiah, T. Beitel, G. Tobar, S.K. Manikandan, I. Pikovski, Stimulated absorption of single gravitons: first light on quantum gravity. arXiv:2407.11929 [gr-qc]
8. D. Kafri, J.M. Taylor, G.J. Milburn, A classical channel model for gravitational decoherence. New J. Phys. **16**, 065020 (2014). https://doi.org/10.1088/1367-2630/16/6/065020, arXiv:1401.0946 [quant-ph]
9. A. Tilloy, Binding quantum matter and space-time, without romanticism. Found. Phys. **48**(12), 1753–1769 (2018). https://doi.org/10.1007/s10701-018-0224-6, arXiv:1802.03291 [physics.hist-ph]
10. J. Oppenheim, Z. Weller-Davies, Covariant path integrals for quantum fields back-reacting on classical space-time. arXiv:2302.07283 [gr-qc]
11. R.J. Glauber, Coherent and incoherent states of the radiation field. Phys. Rev. **131**, 2766–2788 (1963). https://doi.org/10.1103/PhysRev.131.2766
12. E. Sudarshan, Equivalence of semiclassical and quantum mechanical descriptions of statistical light beams. Phys. Rev. Lett. **10**(7), 277 (1963)

13. L. Mandel, E. Wolf, *Optical Coherence and Quantum Optics* (Cambridge University Press, 1995)
14. T. Guerreiro, Quantum effects in gravity waves. Class. Quant. Grav. **37**(15), 155001 (2020). https://doi.org/10.1088/1361-6382/ab9d5d, arXiv:1911.11593 [quant-ph]
15. M. Parikh, F. Wilczek, G. Zahariade, Signatures of the quantization of gravity at gravitational wave detectors. Phys. Rev. D **104**(4), 046021 (2021). https://doi.org/10.1103/PhysRevD.104.046021, arXiv:2010.08208 [hep-th]
16. A. Einstein, Über einen die erzeugung und verwandlung des lichtes betreffenden heuristischen gesichtspunkt. Annalen der Physik **322**(6), (1905) 132–148. https://doi.org/10.1002/andp.19053220607
17. C. Mehta, Diagonal coherent-state representation of quantum operators. Phys. Rev. Lett. **18**(18), 752 (1967)
18. R. Loudon, P.L. Knight, Squeezed light. J. Modern Opt. **34**(6–7), 709–759 (1987)
19. R. Short, L. Mandel, Observation of sub-poissonian photon statistics. Phys. Rev. Lett. **51**, 384–387 (1983). https://doi.org/10.1103/PhysRevLett.51.384
20. W. Jia et al., LIGO operates with quantum noise below the Standard Quantum Limit. arXiv:2404.14569 [gr-qc]
21. J. Beckey, D. Carney, G. Marocco, Quantum measurements in fundamental physics: a user's manual, arXiv:2311.07270 [hep-ph]
22. A.A. Clerk, M.H. Devoret, S.M. Girvin, F. Marquardt, R.J. Schoelkopf, Introduction to quantum noise, measurement, and amplification. Rev. Modern Phys. **82**(2), 1155–1208 (2010)
23. C. Whittle, L. McCuller, V. Sudhir, M. Evans, Unification of thermal and quantum noises in gravitational-wave detectors. Phys. Rev. Lett. **130**(24), 241401 (2023)
24. D. Carney, P.C.E. Stamp, J.M. Taylor, Tabletop experiments for quantum gravity: a user's manual. Class. Quant. Grav. **36**(3), 034001 (2019). https://doi.org/10.1088/1361-6382/aaf9ca, arXiv:1807.11494 [quant-ph]
25. D. Green, R.A. Porto, Signals of a quantum universe. Phys. Rev. Lett. **124**(25), 251302 (2020). https://doi.org/10.1103/PhysRevLett.124.251302, arXiv:2001.09149 [hep-th]
26. D. Carney, L. Chaurette, D. Neuenfeld, G.W. Semenoff, Infrared quantum information. Phys. Rev. Lett. **119**(18), 180502 (2017). https://doi.org/10.1103/PhysRevLett.119.180502, arXiv:1706.03782 [hep-th]

Phantom Fluid Cosmology *or* Ghosts for Gordon

James M. Cline

Abstract The quanta of phantom dark energy models are negative energy particles, whose maximum magnitude of energy $|E|$ must be less than a cutoff $\Lambda \lesssim 20\,\text{MeV}$. They are produced by spontaneous decay of the vacuum into phantoms plus normal particles. I review general cosmological constraints that have been derived from the effects of such phantom fluid production, and a possible application: the generation of boosted dark matter or radiation that could be directly detected. Recent excess events from the DAMIC experiment can be well-fit by such processes.

Keywords Dark energy · Dark matter

1 Happy Birthday, Gordon!

I never had the pleasure of collaborating with Gordon Semenoff, but in 2004 we co-organized a conference at the Banff International Research Station, called "New horizons in string cosmology." The conference photo, Fig. 1, has a peculiar feature, which I asked the audience of the present meeting whether they could identify. Namely, the four participants of this meeting who also attended the BIRS meeting are all grouped together in the bottom left-hand corner. This seems to be a violation of thermodynamics, or causality.[1]

For the present contribution, I tried to think of something that could be remotely of interest to Gordon, considering our rather different research paths, subsequent to our near overlap in the area of string cosmology. What follows is the best I was able to come up with. It is a model of a new kind of dark energy, derived from a rather old

[1] There was another peculiar feature which the audience equally failed to notice: I was wearing exactly the same clothes on both occasions, down to the glasses. It was a little difficult to see anything while giving the talk.

J. M. Cline (✉)
McGill University, Montréal, QC, Canada
e-mail: jcline@physics.mcgill.ca

Trottier Space Institute, Montréal, QC, Canada

M. Paranjape et al. (eds.), *Proceedings of Gravity, Strings and Fields*,
CRM Series in Mathematical Physics,
https://doi.org/10.1007/978-3-031-91266-5_3

Fig. 1 Participants of the 2004 BIRS workshop "New horizons in string cosmology."

one, phantom scalar fields. Unlike the old version, which relies upon the condensate of a scalar field with a wrong-sign kinetic term, our version assumes the condensate is absent, but the corresponding negative-energy particles, that are inevitably produced by the decay of the vacuum, are present [1]. I will start with a review of the original phantom dark energy models by way of motivation.

2 Phantom Dark Energy: A Brief History

I was at CERN in 2003 on my first sabbatical when WMAP (Wilkinson Microwave Anisotropy Probe) made its first data release. Amongst other things, these data led to constraints on the equation of state of the dark energy that drives the acceleration of the Universe [2], $w = p/\rho$, where p is the pressure and ρ the energy density, with allowed regions extending to values of $w < -1$, in violation of the dominant energy condition (DEC). Theorists who had been analyzing various combinations of cosmological data, including pre-WMAP data of the cosmic microwave background, had also found such evidence, for example Refs. [3, 4].

Even before these experimental hints, R. Caldwell had pointed out a simple way of violating the DEC in the context of scalar field theories [5]. Recall that a cosmological constant with stress-energy tensor $T^{\mu\nu} = \Lambda\,\mathrm{diag}(1, -1, -1, -1)$ saturates the DEC, while a homogeneous scalar field has equation of state

$$w = \frac{\frac{1}{2}\dot{\phi}^2 - V}{\frac{1}{2}\dot{\phi}^2 + V} \geq -1\,, \tag{1}$$

where the inequality holds assuming that $V > 0$. This interpolates between cosmological constant, in the limit $\dot{\phi} \to 0$, and $w = 1$ when $V \to 0$. Caldwell's idea was to consider a scalar field with a wrong-sign kinetic term, a "phantom" field, such that $\dot{\phi}^2 \to -\dot{\phi}^2$ in Eq. (1) (but V remains ≥ 0). Then $w < -1$ if $\dot{\phi}^2 < 2V$, and diverges to $-\infty$ as $\dot{\phi}^2 \to 2V$. This will lead to a "big rip" singularity within a finite time, in which the expansion rate of the universe diverges and gravitationally bound structures are ripped apart.[2] This is readily understood from the classical equation of motion of a homogeneous phantom field,

$$\ddot{\phi} + 2H\dot{\phi} = +V'(\phi)\,; \tag{2}$$

the field climbs *up* the potential instead of rolling down.

Currently there is no compelling evidence for violation of the DEC; the Particle Data Group gives $w = -1.028 \pm 0.031$ [6]. However, this has not stopped theorists from continuing to consider phantom dark energy. InSpire database gives 1931 results, showing a steady rate of publication over the last two decades. The vast majority of these papers consider the phantom as a classical field.

But phantoms are not exempt from quantum mechanics, and one should ask what is the quantum field theoretic consequence of a wrong sign kinetic term. The answer depends on the choice of sign for the $i\epsilon$ prescription in the propagator,

$$P(p) = \frac{-i}{p^2 - m^2 \pm i\epsilon}\,. \tag{3}$$

The positive sign leads to negative probability as in the case of Faddeev-Popov ghosts, whereas the negative sign corresponds to positive probability, but with negative energy particles propagating into the future. The latter choice is the lesser of the two evils, and so one must consider phantom particles to carry negative energy if they are to be taken seriously as quantum fields.

Reference [7] understood that the phase space of phantom particles must be cut off at some scale Λ to avoid divergent decay rates, since arbitrarily negative energy phantoms could otherwise appear in the final state. But it was Ref. [8] that pointed out the most stringent irreducible constraint, arising from decay of the vacuum into two phantoms plus two photons, mediated by gravity. The Feynman diagram is shown

[2] It is interesting that Caldwell's paper, although written in 1999, was not published until 2002, when CMB data were hinting at the possibility of $w < -1$.

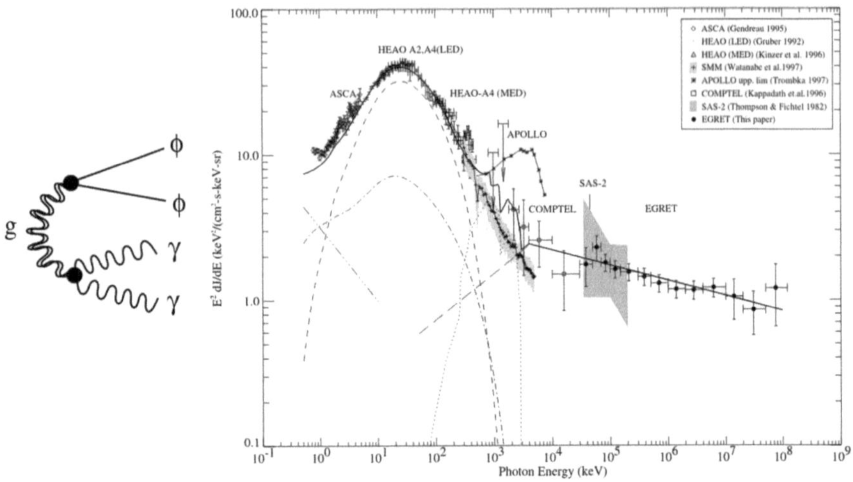

Fig. 2 Left: Feynman diagram for decay of the vacuum to two ghosts plus two photons. Right: schematic illustration of predicted x-ray spectrum (for two values of Λ, in red) versus cosmological diffuse photon background

in Fig. 2. On dimensional grounds, it leads to a constant decay rate per unit volume of order $\Gamma \sim \Lambda^8/m_P^4$, where m_p is the (reduced) Planck mass. The ensuing photon spectrum roughly takes the form

$$\frac{dn}{dE} \sim \frac{\Lambda^7 t_0}{m_p^4}\,\Theta(\Lambda - E)\,, \tag{4}$$

where t_0 is the age of the Universe. It rises steeply in amplitude with Λ, but only linearly in energy. The maximum value of Λ such that the prediction lies below the observed diffuse photon background was estimated to be ~3 MeV, which is rather low for an ultraviolet cutoff. Above this scale, the phantom description must break down and revert to a more normal-looking field theory.

3 Revisiting Phantom Fluids

Since 2003, there have been a few developments motivating a second look at the cosmological production of ghost particles. One is the Hubble tension: the discrepancy between the CMB determination of the Hubble constant, and that coming from more local measurements such as type Ia supernovae; see Ref. [9] for a review. My naive idea was to suppose that the vacuum could decay into massless ghosts plus dark matter. Due to the faster redshifting of the ghost energy density, there would be a net gain in dark matter as a function of time, which could explain why the Universe expands more rapidly at late times than expected. But this should not be a viable resolution, as it has been widely understood that late dark energy solutions do not work for the Hubble tension, while early dark energy may be able to do so. The current mechanism is late-time since the bulk of the accumulated energy is appearing today. The problem with the naive argument is that the appearance of the CMB depends upon the entire expansion history of the Universe, and not just that at early times.

Dark matter produced by vacuum decay to ghosts must be light, with a mass below the ~MeV cutoff. At the same time, it will be produced with an energy of order Λ, hence it is boosted to much higher speeds than ordinary DM, which virializes in the galaxy with speed $\sim 10^{-3}c$. This is reminiscent of another recent development, the interest in boosted light DM [10]. Such light DM particles could not be directly detected if moving slowly, being below the energy threshold of conventional detectors, but boosted DM is able to deposit sufficient energy. Interest in such models was further stimulated by excess events (now attributed to background) observed by the XENON1T experiment in 2010 [11]. These were the motivations for our recent work on the subject, Refs. [1, 12].

4 Revised Limit on Λ

One of our goals was to better quantify the limit on Λ from vacuum decay into ghosts plus photons, improving on the original order-of-magnitude estimate. First, we used the exact expression for the matrix element, which takes the rather aesthetic form $|\mathcal{M}|^2 = (tu/s)^2/8m_p^4$ [13]. Second, we computed the phase space integral accurately to obtain the full photon spectrum. Third, we accounted for reshifting of the spectrum using an appropriate Boltzmann equation, with the result shown in Fig. 3. Finally, we used more recent observational data from the COMPTEL experiment. The new limit on the cutoff is

$$\Lambda < 18\,\text{MeV}\,, \tag{5}$$

six times less stringent than the original estimate.

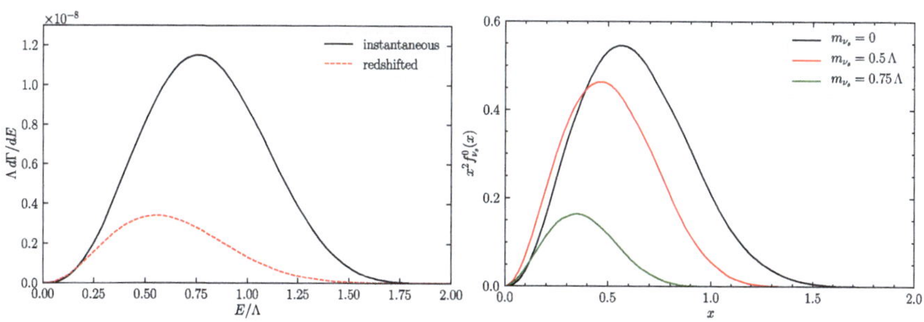

Fig. 3 Left: spectrum of photons from decay of the vacuum, to $\gamma\gamma\phi\phi$. Right: spectra of sterile neutrinos from vacuum decay to $\nu_s\bar{\nu}_s\phi\phi$

5 Constraints on Phantom Fluid Production

For definiteness, we consider a sterile neutrino ν_s as the light DM candidate coupled to phantoms, but similar results would hold for scalar DM. However the gravitational coupling, analogous to Fig. 2, turns out to be too weak (by some 9 orders of magnitude) to be cosmologically relevant. Therefore we consider the possibility that the decay to $\phi\phi\nu_s\bar{\nu_s}$ is mediated by a more strongly coupled intermediate particle, such as a heavy vector or scalar boson. For definiteness I will assume the vector case here (scalar exchange gives similar results); by integrating out the heavy particle, it can be described by an effective interaction,

$$\mathcal{L} \quad \ni \quad \frac{i}{M_v^2}(\phi^* \overleftrightarrow{\partial}_\mu \phi)\, \bar{\nu_s}\gamma^\mu \nu_s\,, \tag{6}$$

where the ghost is complex and the neutrino is Dirac, consistent with a U(1) gauge interaction. We will show that M_v as large as the GUT scale can lead to observable effects, if Λ saturates the bound (5).

For simplicity we take the phantoms to be massless. Then there are two qualitatively different regimes: ν_s is massive or (nearly) massless. The first case is straightforward since it doesn't require cosmological perturbation theory. Instead the main effects are via the time-dependent energy densities of the ghosts and ν_ss produced from vacuum decay. They are described by coupled Boltzmann equations,

$$\frac{d\rho_\phi}{dt} + 4H\rho_\phi = -\Gamma_\rho, \quad \frac{d\rho_{\nu_s}}{dt} + 3H(1+w_{\nu_s})\rho_{\nu_s} = +\Gamma_\rho\,, \tag{7}$$

where the rate of energy density production of ν_s is

$$\Gamma_\rho = \int \prod_i \frac{d^3 p_i}{(2\pi)^3 2|E_i|}(2\pi)^4 \delta^{(4)}\Big(\sum_i p_i^\mu\Big)|\mathcal{M}|^2 \cdot (E_{\nu_s} + E_{\bar{\nu_s}}) \tag{8}$$

with matrix element $|\mathcal{M}|^2 = (2/M_v^2)(s^2 - (t-u)^2)$, and the Hubble parameter is

$$H^2 = H_0^2 \times (\rho_{\rm rad} + \rho_m + \rho_\Lambda + \rho_\phi + \rho_{\nu_s})/\rho_c\,, \tag{9}$$

where ρ_c is the critical density. Because of the terms in red, the Boltzmann equations (7) are nonlinear, but in practice, their sum must remain relatively small to avoid undue modifications of standard cosmology, and then simple analytic approximate solutions can be found.

One can see that if ν_s is massless, $w_{\nu_s} = 1/3$, then the phantom fluid energy density $\rho_g \equiv \rho_\phi + \rho_{\nu_s} = 0$ is an exact, rather uninteresting solution. But if $m_{\nu_s} > 0$, the two species redshift differently, leading to net production of ρ_g. A potential complication is that w_{ν_s} could be time-dependent, since the ν_ss are being produced at a constant rate:

$$w_{\nu_s} = \frac{p_{\nu_s}}{\rho_{\nu_s}} = \frac{\int d^3p\, f_{\nu_s}(p,t)\, p^2/3E}{\int d^3p\, f_{\nu_s}(p,t)\, E} \sim \frac{1}{3} - \left(\frac{m_{\nu_s}}{\Lambda}\right)^2 + \cdots \qquad (10)$$

hence their distribution function f_{ν_s} is time-dependent. However we find that this dependence cancels nearly exactly in the ratio, so that w_{ν_s} is only a function of m_{ν_s}/Λ, which greatly simplifies the analysis.

In effect, ρ_g behaves as an exotic form of dark energy, different from that of the classical phantom condensate, with effective equation of state $w_{\text{eff}} = (p_\phi + p_{\nu_s})/(\rho_\phi + \rho_\nu)$ as shown in Fig. 4 (left). It violates the DEC, $-1.5 \leq w_{\text{eff}} \lesssim -1.3$. Hence, unlike matter, ρ_g grows with time, as shown in Fig. 4 (right).

For the massless ν_s case, no constraint arises at zeroth order in cosmological perturbation theory, even with arbitrarily fast growth of ρ_ϕ or ρ_{ν_s} individually, since they cancel exactly. However, their perturbations do not exactly cancel, since their distribution functions are different (Bose-Einstein versus Fermi-Dirac) and their phase space limits differ (only ghosts are restricted to have $|\mathbf{p}| < \Lambda$). Hence their first-order perturbations can grow and impact the integrated Sachs-Wolfe effect and lead to constraints from the cosmic microwave background (CMB).

For both massive and massless ν_s, we performed a Monte Carlo search of the parameter space using CosmoMC [14] to derive limits from combined cosmological data sets on the new physics parameter M_υ, as well as on the phenomenological quantity Ω_g (the present contribution of the ghost fluid to the cosmological density) as functions of the cutoff scale Λ. Ω_g cannot exceed 0.2, so disappointingly it is not a viable replacement for the cosmological constant $\Omega_\Lambda \gtrsim 0.5$, and could only provide a fraction of the total dark energy. The ensuing lower limits on M_υ versus Λ are shown in Fig. 5 for massive m_{ν_s}. For $m_{\nu_s} = 0$, the density of ν_s can grow to enormous values, $\Omega_{\nu_s} \sim 10^{25}$ since it is canceled by the ghosts. The limit is shown in Fig. 6 (left).

We investigated the extent to which the phantom fluid model can ameliorate the Hubble tension, as well as the S_8 tension (S_8 being the parameter describing the amplitude of fluctuations at the scale $8h^{-1}$ Mpc). As expected on general grounds,

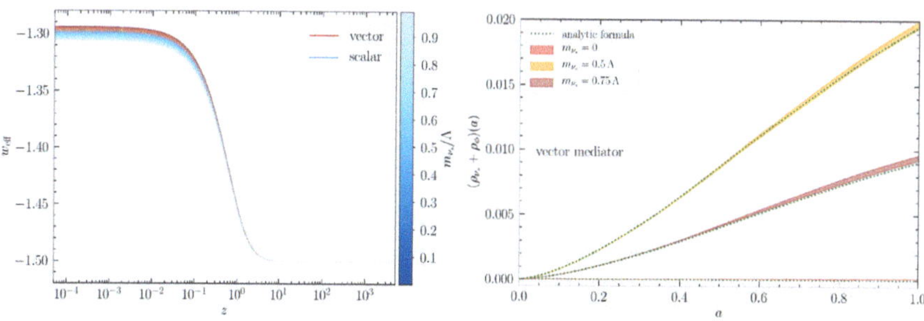

Fig. 4 Left: equation of state of the ghost fluid as a function of redshift. Band indicates variation with respect to m_{ν_s}/Λ for vector and scalar mediator models. Right: growth of the phantom fluid energy density with scale factor, for several m_{ν_s}/Λ values

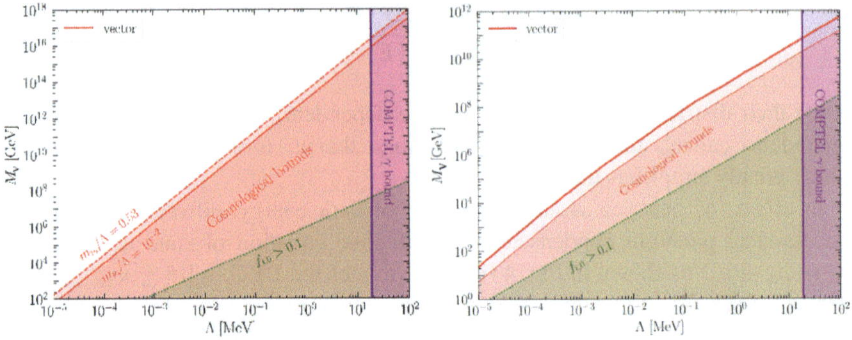

Fig. 5 Left: Lower limit on the new physics scale M_v versus Λ from phantom fluid production, for $m_{\nu_s} > 0$. Right: same for $m_{\nu_s} = 0$. Brown regions are excluded by the technical assumption that the fluids do not become degenerate (for which our analysis is not valid)

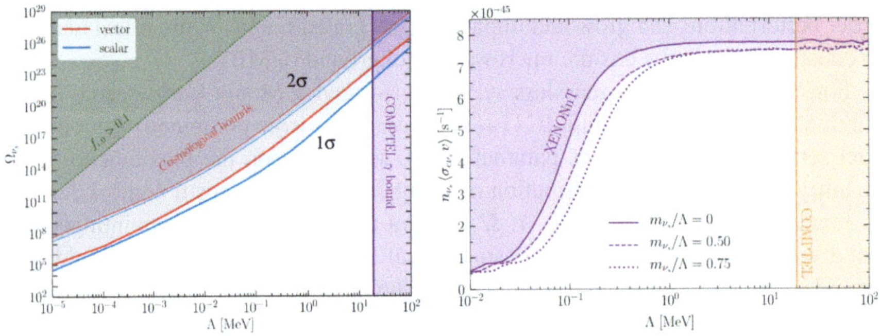

Fig. 6 Left: upper limit on Ω_{ν_s} versus Λ for the case of massless ν_s. Right: direct detection upper limit on cross section for ν_s-e scattering times ν_s density versus Λ, from XENONnT null results

this being a late-dark energy model, it does not provide a full resolution, but it reduces the discrepancy to the 3σ level. The main source of the tension is the baryon acoustic oscillation (BAO) data. If one omits these sets, excellent consistency can be obtained.

6 Direct Detection

With the possibility of such a huge exotic dark radiation component, one is led to wonder if it could have observable consequences. Suppose it was coupled to electrons in a similar manner to Eq. (6),

$$\mathcal{L} \ \ni \ \frac{i}{M'^2}(\bar{e}\gamma_\mu e)(\bar{\nu_s}\gamma^\mu \nu_s)\,, \tag{11}$$

where we allow for a different interaction strength M'. We choose electrons since recent experimental searches for light dark matter have been taking advantage of e-DM scattering: it is easier for a light particle to impart energy to electrons than to nuclei. The dark radiation of massless ν_s carries energy $\sim \Lambda$ which can be far greater than that of thermal radiation, and easily above the energy thresholds $\sim 20\,\text{eV}$ of detectors such as DAMIC [15]. In Ref. [12] we calculated the event rate for such interactions in xenon and silicon detectors. A model-independent constraint on the scattering cross section of ν_s on electrons times the ν_s density can be derived from the null results of the XENONnT experiment [16], which exclude the putative signal that was previously observed by XENON1T [11]. It is shown in Fig. 6 (right).

Translating the XENONnT constraint into model parameters, we find an upper bound on the product $M_\upsilon M'$, corresponding to the product $\sigma_{\nu_s e} n_{\nu_s}$. The signal turns out to be undetectably small for cosmologically allowed values if we assume $M' = M_\upsilon$; one needs $M' > M_\upsilon$ to saturate the bound. This is illustrated in Fig. 7 (left). The DAMIC experiment in SNOLAB has observed excess events that can be interpreted as scattering on electrons, with a spectrum shown in Fig. 7 (right).

The phantom fluid model has essentially two parameters relevant for direct detection: Λ, determining the shape of the spectrum and $\sqrt{M_\upsilon M'}$, determining its normalization; the value of m_{ν_s}/Λ has a relatively small effect. We find an excellent fit to the observed data by taking $\Lambda = 10\,\text{keV}$ and $\sqrt{M_\upsilon M'} = 1.3 \times 10^5\,\text{GeV}$, which is not far from the XENONnT lower limit. It must be raised to the 8th power to get $\sigma_{\nu_s e} n_{\nu_s}$, implying the signal is $1.3^8 \sim 8$ times lower than the current XENONnT sensitivity. There is a data bin at 0.05 keV that is essentially empty. We confirm that the predicted signal falls to negligible values in this bin, due to the binding energy of silicon preventing excitations.

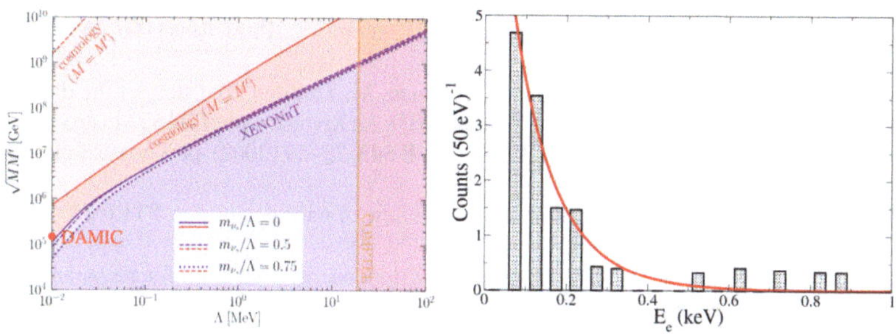

Fig. 7 Left: XENONnT limit on $\sqrt{M_\upsilon M'}$, compared to previous cosmological limits assuming $M' = M_\upsilon$. Right: massless ν_s fit to the DAMIC excess signal, with $\Lambda = 10\,\text{keV}$ and $\sqrt{M' M_\upsilon} = 1.3 \times 10^5\,\text{GeV}$

7 Conclusions

Phantom fluids through vacuum decay are an inevitable consequence of phantom dark energy models, although it is necessary to couple the ghosts more strongly than gravity to make them cosmologically relevant. If one does this, for (massless) ghosts and light dark matter of mass $\lesssim 18\,\text{MeV}$, it gives an exotic dark energy component with equation of state $-1.5 < w < -1.3$ that violates the dominant energy condition. Sadly, cosmological data prevents it from contributing more than $\Omega_g \sim 0.2$ to the total energy budget. Nevertheless, even a small amount will lead to a regularized big rip in the future—regularized, since the theory must revert to being normal at scales above the cutoff. $\Lambda \sim 10\,\text{keV}$ is sufficient to rip apart bound objects like the sun in a future epoch. Apart from these annoying features, phantom fluids can also do something useful by providing a new target for detection of boosted light dark matter or radiation.

This work was supported by NSERC (Natural Sciences and Engineering Research Council, Canada).

References

1. J.M. Cline, M. Puel, T. Toma, Q.S. Wang, Phantom fluid cosmology: impact of a phantom hidden sector on cosmological observables. Phys. Rev. D **108**(9), 095033 (2023). [arXiv:2308.12989 [hep-ph]]
2. D.N. Spergel et al., [WMAP], First year Wilkinson Microwave Anisotropy Probe (WMAP) observations: determination of cosmological parameters. Astrophys. J. Suppl. **148**, 175–194 (2003). [arXiv:astro-ph/0302209 [astro-ph]]
3. S. Hannestad, E. Mortsell, Probing the dark side: constraints on the dark energy equation of state from CMB, large scale structure and Type Ia supernovae. Phys. Rev. D **66**, 063508 (2002). [arXiv:astro-ph/0205096 [astro-ph]]
4. A. Melchiorri, L. Mersini-Houghton, C.J. Odman, M. Trodden, The state of the dark energy equation of state. Phys. Rev. D **68**, 043509 (2003). [arXiv:astro-ph/0211522 [astro-ph]]
5. R.R. Caldwell, A phantom menace?. Phys. Lett. B **545**, 23–29 (2002). [arXiv:astro-ph/9908168 [astro-ph]]
6. R.L. Workman et al., [Particle Data Group], Review of particle physics. PTEP **2022**, 083C01 (2022)
7. S.M. Carroll, M. Hoffman, M. Trodden, Can the dark energy equation-of-state parameter w be less than -1?. Phys. Rev. D **68**, 023509 (2003). [arXiv:astro-ph/0301273 [astro-ph]]
8. J.M. Cline, S. Jeon, G.D. Moore, The phantom menaced: constraints on low-energy effective ghosts. Phys. Rev. D **70**, 043543 (2004). [arXiv:hep-ph/0311312 [hep-ph]]
9. E. Di Valentino, O. Mena, S. Pan, L. Visinelli, W. Yang, A. Melchiorri, D.F. Mota, A.G. Riess, J. Silk, In the realm of the Hubble tension—a review of solutions. Class. Quant. Grav. **38**(15), 153001 (2021). [arXiv:2103.01183 [astro-ph.CO]]
10. K. Agashe, Y. Cui, L. Necib, J. Thaler, (In)direct detection of boosted dark matter. JCAP **10**, 062 (2014). [arXiv:1405.7370 [hep-ph]]
11. E. Aprile et al. [XENON], Excess electronic recoil events in XENON1T. Phys. Rev. D **102**(7), 072004 (2020). [arXiv:2006.09721 [hep-ex]]
12. J.M. Cline, M. Puel, T. Toma, Boosted dark matter from a phantom fluid. Phys. Lett. B **848**, 138377 (2024). [arXiv:2308.01333 [hep-ph]]

13. N. Bernal, M. Dutra, Y. Mambrini, K. Olive, M. Peloso, M. Pierre, Spin-2 portal dark matter. Phys. Rev. D **97**(11), 115020 (2018). [arXiv:1803.01866 [hep-ph]]
14. A. Lewis, S. Bridle, Cosmological parameters from CMB and other data: a Monte Carlo approach. Phys. Rev. D **66**, 103511 (2002). [arXiv:astro-ph/0205436 [astro-ph]]
15. A. Aguilar-Arevalo et al. [DAMIC, DAMIC-M and SENSEI], Confirmation of the spectral excess in DAMIC at SNOLAB with skipper CCDs. [arXiv:2306.01717 [astro-ph.CO]]
16. E. Aprile et al. [XENON], Search for new physics in electronic recoil data from XENONnT. Phys. Rev. Lett. **129**(16), 161805 (2022). [arXiv:2207.11330 [hep-ex]]

Superinsulators: The Discovery of Electric Confinement in Condensed Matter Systems

Maria Cristina Diamantini

Abstract Superinsulators are a new topological state of matter, predicted by our collaboration and experimentally observed in the critical vicinity of the superconductor-insulator transition (SIT). Superinsulators are dual to superconductors and realise electric-magnetic (S) duality. In superinsulators, Cooper pairs are linearly bound by electric fields squeezed into strings (dual Meissner effect) by a monopole condensate (instantons plasma in (2+1) dimensions), in analogy to quarks in hadrons. Superinsulators realise, thus, the electric version of the dual superconductor mechanism proposed to explain confinement in QCD.

Keywords Superinsulation · Monopoles · Confinement

1 Introduction

Quark confinement is one the most fascinating and still open problem of particle physics. Based on the analogy with the Meissner effect in type-II superconductors, 't Hooft [1] described quark confinement inside mesons as the formation of thin chromo-electric flux tubes with quarks at their ends, squeezed in a condensate of magnetic monopoles. 't Hooft identified this phase with the "extreme opposite" of a superconductor, a "superinsulator". In fact the zero quark mobility in this phase can be interpreted as an infinite chromo-electric resistance.

A crucial role in the dual superconductivity mechanism is played by chromomagnetic monopoles. It is not, however, straightforward to identify monopoles in QCD. To accommodate this scenario, 't Hooft proposed a partial gauge fixing, the Abelian projection, e.g. $SU(3) \to U(1) \times U(1)$, where Abelian magnetic monopoles are obtained by breaking the original gauge group to its maximal Abelian subgroup.

Superinsulators are the realization of Abelian electric confinement through the dual superconductivity mechanism in condensed matter systems: Cooper pairs are

M. C. Diamantini (✉)
NiPS Laboratory, INFN and Dipartimento di Fisica e Geologia, University of Perugia, Perugia, Italy
e-mail: cristina.diamantini@pg.infn.it

M. Paranjape et al. (eds.), *Proceedings of Gravity, Strings and Fields*,
CRM Series in Mathematical Physics,
https://doi.org/10.1007/978-3-031-91266-5_4

bound by electric flux tubes in a condensate of magnetic monopoles (instantons plasma in (2+1) dimensions). In this one-color version of QCD , Cooper pairs play the role of quarks. This new state of matter was originally predicted in Josephson junction arrays (JJA) [2] and then experimentally detected in InO superconducting films [3], Tin films [4, 5] and NbTiN films [6]. The final form of the theory was developed in [7].

Superinsulation appears on the insulating side of materials which harbour the SIT. At the SIT, the relevant degrees of freedom are Cooper pairs and vortices, so the phases that it harbours are determined by the competition between two quantum orders embodied in the topological Aharonov-Bohm/Aharonov-Casher (ABC) interactions between the two degrees of freedom. ABC interactions are non-local, but Wilczek [8] as shown that a local formulation is possible at the price of introducing two emergent gauge fields a_μ and b_μ (a tensor field $b_{\mu\nu}$ in (3+1) dimensions) coupled to the conserved charge and vortex currents, respectively. The emergent gauge fields are coupled through a mixed Chern-Simons (CS) term in (2+1) dimensions and a BF term in (3+1) dimensions.

Superinsulators are characterised by an infinite resistance due to linear confinement of charges [7], which persists at finite temperatures: charges cannot move even if a voltage (below a critical threshold) is applied. Like in QED, [9] the monopoles Bose condensate constricts electric fields into electric flux tubes that bind Cooper pairs and anti-Copper pairs. Magnetic monopoles, which are elusive as elementary particles, appear in these condensed matter systems as emergent quasiparticle excitations and play a crucial role in the formation and properties of the superinsulating state [10].

In what follows we will concentrate on the (2+1)-dimensional case and we will present the effective gauge theory description of the SIT, derive the phase diagram and show how the BKT transition arises. We then present experimental evidences of superinsulation. We will use natural units $c = 1, \hbar = 1$, $\epsilon_0 = 1$ but restore physical units when necessary.

2 Effective Gauge Theory in (2+1) Dimensions

In the vicinity of the SIT the relevant degrees of freedom are charges and vortices, whose world-lines are given, respectively, by:

$$
\begin{aligned}
Q_\mu &= \sum_{\rm i} \int_{{\rm x}_q^{(i)}} d\tau \frac{dx_{{\rm q}\mu}^{\rm (i)}(\tau)}{d\tau}\, \delta^3 \left(x - x_{\rm q}^{\rm (i)}(\tau)\right) , \\
M_\mu &= \sum_{\rm i} \int_{{\rm x}_m^{(i)}} d\tau \frac{dx_{{\rm m}\mu}^{\rm (i)}(\tau)}{d\tau}\, \delta^3 \left(x - x_{\rm m}^{(i)}(\tau)\right) ,
\end{aligned}
\qquad (1)
$$

where the index i labels the elementary charges and vortices, parametrized by the coordinates $x_{\rm q}^{(i)}$ and $x_{\rm m}^{(i)}$, respectively. Greek subscripts run over the Euclidean three dimensional space encompassing the 2D space coordinates and the Wick rotated time coordinate.

Charges and vortices are subject to the infinite-range topological ABC interaction, embodying the quantum phase acquired either by a charge encircling a vortex or by a vortex encircling a charge, which is given by the Gauss linking number between the two curves Eq. (1):

$$S_{\text{linking}} = i \int d^3x \, Q_\mu \epsilon_{\mu\alpha\nu} \frac{\partial_\alpha}{-\nabla^2} M_\nu \,, \tag{2}$$

where $\epsilon_{\mu\alpha\nu}$ is the completely antisymmetric tensor. As shown by Wilczek [8], Eq. (2) admits a local formulation at the price of introducing two emergent gauge fields, a_μ and b_μ interacting through a mixed CS term [11]:

$$S^{\text{CS}} = \int d^3x \left[i \frac{1}{2\pi} a_\mu \epsilon_{\mu\alpha\nu} \partial_\alpha b_\nu + i a_\mu Q_\mu + i b_\mu M_\mu \right] . \tag{3}$$

The CS term contains only one field derivative and represents, thus, the dominant contribution to the action at long distances. It is invariant under the gauge transformations $a_\mu \to a_\mu + \partial_\mu \lambda$ and $b_\mu \to b_\mu + \partial_\mu \chi$.

The CS kernel has a zero mode [11] and needs a regularisation. To this end we will use the next-order terms in the effective action of the SIT that contain two field derivatives and are gauge invariant. Introducing the dual field strengths $f_\mu = \epsilon_{\mu\alpha\nu} \partial_\alpha b_\nu$ and $g_\mu = \epsilon_{\mu\alpha\nu} \partial_\alpha a_\nu$ we obtain the action

$$S_{2\text{D}} = \int d^3x \, i \frac{1}{2\pi} a_\mu \epsilon_{\mu\alpha\nu} \partial_\alpha b_\nu + \frac{1}{2e_v^2 \mu} f_0^2 + \frac{\varepsilon}{2e_v^2} f_i^2 + \frac{1}{2e_q^2 \mu} g_0^2 + \frac{\varepsilon}{2e_q^2} g_i^2 + i a_\mu Q_\mu + i b_\mu M_\mu \,, \tag{4}$$

where f_0 and g_0 are the magnetic fields, f_i and g_i the electric fields μ is the magnetic permeability and ε is the electric permittivity which define the speed of light $v = 1/\sqrt{\mu\varepsilon}$ in the material. Note that the introduction of these terms is required in order to correctly define the effective theory which describes this physical system and they will be crucial to determine the phase structure of the theory. In fact, although the two coupling constants e_q^2 and e_v^2 have canonical dimension [1/length] and the two kinetic terms are, naively, infrared-irrelevant, they are necessary to correctly define the pure CS limit in which the topological mass $m = e_q e_v / 2\pi v \to \infty$ [12, 13]. This is due to the fact that, with the two energy scales e_q^2 and e_v^2 we can define a dimensionless coupling constant $g = e_v / e_q$. The electric-magnetic duality (charge–vortex symmetry) is given by the action symmetry with respect to the transformation $g \equiv e_v / e_q \leftrightarrow 1/g$. Thus, g is a tuning parameter driving the system across the SIT, and the SIT itself corresponds to $g = g_c = 1$. To make the connection with the films parameters we notice that $g = O(d/(\alpha \lambda_L))$ where λ_L is the London penetration depth in the bulk material, d is the thickness of the film and $\alpha = e^2/4\pi$ is the fine structure constant.

The two emergent gauge fields must be compact in order to describe the Gauss linking number. A way to formulate compact U(1) gauge symmetries is to use a lattice regularisation introducing a lattice of spacing ℓ. In what follows, we will not present the details of the lattice calculation that can be found in [2], but we will review the main results.

In the lattice formulation Q_μ and M_μ represent integer point charges and vortices and play the role of topological excitations due to the compactness of the gauge symmetries. Integrating out the fictitious gauge fields we obtain an action for the topological excitations alone. Following the standard lattice gauge field theory arguments of [14] we retain only the self-interaction terms between the topological excitations, obtaining:

$$S_{\text{top}} = 2\pi m\ell v G(m\ell v) \left[\frac{e_q}{e_v} \, Q^2 + \frac{e_v}{e_q} \, M^2 \right] N \, , \tag{5}$$

where we consider strings made of N bonds with integer electric and magnetic quantum numbers Q and M, and $G(m\ell v)$ a lattice parameter [2]. The entropy of a string is proportional to its length, and is given by μN with $\mu \approx \ln(5)$ since, at each step, the non backtracking string can choose among 5 possible directions on how to continue. The main contribution to the free energy is thus:

$$F = 2\pi m\ell v G(m\ell v) \left[\frac{e_q}{e_v} \, Q^2 + \frac{e_v}{e_q} \, M^2 - \frac{1}{\eta} \right] N \, , \tag{6}$$

with the dimensionless parameter η given by:

$$\eta = \frac{2\pi m\ell v G(m\ell v)}{\mu} \, . \tag{7}$$

η and $g = e_v/e_q$ determine the condensation (or lack thereof) of topological defects, and, thus, the quantum phase structure. When the energy term in (6) dominates, the free energy is positive and minimized by short closed loop configurations while, when the entropy dominates, the free energy is negative and minimised by large strings and long closed loops giving the following condensation conditions for long strings with integer quantum numbers Q and M:

$$\eta \frac{e_q}{e_v} Q^2 + \eta \frac{e_v}{e_q} M^2 < 1 \, . \tag{8}$$

If two or more condensations are allowed, one has to choose the one with the lowest free energy. This condition describes the interior of an ellipse with semi-axes

$$\begin{aligned} r_Q &= \sqrt{\tfrac{e_v}{e_q}} \sqrt{\tfrac{1}{\eta}} \, , \\ r_M &= \sqrt{\tfrac{e_q}{e_v}} \sqrt{\tfrac{1}{\eta}} \, , \end{aligned} \tag{9}$$

on a square lattice of integer electric and magnetic charges. The ratio $g = e_v/e_q$ determines the ratio of the semi-axes while the parameter η sets the overall scale of the ellipse. The quantum phase diagram is found by noting which integer charges lie within the ellipse when the semi-axes and the overall scale are varied:

$$\eta < 1 \rightarrow \begin{cases} g < 1 \text{ , electric condensation } = \text{ type III superconductor ,} \\ g > 1 \text{ , magnetic condensation } = \text{ superinsulator ,} \end{cases}$$

$$\eta > 1 \rightarrow \begin{cases} g < \frac{1}{\eta} \text{ , electric condensation } = \text{ type III superconductor ,} \\ \frac{1}{\eta} < g < \eta \text{ , no condensation } = \text{ bosonic topological insulator/Bose metal ,} \\ g > \eta \text{ , magnetic condensation } = \text{ superinsulator .} \end{cases}$$

For $\eta > 1$, at the SIT, an intermediate phase, in which both Q_μ and M_μ are diluted, appears with an effective action given by:

$$S = \sum_x i2\pi \; a_\mu K_{\mu\nu} b_\nu \; , \tag{10}$$

which describes a bosonic topological insulator. The existence of this phase was first predicted in [2].

This topological state is often referred to as a Bose metal because it hosts symmetry-protected metallic edge states [15]. It forms due to the same competition of two quantum orders, the charge condensate and the vortex condensate that leads to the universal SIT point. This intermediate phase is separated by two quantum BKT transition from the superinsulating phase for $g < 1$ and from the superconducting phase for $g > 1$ [15, 16].

The superconducting phase is the phase in which the electric topological defects proliferate, and it is this new type of superconductivity that we called type III [17]. It is characterized by a topological gap and superconductivity is lost through a BKT transition both at $T = 0$ and at finite T, as discussed in the introduction.

For $\eta < 1$, instead, when g lies between η and $1/\eta$ there is coexistence of electric and magnetic excitations indicating a direct first order phase transition between the superinsulator and the superconductor.

3 Superinsulators in (2+1) Dimensions

In this section we will use the lattice notion. We will consider a square lattice of lattice spacing l.

In the superinsulating state charges are diluted [2] and we set $Q_\mu = 0$. To compute the electromagnetic response of an ensemble of charges in a superinsulator we couple the charge current j_μ to the physical electromagnetic gauge field A_μ by adding to the action the minimal coupling term $2eA_\mu j_\mu$. We, then, integrate out the gauge fields a_μ and b_μ, and we sum over the condensed vortices M_μ. The effective action $S_{\text{eff}}(A_\mu)$ is:

$$S_{\text{top}}\left(M_\mu, A_\mu\right) = \sum_{x,i} \frac{1}{2e_{\text{eff}}^2} \left(\mathcal{F}_i + 2\pi M_i\right)^2 \,, \tag{11}$$

where e_{eff}^2 is the effective coupling constant

$$e_{\text{eff}}^2 = \frac{2\pi^2}{\mu} \frac{1}{\eta g} = e^2 \frac{\pi}{2\mu\eta} \frac{\lambda_L}{d} = e^2 O\left(\frac{\lambda_L}{d}\right) . \tag{12}$$

In Eq. (11) $\mathcal{F}_i$ is the electric field and we recognize, thus, the deep non-relativistic version of Polyakov's compact QED action [9] in which only the electric fields survive. We now notice that the gauge invariance of the b_μ fields in Eq. (4) enforces the constraint $d_\mu M_\mu = 0$. This condition is satisfied by choosing $d_i M_i = m$ and consequently $d_t M_0 = -m$. The m represent the monopoles, instanton quantum tunnelling events in (2+1) dimensions in which vortices on the film appear and disappear and their magnetic flux flows in and out isotropically in the four available spatial directions. The partition function we obtain is:

$$Z = Z_0 \cdot Z_{\text{inst.}} = \int_{-\infty}^{+\infty} \mathcal{D}\mathcal{A}_\mu \, \mathrm{e}^{-\frac{1}{2e_{\text{eff}}^2} \sum_{x,i} \mathcal{F}_i^2} \cdot \sum_{\{m\}} \mathrm{e}^{-\frac{2\pi^2}{e_{\text{eff}}^2} \sum_x m \frac{1}{-\nabla_2^2} m} \,, \tag{13}$$

where ∇_2^2 is the spatial lattice Laplacian and the interaction of the monopoles near the SIT is is logarithmic, $(e_{\text{eff}}^2/2\pi)\ln|\mathbf{x}|$ contrary to the inverse linear power of the relativistic model.

Another important effect of instantons is that they disorder the system and generate a mass for the photon given by [9]

$$m_\gamma = \frac{8\pi^2}{e_{\text{eff}}^2} z \,, \tag{14}$$

rendering thus the Coulomb potential screened with a a screening length $\lambda_{\text{el}} = 1/m_\gamma$.

To probe Cooper pair confinement we compute the expectation value of the Wilson loop operator $W(C)$, where C is a closed loop in 3D Euclidean space-time (a factor ℓ is absorbed into the gauge field A_μ to make it dimensionless),

$$\langle W(C) \rangle = \frac{1}{Z_{A_\mu, M_i}} \sum_{\{M_i\}} \int_{-\pi}^{+\pi} \mathcal{D}A_\mu \, \mathrm{e}^{-\frac{1}{2e_{eff}^2} \sum_x (\mathcal{F}_i - 2\pi M_i)^2} \mathrm{e}^{i q_{\text{ext}} \sum_C A_\mu} \,. \tag{15}$$

In [7] we have shown that, for $q_{\text{ext}} = 1$, i.e., Cooper pair probes, the sum over instantons gives rise to an area law for the expectation value of the Wilson loop (15) with a string tension given by [18]

$$\sigma = \frac{\sqrt{8}}{\ell_0 \ell} \frac{e_{\text{eff}}}{\pi} \mathrm{e}^{-\frac{\pi^2}{e_{\text{eff}}^2} G_2(0)} \,, \tag{16}$$

where $G_2(0)$ is now the infrared-regularized 2D lattice Coulomb potential at coinciding points. We have, thus, the formation of an electric string of typical width λ_{el} [19] and typical length $d_s = 1/\sqrt{\sigma}$ which bounds together Cooper pairs and Cooper holes in an "electric pion". When one pulls this string by, say, an external voltage, Cooper pairs and Cooper holes start moving apart but there comes a moment where it becomes energetically favourable for the system to pop out a Cooper pair-Cooper hole pair in some intermediate island and to form two short strings. Only neutral states exist asymptotically in this phase of the system and the resistance becomes infinite since charges cannot move anymore.

The logarithmic interactions between monopoles have a very important consequence: contrary to relativistic compact QED, in which monopoles are always in a plasma phase, in superinsulators near the SIT they can undergo a confining quantum BKT transition [20, 21] for sufficiently strong coupling constants g. Summing over m, the partition function Eq. (13) can be rewritten as

$$Z_{\text{inst}} = \int_{-\infty}^{+\infty} \mathcal{D}\chi \; e^{-\sum_{\mathbf{x},i} \frac{e_{eff}^2}{8\pi^2}(\Delta_i \chi_{\mathbf{x}})^2 + 2z(1-\cos(\chi_{\mathbf{x}}))} , \tag{17}$$

which is nothing else than the partition function of the sine-Gordon model (for a review see [22]) which describes the physics of the planar XY model. The parameter z, given by:

$$z = e^{\frac{-2\pi^2 G(0)}{e_{eff}^2}} , \tag{18}$$

is the instanton fugacity. Following the results for the XY model [22] we have that, for low values of g, instantons are free and charges are confined, while instantons undergo a confining transition and become logarithmically confined at $g = g_{\text{cr}} = (4\pi/\mu)(1/\eta)$. This quantum BKT transition represents the SIT itself, with a transition between the superinsulating phase and the intermediate Bose metal, the bosonic topological insulator phase.

4 Experimental Evidence of Superinsulation

Using the renormalisation group flow properties of the sine-Gordon model [22], we see that the confining interaction decreases when flowing towards short scales showing what is called *asymptotic freedom*. This phenomenon is typically associated with non-Abelian gauge theories, where it characterises their UV fixed point . Here it is associated with the sine-Gordon model (and not the compact QED) and describes an IR fixed point so it should be called *asymptotic safety*, but here will use the more familiar term asymptotic freedom. Wath is the experimental signature of this phenomenon in superinsulating materials? While it is impossible to "look inside hadrons" to study the UV to IR confining transition, it is, instead, possible in superinsulators to look inside the "electric pion", since the electric interaction is much weaker than the strong interaction. In superinsulators, if the string length scale d_s is large enough so that $\lambda_{el} < d_s$, we can probe the systems on distances r such that $\lambda_{el} < r < d_s$. In this regime, which corresponds to samples of size L in the range $\lambda_{el} < L < d_s$, Cooper pairs do not feel the string tension, since the string is loose on these scales, and the Coulomb interaction is screened on the scale λ_{el}. We expect thus to observe a transition from hyperactivated resistance behaviour to a metallic saturation at the lowest temperatures when the sample size is decreased. This is exactly what has been observed in a NbTiN superinsulating film by varying the bridge length on which the external voltage is applied [23], as shown in Fig. 1. For large bridge lengths the film displays hyperactivated resistances, for the smallest bridge length 0.2 mm, however, we see metallic saturated behaviour at low temperatures. The crossover from hyperactivation to metallic behaviour should take place around a bridge length $L \approx d_s$.

In [23] we also derived and measured the I(V) characteristics of a superinsulator as shown in Fig. 2. Without going into details, let us try to understand the results. A string of length d_s has the energy $\sqrt{v\sigma}$. Hence the energy to form a string chain spanning the entire system is $2eV_{c1} = \sigma L$ where L is the size of the sample. Accordingly, the critical electric field

$$E_{c1} = \frac{\sigma}{2e}\,, \tag{19}$$

corresponds to the lower critical field H_{c1} in superconductors and is the minimum electric field necessary for a string to penetrate the sample. Thus, at $E < E_{c1}$ or an applied voltage $V < V_{c1}$ only neutral electric pions are present; we called this the electric Meissner state. When V reaches the critical value $V = V_{c1}$, enough electric pions can be created such that a chain of them reaches from one end of the sample to the other, an electric string can traverse the sample end-to-end and the Meissner state is destroyed. For $V = V_{c1}$ electric fields penetrate the superinsulator in form of flux tubes of typical width λ_{el}, like Abrikosov vortices in type-II superconductors, and the mixed state of the superinsulator sets in. Increasing the applied voltage we reach

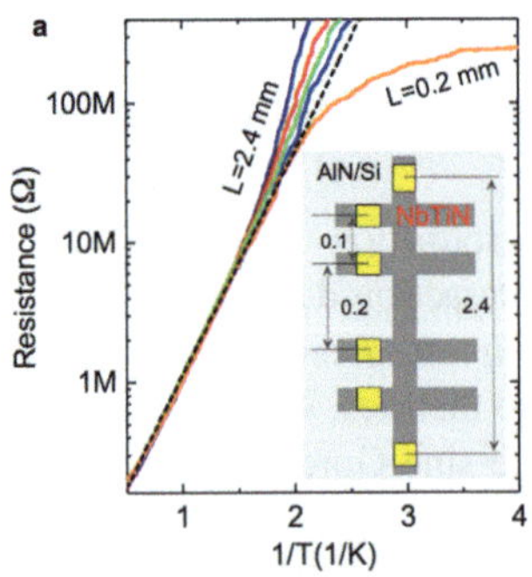

Fig. 1 Sheet resistance of a NbTiN superinsulating film as a function of the effective sample size (bridge length). Logarithmic plot of sheet resistance $R_\square$ vs. inverse temperature $1/T$ for bridges of various length L. The dashed straight line shows the Arrhenius behaviour $R \propto \exp(1/T)$

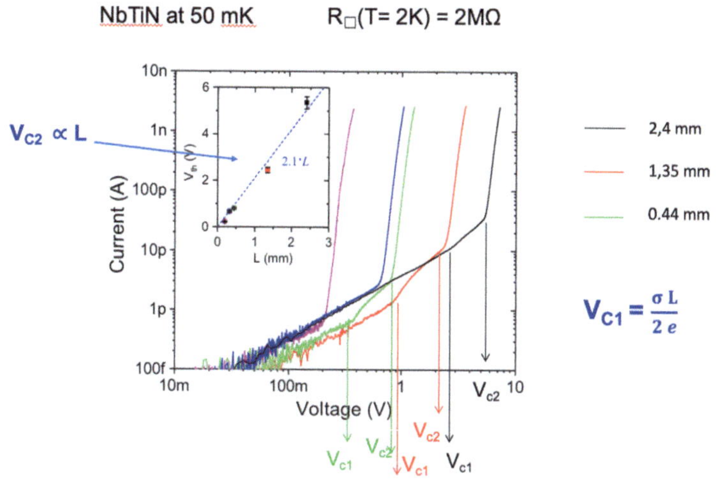

Fig. 2 Threshold voltage of bridges of different length. The I-V curves of two different NbTiN samples with $R_\square$ (T =2 K) = 2 M Ω taken at the same temperature T = 50 mK. Different colors correspond to different length of the bridge L (distance between electrodes)

a critical value V_{c2} at which superinsulation breaks down and the system switches into a normal insulator. In [23] we have shown that the critical value V_{c2} has a linear dependence on the sample size L.

Cooper pairs in superinsulators are bound by a linear potential and current starts to pass when the applied potential reaches the critical value $V_c = \sigma L/2e$. As shown in [24], in the center of mass coordinates, we can reduce the problem to one particle of reduced mass $m/2$ at the origin, subject to the potential energy $U(r) = 2e\sigma r$, with r denoting the separation distance. The binding energy of the $\pm$ charge pair is $2eV_{c1} \approx 0.3$ eV, where we used the average value of the threshold voltages in Fig. 3 .Since this is far smaller than the reduced mass 0.5 MeV of a Cooper pair we can use classical considerations to compute the behaviour of the centre of mass coordinate. A positive linear potential results in an attractive constant force $F_a = 2e\sigma = 2eV_{c1}/L$. Let us now apply an external potential V. This amounts to an additional repulsive force $F_r = 2eV/L$. The total acceleration of separation is thus $a = (2/m)F_{\text{tot}} = (4e/mL)(V - V_{c1})$. The corresponding equation of motion separating the charge-hole system is

$$r(t) = \frac{2e}{mL}(V - V_{c1})t^2 \,. \tag{20}$$

Current starts to pass when the charges of opposite sign reach the sample boundaries, i.e. when $r(t_{\text{cr}}) = L$. The delay for current passage is thus given by inverting Eq. (20):

$$t_{\text{cr}} = \frac{mL^2}{2e}(V - V_{c1})^{-1/2} \,. \tag{21}$$

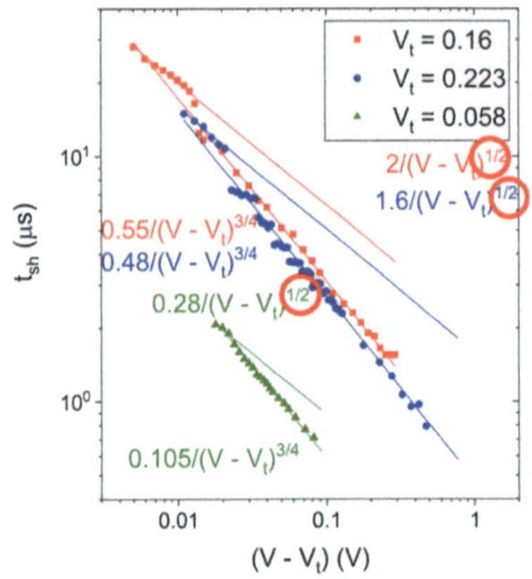

Fig. 3 Scaling of local minima time t_{sh} dependence upon amplitude of rectangular voltage pulse (symbols). The solid lines show the fitting functions

The critical exponent for the current passage $\mu = 1/2$ is a direct consequence of the linear binding potential between charges of the opposite sign. This exponent has been measured in NbTiN films and the results are shown in Fig. 3 with V_P identified with the first critical voltage V_{c1}. This experiment thereby fully confirming the electric strings linear potential confining the charges of opposite signs in the electric Meissner state and effectively rules out disorder-induced localization as a mechanism for superinsulation.

The $\mu = 3/4$ correspond to jumps from the electric Meissner state to the mixed state and to the superinsulating resistive state with broken charge confinement [24].

References

1. G. ‘t Hooft, On the phase transition towards permanent quark confinement. Nucl. Phys. **B138**, 1–25 (1978)
2. M.C. Diamantini, P. Sodano, C.A. Trugenberger, Gauge theories of Josephson junction arrays. Nuclear Phys. B **474**, 641–677 (1996)
3. G. Sambandamurthy, L.W. Engels, A. Johansson, E. Peled, D. Shahar, Experimental evidence for a collective insulating state in two-dimensional superconductors. Phys. Rev. Lett. **94**, 017003 (2005)
4. V.M. Vinokur et al., Superinsulator and quantum synchronization. Nature **452**, 613–615 (2008)
5. T.I. Baturina, V.M. Vinokur, Superinsulator-superconductor duality in two dimensions. Ann. Phys. **331**, 236–257 (2013)
6. A.Yu. Mironov, D.M. Silevitch, T. Proslier, S.V. Postolova, M.V. Burdastyh, A.K. Gutakovskii, T.F. Rosenbaum, V.M. Vinokur, T.I. Baturina, Charge Berezinskii-Kosterlitz-Thouless transition in superconducting NbTiN films. Sci. Rep. **8**, 4082 (2018)
7. M. C. Diamantini, C.A. Trugenberger, V.M. Vinokur, Confinement and asymptotic freedom with Cooper pairs. Comm. Phys. **1:77**, 1 (2018)
8. F. Wilczek, Disassembling anyons. Phys. Rev. Lett. **62**, 132 (1992)
9. A.M. Polyakov, *Gauge Fields and Strings* (Harwood Academic Publisher, Chur (Switzerland), 1987)
10. M.C. Diamantini, C.A. Trugenberger, V.M. Vinokur, Quantum magnetic monopole condensate. Nat. Comm. Phys. **4**, 25 (2021)

11. S. Deser, R. Jackiw, S. Templeton, Topologically massive gauge theories. Ann. Phys. (N.Y.) **140**, 372–411 (1982)
12. G. Dunne, R. Jackiw, C.A. Trugenberger, Topological (Chern-Simons) quantum mechanics. Phys. Rev. D **41**, 661–666 (1990)
13. G. Dunne, R. Jackiw, C.A. Trugenberger, Chern-Simons theory in the Schrödinger representation. Ann. Phys. **194**, 197–223 (1989)
14. T. Banks, R. Myerson, J. Kogut, Phase transitions in Abelian lattice Gauge theories. Nuclear Phys. B **129**, 493–510 (1977)
15. M.C. Diamantini, AYu. Mironov, S.V. Postolova, X. Liu, Z. Hao, D.M. Silevitch, Ya.. Kopelevich, P. Kim, C.A. Trugenberger, V.M. Vinokur, Bosonic topological intermediate state in the superconductor-insulator transition. Phys. Lett. A **384**, 126570 (2020)
16. M.C. Diamantini, C.A. Trugenberger, V.M. Vinokur, The superconductor-insulator transition in absence of disorder. Phys. Rev. B **103**, 174516 (2021)
17. M.C. Diamantini, C.A. Trugenberger, V.M. Vinokur, Type III superconductivity. Adv. Sci. 2206523 (2023)
18. C.A. Trugenberger, M.C. Diamantini, N. Poccia, F.S. Nogueira, V.M. Vinokur, Magnetic monopoles and superinsulation in Josephson junction arrays. Quantum Rep. **2**, 388–399 (2020)
19. M. Caselle, M. Panero, D. Vadacchino, Width of the flux tube in compact U(1) gauge theory in three dimensions. JHEP **02**, 180 (2016)
20. V.L. Berezinskii, Destruction of long-range order in one-dimensional and two-dimensional systems having a continuous symmetry group I. Classical systems. Sov. Phys.–JETP **32**, 493–500 (1970)
21. J.M. Kosterlitz, D.J. Thouless, Ordering, metastability and phase transitions in two-dimensional systems. J. Phys. C: Solid State Phys. **6**, 1181–1203 (1973)
22. J. Zinn-Justin, *Quantum Field Theory and Critical Phenomena* (Clarendon Press, Oxford, 1989)
23. M.C. Diamantini, L. Gammaitoni, C. Strunk, S.V. Postolova, A.Yu. Mironov, C.A. Trugenberger, V.M. Vinokur, Direct probe of the interior of an electric pion in a Cooper pair superinsulator. Nat. Comm. Phys. **3**, 142 (2020)
24. A.Yu. Mironov, M.C. Diamantini, C.A. Trugenberger, V.M. Vinokur, Relaxation electrodynamics of superinsulators. Sci. Rep. **12**, 19918 (2022)

Gravitational Dynamics of Binary Black Holes Near Supermassive Black Holes

Gianluca Grignani, Filippo Camilloni, Troels Harmark, Marta Orselli, and Daniele Pica

Abstract We explore the mergers of binary-black-hole (BBH) systems in close proximity to a supermassive black hole (SMBH) bound in an orbit. Our findings reveal that incorporating the strong gravity effects of the SMBH using a Kerr metric leads to significant modifications in the system's dynamics, deviating from the simplified Newtonian point particle model for the SMBH. The study involves a detailed analysis of the system's dynamics, incorporating a quadrupole approximation to account for tidal forces from the SMBH. The secular Hamiltonian is derived by averaging over both inner and outer orbits. The strong gravity effects of the SMBH play a crucial role in amplifying the von Zeipel-Lidov-Kozai mechanism, resulting in an increased number of cycles, higher eccentricity, and a shortened merger time, especially when the binary is close to or at the SMBH's innermost stable orbit.

1 Introduction

Gravitational waves (GWs) from merging binary black holes (BBHs) provide a unique observational opportunity [1, 2]. Galactic centers are believed to host numerous BBHs near supermassive black holes (SMBHs), prompting a study of GW signatures. Proximity to SMBHs can lead to shorter merger times and high eccentricities, influencing LIGO-Virgo-KAGRA observations [3–8].

The common approach approximates BBHs as particles near an SMBH, assuming weak gravity and small relative velocities. However, when BBH mergers occur close to the SMBH, the particle/PN approximation breaks down. In this case, a metric for

G. Grignani (✉) · M. Orselli · D. Pica
Dipartimento di Fisica e Geologia, Università di Perugia, INFN Sezione di Perugia, Perugia, Italy
e-mail: gianluca.grignani@unipg.it

F. Camilloni
Institut für Theoretische Physik, Goethe Universität, Frankfurt am Main, Germany

T. Harmark · M. Orselli · D. Pica
Niels Bohr International Academy, Niels Bohr Institute, Copenhagen University, Copenhagen, Denmark

M. Paranjape et al. (eds.), *Proceedings of Gravity, Strings and Fields*,
CRM Series in Mathematical Physics,
https://doi.org/10.1007/978-3-031-91266-5_5

the SMBH is necessary, exposing the BBH to non-perturbative gravitational effects. This paper studies a BBH near an SMBH described by the Kerr metric, assuming small binary size compared to the SMBH curvature scale.

This paper begins by deriving the Hamiltonian for a BBH near an SMBH with the Kerr metric. The center of mass motion is approximated as a geodesic on a circular equatorial orbit. Quadrupole tidal forces from the Kerr metric are considered, applying the Newtonian approximation for the internal binary dynamics. This leads to the derivation of the secular Hamiltonian, summarizing the long-time-scale dynamics. The frame of reference where the binary system is located is the local inertial frame introduced by Marck [9] for the Kerr black hole. Section 4 introduces Euler angles and action-angle variables in the Marck's frame, setting the stage for the dynamics study. In Sect. 5 we derive the secular Hamiltonian, describing the long-time-scale dynamics of the BBH-SMBH system. This includes averaging over the inner and outer orbits. Then in Sect. 6 we will focus on the ZLK mechanism's. The ZLK frequency, affected by the full SMBH metric, is compared to Newtonian approximations.

These proceedings are based on the recent work by the authors [10] and is dedicated to Gordon W. Semenoff a master for some of them.

2 Binary System Dynamics in Tidal Force Background

In this section, we review the dynamics of a binary system of black holes in a general tidal force background up to the quadrupole approximation. The tidal forces arise from the curvature of the space-time that the binary system moves in. We simplify the problem by using the Newtonian approximation for the black holes, treating them as particles. The quadrupole tidal force background is kept completely general. We assume the binary system is freely falling, and to leading order, its center of mass moves on a geodesic of the background space-time. The metric near a geodesic in Thorne-Hartle Fermi-normal coordinates, is given by:

$$
\begin{aligned}
g_{00} &= -1 - \mathcal{E}_{ij}x^i x^j + \mathcal{O}((x/\mathcal{R})^3) \ , \\
g_{0i} &= -\tfrac{2}{3}\epsilon_{ijk}\mathcal{B}^j{}_l x^k x^l + \mathcal{O}((x/\mathcal{R})^3) \ , \\
g_{ij} &= \delta_{ij}\left(1 - \mathcal{E}_{kl}x^k x^l\right) + \mathcal{O}((x/\mathcal{R})^3) \ ,
\end{aligned}
\tag{2.1}
$$

Here, $i, j = 1, 2, 3$, $x^2 = x^i x^i$, and $\mathcal{R}$ is the curvature length scale. The "small-tide approximation" requires $x \ll \mathcal{R}$. The electric and magnetic tidal moments $\mathcal{E}_{ij}$ and $\mathcal{B}_{ij}$ are related to the Riemann curvature tensor. They depend on the proper time τ of the geodesic. Assuming $R_{\mu\nu} = 0$ for the background space-time (true for the Kerr space-time), we keep the expressions general. In Sect. 3, we will focus on tidal moments around a Kerr black hole. Modeling the two black holes (BHs) as particles moving in the background metric (2.1), with the center of mass of the binary located at the geodesic, the Lagrangian for particle 1 is:

$$L_{(1)} = -m_1 c^2 \sqrt{-G_{00} - 2G_{0i}\frac{v^i_{(1)}}{c} - G_{ij}\frac{v^i_{(1)}v^j_{(1)}}{c^2}}\ . \qquad (2.2)$$

The metric $G_{\mu\nu}$ includes the background metric with tidal forces (2.1) and gravitational potentials generated by particle 2, expanded up to 1PN.

$$G_{00} = g_{00} - 2\phi_{(2)} + \cdots\ , \quad G_{0i} = g_{0i} + \cdots\ , \quad G_{ij} = g_{ij} - 2\delta_{ij}\phi_{(2)} + \cdots \qquad (2.3)$$

Performing a PN expansion of the Lagrangian for particle 1, considering tidal moments as small perturbations up to first order, gives:

$$L_{(1)} = L_{(1)}|_{\mathscr{E}=\mathscr{B}=0} - \frac{1}{2}m_1 c^2 x^i_{(1)}x^j_{(1)}\mathscr{E}_{ij} - \frac{2c}{3}m_1 v^i_{(1)}x^k_{(1)}x^l_{(1)}\epsilon_{ijk}\mathscr{B}^j{}_l + \cdots \qquad (2.4)$$

The terms in (2.4) include the Lagrangian solely from the gravitational interaction with particle 2 and the kinetic energy of particle 1. The second term couples to $\mathscr{E}_{ij}$, giving the leading coupling to quadrupole tidal moments, while the third term couples to $\mathscr{B}_{ij}$ and is subleading. In the Newtonian limit, after adding the analogous contribution for particle 2, the Lagrangian of the binary system is:

$$\begin{aligned} L_{\text{Newton}} &= -(m_1 + m_2)c^2 + \frac{1}{2}m_1 v^2_{(1)} + \frac{1}{2}m_2 v^2_{(2)} \\ &+ \frac{Gm_1m_2}{r} - \frac{c^2}{2}\left[m_1 x^i_{(1)}x^j_{(1)} + m_2 x^i_{(2)}x^j_{(2)}\right]\mathscr{E}_{ij}. \end{aligned} \qquad (2.5)$$

Introducing center of mass quantities and making the center of mass motion decoupled from the relative motion, the Lagrangian becomes:

$$L_{\text{Newton}} = -Mc^2 + \frac{1}{2}MV^2 + \frac{1}{2}\mu v^2 + \frac{GM\mu}{r} - \frac{c^2}{2}\left[MX^iX^j + \mu x^i x^j\right]\mathscr{E}_{ij}. \quad (2.6)$$

Setting $\mathbf{X} = 0$ and $\mathbf{V} = 0$, putting the center of mass of the binary system at the geodesic, and using the general formula [11]:

$$x^i x^j \mathscr{E}_{ij} = r^2 \mathscr{E}^q\ , \qquad (2.7)$$

The binary system Lagrangian is obtained:

$$L_{\text{Newton}} = \frac{1}{2}\mu v^2 + \frac{GM\mu}{r} - \frac{c^2}{2}\mu r^2 \mathscr{E}^q\ . \qquad (2.8)$$

Finally, performing a Legendre transform to obtain the Hamiltonian:

$$H_{\text{Newton}} = \frac{1}{2\mu}p^2 - \frac{GM\mu}{r} + \frac{c^2}{2}\mu r^2 \mathcal{E}^q \ , \tag{2.9}$$

This Hamiltonian describes the dynamics of a binary system of particles moving on a geodesic of a background space-time, subject to quadrupole tidal forces.

3 Three-Body Black Hole System

Consider a system of three black holes (BHs) with masses m_1, m_2, and m_3. The BHs with masses m_1 and m_2 are in a bound motion, forming a binary black hole system, referred to as the inner binary. The third BH, with mass m_3, is much larger, identified as a supermassive black hole.

We assume two independent separations of scales for analytical control of the dynamics. Firstly, the inner binary is in the particle/Post-Newtonian (PN) regime, where their Schwarzschild radii are much smaller than their separation r:

$$r \gg \frac{2Gm_1}{c^2} \ , \ \frac{2Gm_2}{c^2} \ . \tag{3.1}$$

Secondly, the small-tide approximation is assumed, meaning the size of the binary system r is much smaller than the curvature length scale $\mathcal{R}$ of the SMBH:

$$r \ll \hat{r}\sqrt{\frac{c^2\hat{r}}{Gm_3}} \ , \tag{3.2}$$

with $\hat{r}$ the radial coordinate of the Kerr metric.

This allows us to treat the influence of the SMBH on the BBH via a tidal approximation, even when the BBH is close to the SMBH, and to focus on the leading quadrupole effect from the Kerr curvature tensor for the SMBH.

3.1 Kerr Black Hole Background

The SMBH is modeled as a Kerr black hole with the line element in Boyer-Lindquist coordinates:

$$d\hat{s}^2 = -\Big(1 - \frac{2Gm_3\hat{r}}{c^2\Sigma}\Big)c^2 d\hat{t}^2 - \frac{4Gm_3\hat{r}}{c^2\Sigma} s_3 \sin^2\hat{\theta}\, c\, d\hat{t}\, d\hat{\phi} + \frac{\mathcal{A}}{\Sigma}\sin^2\hat{\theta}\, d\hat{\phi}^2 + \frac{\Sigma}{\Delta}d\hat{r}^2 + \Sigma d\hat{\theta}^2 \ . \tag{3.3}$$

Here, m_3 is the black hole mass, $s_3 = J_3/(c\, m_3)$ is the specific angular momentum, and Σ, Δ, $\mathcal{A}$ are defined in terms of $\hat{r}$ and $\hat{\theta}$. The dimensionless spin parameter χ

is introduced as $\chi = s_3c^2/(Gm_3)$. The geodesic motion in the Kerr space-time is parametrized by constants of motion: energy $\hat{E}$, angular momentum $\hat{L}$, and Carter constant K per unit of rest energy.

For circular and equatorial $\hat{\theta} = \pi/2$ geodesics, the tangent vector components $u^t = d\hat{t}/d\tau$ and $\Omega = d\hat{\phi}/d\hat{t}$ are defined, with Ω being the coordinate angular velocity. The constants of motion in terms of the orbital radius $\hat{r}$, black hole mass m_3, and spin s_3 for a circular and equatorial geodesic can be found in [12], and more generically in [13].

3.2 Binary System on a Kerr Geodesic

We can regard the black holes of the BBH system as particles in a Newtonian limit since we assume the condition (3.1) holds. The BBH system moves along a circular equatorial orbit in the background of a Kerr black hole describing the SMBH. As seen in Sect. 2, for a sufficiently small binary system (3.2), we can approximate the influence of the SMBH through quadrupole tidal forces acting on the BBH, while the center of mass of the BBH moves approximately on a geodesic of the Kerr metric. This means we can employ the results of Sect. 2.

For the Kerr metric (3.3), we can use the *Marck's tetrad* [9] to describe the local Fermi-normal coordinates (2.1) of Sect. 2. Marck's tetrad is given by the four vectors λ^μ_A which provide an orthonormal basis for the vector space at each point of the Kerr-geodesic since $\lambda^\mu_A\lambda^\nu_B g_{\mu\nu} = \eta_{AB}$ and $\lambda^\mu_A\lambda^\nu_B\eta^{AB} = g_{\mu\nu}$, where in particular $\lambda^\mu_0 = u^\mu$ is the four-velocity. Note that $A = 0, 1, 2, 3$ are the flat tetrad indices. One can equivalently represent Marck's tetrad as the orthonormal one-forms $\lambda^A_\mu = \eta^{AB} g_{\mu\nu}\lambda^\nu_B$.

One can now employ the standard map of the Fermi-Normal coordinates between vectors on the Kerr geodesic and events in the neighborhood of the geodesic. For Marck's tetrad, we have in addition to the time coordinate $x^0 = \tau$, which is the proper time on the geodesic, also the spatial coordinates x^i parametrizing an orthogonal vector $\sum_{i=1}^3 x^i\lambda^\mu_i$ at the geodesic. In this way, the coordinates x^A describe a neighborhood of the geodesic. We will call this coordinate system the *Marck's frame of reference*. Since Marck's tetrad is parallel-transported along the geodesic, it provides an inertial frame, meaning it is characterized by a vanishing acceleration and vanishing angular velocity of rotation of spatial basis vectors [14]: $a^i \equiv \lambda^i_\mu \frac{D\lambda^\mu_0}{d\tau} = 0\ ,\ \ \omega^i \equiv -\frac{1}{2}\epsilon^{ij}{}_k\lambda^\mu_j\frac{D\lambda^k_\mu}{d\tau} = 0.$

For circular geodesics in the equatorial plane ($\hat{\theta} = \pi/2$), Marck's tetrad can be conveniently written in the one-forms basis

$$
\begin{aligned}
&\lambda^0_\mu = \left(\hat{E}, 0, 0, \hat{L}\right)\ ,\ \ \lambda^1_\mu = \cos\Psi\ \tilde{\lambda}^1_\mu - \sin\Psi\ \tilde{\lambda}^2_\mu\ , \\
&\lambda^2_\mu = \sin\Psi\ \tilde{\lambda}^1_\mu + \cos\Psi\ \tilde{\lambda}^2_\mu\ ,\ \ \lambda^3_\mu = \left(0, 0, -\hat{r}, 0\right)\ ,
\end{aligned}
\tag{3.4}
$$

with

$$\tilde{\lambda}^1_\mu = \left(0, \sqrt{\frac{\hat{r}^2}{K+\hat{r}^2}} \frac{(\sqrt{K} s_3 + \hat{E}\hat{r}^2)}{\Delta}, 0, 0\right) ,$$
$$\tilde{\lambda}^2_\mu = \left(\frac{(s_3 - \hat{E}\sqrt{K})}{\sqrt{K+\hat{r}^2}}, 0, 0, \frac{\hat{E}\sqrt{K} s_3 - s_3^2 - \hat{r}^2 - K}{\sqrt{K+\hat{r}^2}}\right) , \tag{3.5}$$

where $\hat{r}$ is the constant radius of the circular equatorial geodesic and the angle Ψ is introduced to ensure that the tetrad is parallel transported along the geodesic. The explicit expression for Ψ in terms of the geodesic's proper time τ is given by $\Psi = \Omega_\Psi \tau$, with

$$\Omega_\Psi \equiv \sigma \sqrt{\frac{G m_3}{\hat{r}^3}} . \tag{3.6}$$

To use the results for the Lagrangian and Hamiltonian of Sect. 2, given in Eqs. (2.8) and (2.9), respectively, we record that the electric tidal moments in the equatorial plane are [9, 12]

$$\mathcal{E}_{11} = \left[1 - 3\left(1 + \frac{K}{\hat{r}^2}\right)\cos^2\Psi\right]\frac{G m_3}{c^2 \hat{r}^3} , \quad \mathcal{E}_{22} = \left[1 - 3\left(1 + \frac{K}{\hat{r}^2}\right)\sin^2\Psi\right]\frac{G m_3}{c^2 \hat{r}^3} ,$$
$$\mathcal{E}_{33} = \left(1 + 3\frac{K}{\hat{r}^2}\right)\frac{G m_3}{c^2 \hat{r}^3} , \qquad \mathcal{E}_{12} = -3\left(1 + \frac{K}{\hat{r}^2}\right)\frac{G m_3}{c^2 \hat{r}^3}\cos\Psi \sin\Psi . \tag{3.7}$$

Using these with Eq. (2.7), one gets an explicit expression for the scalar quadrupole electric tidal moment induced by the Kerr black hole, as measured by an observer using Fermi-normal coordinates, namely

$$r^2 \mathcal{E}^q = \frac{G m_3}{c^2 \hat{r}^3}\left[r^2 + 3(x^3)^2 \frac{K}{\hat{r}^2} - 3\left(1 + \frac{K}{\hat{r}^2}\right)(x^1 \cos\Psi + x^2 \sin\Psi)^2\right] . \tag{3.8}$$

With this, we can describe the dynamics of the BBH system in the approximations (3.1) and (3.2) via the Lagrangian (2.8) and Hamiltonian (2.9).

4 Euler Angles and Action-Angle Variables

4.1 Inner BBH Configuration

While the center of mass for the inner Newtonian BBH system moves along the circular equatorial geodesic in the Kerr space-time, the orientation of its inner orbital plane can vary with respect to the outer orbital plane of the Kerr SMBH.

The vector describing the relative position of a body in a Newtonian elliptic orbit of arbitrary orientation can be written as $\boldsymbol{r} = r(\cos\psi\, \hat{\boldsymbol{u}} + \sin\psi\, \hat{\boldsymbol{v}})$, with

$$r = \frac{a(1-e^2)}{1+e\cos\psi}\ , \tag{4.1}$$

where a is the *semi-major axis*, e the *eccentricity* of the orbit and $\psi \in [0, 2\pi]$ the *true anomaly*, keeping track of the relative motion along the orbit trajectory. $\psi = 0$ identifies the *periapsis direction*, whereas $\psi = \pi/2$ gives the *direction of the ascending nodes*. For an inner binary in a Newtonian regime, it is possible to unambiguously introduce the *eccentric anomaly* ζ and the *mean anomaly* β, defined by means of the Kepler's equation

$$\cos\psi = \frac{\cos\zeta - e}{1 - e\cos\zeta}\ , \quad \beta = \zeta - e\sin\zeta\ . \tag{4.2}$$

The mean anomaly $\beta = \sqrt{GM/a^3}\tau$ is the angle that a fictitious body would have if it was moving uniformly along a circular orbit with the same orbital period that the actual body has along the elliptic orbit. By definition $\beta \in [0, 2\pi]$.

In our setup $m_3 \gg (m_1 + m_2)$, we assume that the equatorial plane of the Kerr black hole represents the outer orbital plane $\hat{\theta} = \pi/2$, where the inner BBH system moves following a fully relativistic circular geodesic. The equatorial plane is characterized by the two normal directions $\hat{\boldsymbol{x}}$ and $\hat{\boldsymbol{y}}$, whereas the outer orbital momentum direction specifies the third direction $\hat{\boldsymbol{z}} = \hat{\boldsymbol{L}}_{\rm out}$. It is then easy to check that the arbitrary orientation of a binary system is characterised by

$$\begin{aligned}\hat{\boldsymbol{u}} =&(\cos\gamma\cos\theta - \cos I\sin\gamma\sin\theta)\,\hat{\boldsymbol{x}} + (\cos\gamma\sin\theta + \cos I\sin\gamma\cos\theta)\,\hat{\boldsymbol{y}}\\ &+ \sin I\sin\gamma\,\hat{\boldsymbol{z}}\ ,\\ \hat{\boldsymbol{v}} =&(-\sin\gamma\cos\theta - \cos I\cos\gamma\sin\theta)\,\hat{\boldsymbol{x}} + (-\sin\gamma\sin\theta + \cos I\cos\gamma\cos\theta)\,\hat{\boldsymbol{y}}\\ &+ \sin I\cos\gamma\,\hat{\boldsymbol{z}}\ .\end{aligned} \tag{4.3}$$

where the Euler's angles are the *longitude of ascending nodes* $\theta \in [0, 2\pi]$, is the *orbital inclination* $I \in [0, \pi/2]$, and the *argument of the periapsis* $\gamma \in [0, 2\pi]$. This two directions span the inner orbital plane of the inner BBH system, whereas its inner orbital angular momentum direction is

$$\hat{\boldsymbol{L}}_{\rm in} = \hat{\boldsymbol{u}} \times \hat{\boldsymbol{v}} = \sin I\sin\theta\,\hat{\boldsymbol{x}} - \sin I\cos\theta\,\hat{\boldsymbol{y}} + \cos I\,\hat{\boldsymbol{z}}\ . \tag{4.4}$$

4.2 Action-Angle Variables in Marck's Tetrad

It is customary in celestial mechanics to introduce angle-action variables in order to simplify the dynamics of binary systems. The advantage is that it is possible to express the non-tidal part of the Hamiltonian in terms of constants of motion. We introduce the *Delauney variables*, in which the set of angle variables (β, γ, θ) is used to specify the relative position of a body along an arbitrarily-oriented Newtonian orbit, whereas the three corresponding action variables

$$J_\beta = \mu\sqrt{GMa}\ , \quad J_\gamma = \mu\sqrt{GMa(1-e^2)}\ , \\ J_\theta = \mu\sqrt{GMa(1-e^2)}\cos I\ , \tag{4.5}$$

with M the total mass and μ the reduced mass of the BBH system, are related to the magnitude and the orientation of the angular momentum of the inner binary with respect to the outer orbital plane, according to $J_\gamma = |\boldsymbol{L}_{\rm in}|$ and $J_\theta = \boldsymbol{L}_{\rm in} \cdot \hat{\boldsymbol{L}}_{\rm out}$. The total Hamiltonian of the inner binary becomes $H = -\left(\frac{GM}{J_\beta}\right)^2 + H_{\rm q}$, where H_q is the quadrupole tidal contribution given by (3.8), which in the Marck's tetrad explicitly reads

$$H_{\rm q} = \frac{\mu}{2}\frac{Gm_3 r^2}{\hat{r}^3}\left[1 + 3\frac{K}{\hat{r}^2}\sin^2(\gamma+\psi)\sin^2 I \right. \\ \left. -3\left(1+\frac{K}{\hat{r}^2}\right)\left(\cos(\Psi-\theta)\cos(\gamma+\psi) + \sin(\Psi-\theta)\sin(\gamma+\psi)\cos I\right)^2\right]\ , \tag{4.6}$$

where we used the projections $x^1 \equiv \boldsymbol{r}\cdot\hat{\boldsymbol{x}}$, $x^2 \equiv \boldsymbol{r}\cdot\hat{\boldsymbol{y}}$ and $x^3 \equiv \boldsymbol{r}\cdot\hat{\boldsymbol{z}}$ with $\boldsymbol{r} = r(\cos\psi\ \hat{\boldsymbol{u}} + \sin\psi\ \hat{\boldsymbol{v}})$ and Eq. (4.3). The same construction in alternative relativistic frames which are not parallel transported like Marck's tetrad would produce additional non-tidal terms, associated for instance to the presence of the gyroscope precession of GR and additional fictitious forces. This was explored in details in the main article [10].

5 Secular Hamiltonian

Triple systems display various dynamical effects governed by specific time scales, with the von Zeipel-Lidov-Kozai (ZLK) mechanism being a prominent example. Secular effects, unfolding over much longer periods than other system time scales, include internal motion within the binary black hole system and outer geodesic motion around the Kerr black hole. To focus on these longer time scales, an averaging procedure is crucial, revealing the ZLK mechanism after applying secular averaging to the tidal interaction term H_q in the inner BBH Hamiltonian. Analyzing the time scales in the triple system requires examining the inner motion time scale, where a uniformly growing angle (mean anomaly β) is needed for the elliptic motion. The Hamiltonian's tidal part is a function of the true anomaly ψ, so the averaging involves translating $(2\pi)^{-1}\int_0^{2\pi} d\beta$ into an integral in the true anomaly ψ using Kepler's relations (4.2).

Regarding the outer orbit, it is a fully relativistic circular geodesic in a curved background, making the averaging procedure more subtle. A significant simplification occurs in local frames associated with freely-falling observers, such as Marck's tetrad (3.4). The inner BBH Hamiltonian (4.6), expressed using Delauney action-angle variables in Marck's local frame, is periodic in the angle Ψ, ensuring the Marck's tetrad is parallel transported along the geodesic. The Marck's angle Ψ has a period of 2π and a constant frequency $d\Psi/d\tau = \Omega_\Psi$, making the outer orbit average straightforward as $(2\pi)^{-1}\int_0^{2\pi} d\Psi$. The secular dynamics then emerges from the double-averaged Hamiltonian (4.6) in Marck's frame of reference.

$$\langle H\rangle = -\left(\frac{GM}{J_\beta}\right)^2 + \langle H_{\rm q}\rangle \ , \tag{5.1}$$

with

$$\langle H_{\rm q}\rangle \equiv \frac{1}{(2\pi)^2}\int_0^{2\pi} d\Psi \int_0^{2\pi} d\psi \frac{(1-e^2)^{3/2}}{(1+e\cos\psi)^2} H_{\rm q} \ , \tag{5.2}$$

and H_q defined in Eq. (4.6). The result obtained after averaging the tidal part in Marck's tetrad is consistent with the analogous procedure in a frame for an observer at rest with the distant stars if one averages over the Kerr geodesic azimuthal coordinate $\hat{\phi}$ [10]. The secularly-averaged quadrupole tidal part of the inner BBH system Hamiltonian is written as

$$\begin{aligned} \langle H_{\rm q}\rangle &= -\Omega_{\rm ZLK}^{\rm (GR)} J_\gamma \left((1-e^2)(\cos^2 I - 2) - 5e^2\sin^2 I \sin^2\gamma + \frac{5}{3}\right) \ , \\ \Omega_{\rm ZLK}^{\rm (GR)} &= \Omega_{\rm ZLK}^{\rm (N)}\left(1 + 3\frac{K}{\hat{r}^2}\right) \ , \quad \Omega_{\rm ZLK}^{\rm (N)} = \frac{3}{8J_\gamma}\left(\frac{Gm_3\mu}{\hat{r}}\right)\left(\frac{a}{\hat{r}}\right)^2 \ , \end{aligned} \tag{5.3}$$

where we factorized the characteristic frequency $\Omega_{\rm ZLK}^{\rm (GR)}$ of the ZLK mechanism, which in our treatment takes into account GR effects associated to the presence of the Kerr SMBH companion by means of the presence of the Carter constant K. In the weak field limit $\hat{r} \to \infty$ one recovers the Newtonian secular Hamiltonian (see for instance [5]), and $\Omega_{\rm ZLK}^{\rm (GR)} \to \Omega_{\rm ZLK}^{\rm (N)}$.

6 ZLK Mechanism in a Strong GR Background

Having the secular Hamiltonian, it is possible to derive the equations of motion describing the dynamics of the inner BBH system perturbed by the presence of the Kerr companion. The non-tidal part of $\langle H\rangle$ only contributes to the evolution of the mean anomaly β according to the usual Keplerian frequency. From the quadrupole tidal part $\langle H_q\rangle$ one derives the equations for the remaining angle variables

$$\frac{d\gamma}{d\tau} \equiv \frac{\partial\langle H\rangle}{\partial J_\gamma} = 2\Omega_{\rm ZLK}^{\rm (GR)}[2(1-e^2) - 5(1-e^2-\cos^2 I)\sin^2\gamma] \ , \tag{6.1}$$

$$\frac{d\theta}{d\tau} \equiv \frac{\partial \langle H \rangle}{\partial J_\theta} = -2\Omega^{(\mathrm{GR})}_{\mathrm{ZLK}} \cos I \,(1 - e^2 + 5e^2 \sin^2 \gamma) \ . \tag{6.2}$$

The equation of motion for the action variable J_γ, instead, provides the evolution equation for the inner angular momentum magnitude

$$\frac{dJ_\gamma}{d\tau} \equiv -\frac{\partial \langle H \rangle}{\partial \gamma} = -5\Omega^{(\mathrm{GR})}_{\mathrm{ZLK}} J_\gamma e^2 \sin^2 I \sin 2\gamma \ , \tag{6.3}$$

all the other equations being trivial as far as the gravitational backreaction is neglected, so that $J_\theta = \boldsymbol{L}_{\mathrm{in}} \cdot \hat{\boldsymbol{L}}_{\mathrm{out}} = \mu\sqrt{GMa(1-e^2)} \cos I$ is conserved. With these equations it is possible to work-out the ZLK mechanism, for which a binary system under the influence of the tidal forces generated by an outer third body exhibits a periodic exchange of eccentricity and orbital inclination, over a timescale much larger than both the inner and outer binaries orbital periods [15–17]. By exploiting the definitions (4.5) it is easy to derive the equations describing the evolution of the orbital inclination I and of the orbital eccentricity e, explicitly

$$\frac{dI}{d\tau} = -\frac{5}{2}\Omega^{(\mathrm{GR})}_{\mathrm{ZLK}} e^2 \sin 2I \sin 2\gamma \ , \qquad \frac{de}{d\tau} = 5\,\Omega^{(\mathrm{GR})}_{\mathrm{ZLK}} e(1-e^2) \sin^2 I \sin 2\gamma \ . \tag{6.4}$$

The stationary points for minimum and maximum inclination and eccentricity correspond to $\gamma = 0, \pi/2$. The orbital inclination has a maximum I_{max} when the eccentricity is minimal e_{min} (and viceversa) as a consequence of the conservation of $J_\theta = \boldsymbol{L}_{\mathrm{in}} \cdot \hat{\boldsymbol{L}}_{\mathrm{out}}$. Hence, the pair $(I_{\mathrm{max}}, e_{\mathrm{min}})$ occurs for $\gamma = 0$, whereas $(I_{\mathrm{min}}, e_{\mathrm{max}})$ corresponds to $\gamma = \pi/2$. To understand what triggers the ZLK mechanism one can restrict to the case in which the inner orbit is initially circular, $e_0 = 0$, and from the energy and angular momentum conservation ($(\langle \mathcal{H} \rangle_0 = \langle H \rangle^{I_{\mathrm{min}}, e_{\mathrm{max}}}_{\gamma=\pi/2}$ and $(J_\theta)_0 = (J_\theta)^{I_{\mathrm{min}}, e_{\mathrm{max}}}_{\gamma=\pi/2}$) derive that $e_{\mathrm{max}} = \sqrt{1 - \frac{5}{3}\cos^2 I_0}$ and $\cos I_{\mathrm{min}} = \pm\sqrt{\frac{3}{5}}$. For the ZLK mechanism to be triggered, thus, the initial inclination I_0 of the binary should obey $|\cos I_0| < \sqrt{3/5}$. For very high initial inclination $I_0 \approx \pi/2$ and for the binary initially on a circular orbit ($e_0 = 0$), the BBH system can reach extremely high eccentricity $e_{\mathrm{max}} \approx 1$ [17].

Incorporating full general relativistic effects into the outer orbit does not change the onset of the von Zeipel-Lidov-Kozai (ZLK) mechanism. However, the frequency of eccentricity/inclination oscillations is influenced by $\Omega^{(\mathrm{GR})}_{\mathrm{ZLK}}$, defined in Eq. (5.3), which incorporates GR inputs like the Kerr Carter constant K and parameters characterizing the outer orbit. This GR-influenced ZLK frequency significantly differs from its Newtonian counterpart when using a general relativistic geodesic description for the outer orbit. Particularly, when the binary black hole system approaches a supermassive black hole closely enough to experience strong gravity effects, notable deviations in the ZLK oscillation frequency occur compared to the frequency obtained with a Newtonian point particle approximation. One can compare $\Omega^{(\mathrm{GR})}_{\mathrm{ZLK}}$ to the Newtonian frequency $\Omega^{(\mathrm{N})}_{\mathrm{ZLK}}$, *i.e.* the frequency that would have resulted from Newtonian quadrupole tidal forces induced by a particle of mass m_3. According to Eq. (5.3), one has that

$$\frac{\Omega_{\rm ZLK}^{\rm (GR)}}{\Omega_{\rm ZLK}^{\rm (N)}} = \left(1 + 3\frac{K}{\hat{r}^2}\right) = \frac{1 + \frac{3\chi^2}{d^2} - \frac{4\sigma\chi}{d^{3/2}}}{1 - \frac{3}{d} + \frac{2\sigma\chi}{d^{3/2}}} \,, \tag{6.5}$$

where in the last equality the explicit expression for the Carter constant K on the equator is written in terms of the dimensionless outer orbit radius $d = \hat{r}c^2(Gm_3)^{-1}$. The behaviour of this ratio is plotted in Fig. 1 as a function of the distance d to highlight how the Kerr SMBH spin can impact on the secular dynamics of the inner BBH system via the ZLK mechanism.

One of the consequences of having a fully general relativistic description for the external companion is that a stable outer circular orbit can only exist up to the limiting minimum value given by the ISCO, the innermost stable circular orbit, as opposed to the Newtonian case. Remarkably, at the ISCO $\hat{r} = \hat{r}^{\sigma}_{\rm ISCO}$ one has the nice relation for the Carter constant $K = \frac{1}{3}\hat{r}^2_{\rm ISCO}$ (see [12]), so that the ratio (6.5) takes an universal value

$$\Omega_{\rm ZLK}^{\rm (GR)} = 2\Omega_{\rm ZLK}^{\rm (N)} \,. \tag{6.6}$$

The frequency of the eccentricity/inclination oscillations induced by the ZLK mechanism can deviate in a significant manner from its Newtonian counterpart, with the deviation being $\sim$ 100% when the BBH is located at the minimum reasonable radius given by the ISCO. It is important to realize that this change in the ZLK frequency has profound implications for the GW signal that one expects to observe from binary

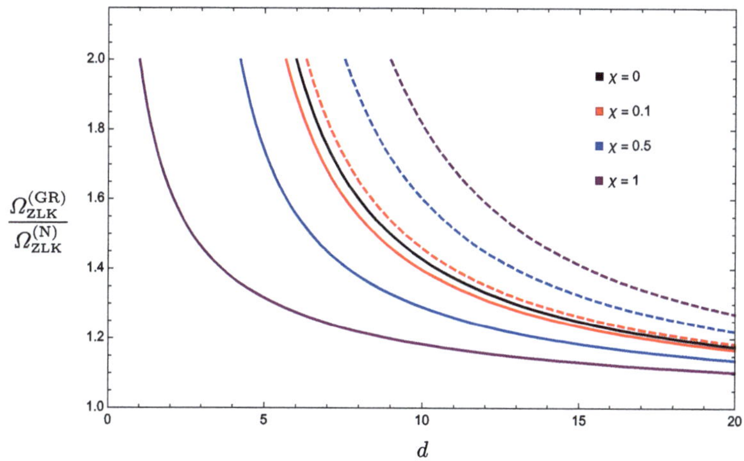

Fig. 1 Diagram of the ratio $\Omega_{\rm ZLK}^{\rm (GR)}/\Omega_{\rm ZLK}^{\rm (N)}$ versus the distance d from the SMBH. Different colors label various values of the black hole spin χ where the solid and dashed lines represent co-rotating ($\sigma = +1$) and counter-rotating ($\sigma = -1$) orbits, respectively. Each curve terminates at the ISCO. Taken from [10]

coalescence taking place close to the galactic centre: the GR-enhanced ZLK mechanism can indeed lead to boosting the BBH merger with modification in the measured waveform. For an exhaustive investigation which takes into account GW emission and periastron precession of the inner BBH system, as well as the redshift factor u^t associated with the outer geodesic around the Kerr SMBH, we refer the reader to [10].

References

1. Abbott et al. (LIGO Scientific, Virgo), Phys. Rev. Lett. **116**, 061102 (2016). https://doi.org/10.1103/PhysRevLett.116.061102. arXiv:1602.03837 [gr-qc]
2. Abbott et al. (LIGO Scientific, Virgo), Astrophys. J. Lett. **900**, L13 (2020). https://doi.org/10.3847/2041-8213/aba493. arXiv:2009.01190 [astro-ph.HE]
3. J. Samsing, Phys. Rev. D **97**, 103014 (2018). https://doi.org/10.1103/PhysRevD.97.103014. arXiv:1711.07452 [astro-ph.HE]
4. J. Samsing, E. Ramirez-Ruiz, Astrophys. J. Lett. **840**, L14 (2017). https://doi.org/10.3847/2041-8213/aa6f0b. arXiv:1703.09703 [astro-ph.HE]
5. L. Randall, Z.-Z. Xianyu, Astrophys. J. **864**, 134 (2018). https://doi.org/10.3847/1538-4357/aad7fe. arXiv:1802.05718 [gr-qc]
6. Samsing et al., Nature **603**, 237 (2022). https://doi.org/10.1038/s41586-021-04333-1. arXiv:2010.09765 [astro-ph.HE]
7. Tagawa et al., Astrophys. J. Lett. **907**, L20 (2021). https://doi.org/10.3847/2041-8213/abd4d3. arXiv:2010.10526 [astro-ph.HE]
8. Rom et al., Formation of merging stellar-mass black hole binaries by gravitational waves emission in agn disks (2023), arXiv:2310.03801 [astro-ph.HE]
9. J.-A. Marck, Proc. R. Soc. Lond. A **385**, 431 (1983). https://doi.org/10.1098/rspa.1983.0021
10. Camilloni, Grignani et al., Mon. Not. Roy. Astron. Soc. **531**, 1884 (2024). https://doi.org/10.1093/mnras/stae1093. arXiv:2310.06894 [gr-qc]
11. E. Poisson, I. Vlasov, Phys. Rev. D **81**, 024029 (2010). https://doi.org/10.1103/PhysRevD.81.024029. arXiv:0910.4311 [gr-qc]
12. Camilloni, Grignani et al., Phys. Rev. D **107**, 084011 (2023). https://doi.org/10.1103/PhysRevD.107.084011. arXiv:2301.04879 [gr-qc]
13. Bardeen et al., Astrophys. J. **178**, 347 (1972). https://doi.org/10.1086/151796
14. Misner et al., Gravitation, ed. by P. U. Press (Princeton University Press, 1973)
15. Y. Kozai, Astrophys. J. **67**, 591 (1962). https://doi.org/10.1086/108790
16. M.L. Lidov, S.L. Ziglin, Celest. Mech. **13**, 471 (1976). https://doi.org/10.1007/BF01229100
17. S. Naoz, Annu. Rev. Astron. Astrophys. **54**, 441 (2016). https://doi.org/10.1146/annurev-astro-081915-023315

Integrable Holographic Defect CFTs

Charlotte Kristjansen and Konstantin Zarembo

Abstract We review a class of integrable, supersymmetric defect conformal field theories which have holographic duals in the form of probe brane models. Our main examples are defect versions of $\mathcal{N}=4$ SYM and ABJM theory, both involving a domain wall with Nahm pole boundary conditions, and the case of a 't Hooft line embedded in $\mathcal{N}=4$ SYM. The field theory defect respectively the probe D-brane can be described as an integrable boundary state of the integrable system underlying the AdS/CFT correspondence.

Keywords Defect CFTs · 't Hooft lines · Holography · Integrability · One-point functions

1 Introduction

Holography and integrability appear to be closely tied to each other with the most well established examples being the AdS_5/CFT_4 and the AdS_4/CFT_3 correspondence where in both cases the planar spectral problem is described in terms of an integrable spin chain [1]. Supersymmetry and conformal symmetry seem to be key prerequisites for this connection. From the original AdS/CFT systems one can engineer set-ups which conserve a subset of the supersymmetries and likewise carry conformal symmetry only in a certain subspace. We will consider configurations where one introduces a flat defect in the field theory and correspondingly a probe brane

C. Kristjansen (✉) · K. Zarembo
Niels Bohr International Academy, Niels Bohr Institute, University of Copenhagen, Copenhagen, Denmark
e-mail: kristjan@nbi.dk

K. Zarembo
e-mail: zarembo@nbi.ku.dk; zarembo@kth.se

K. Zarembo
Nordita, KTH Royal Institute of Technology and Stockholm University, Stockholm, Sweden

M. Paranjape et al. (eds.), *Proceedings of Gravity, Strings and Fields*,
CRM Series in Mathematical Physics,
https://doi.org/10.1007/978-3-031-91266-5_6

in the string theory in such a way that the entire system is 1/2 BPS and conformal symmetry is conserved on the defect, leading to a defect CFT on the field theory side. In particular, we will highlight three models where integrability is preserved after the reduction in conformal- and super-symmetry as a consequence of the defect/probe brane being identified with an integrable boundary state of the underlying spin chain. Two of these models have defects of co-dimension one and are built from respectively $\mathcal{N} = 4$ SYM theory and ABJM theory by the introduction of a domain wall at which Nahm pole boundary conditions are imposed on a part of the field components [2–5]. The third model involves a defect of co-dimension three and is defined by a 't Hooft line embedded in $\mathcal{N} = 4$ SYM theory [6, 7]. The three models are treated in respectively Sects. 2, 3 and 4, and we finish with a summary and outlook in Sect. 5. In addition, we have included an appendix which briefly reviews the concepts needed from the language of spin chain integrability.

2 The Nahm Pole Defect in $\mathcal{N} = 4$ SYM

Supersymmetric boundary conditions in $\mathcal{N} = 4$ SYM have been extensively studied, e.g. in [2, 8], and include Nahm pole boundary conditions [9–11] where three out of the six scalar fields of the theory diverge in a way compatible with conformal symmetry as the boundary is approached. With these boundary conditions as the starting point one can construct a defect CFT version of $\mathcal{N} = 4$ SYM theory which has an explicit holographic dual. More precisely one glues together two copies of $\mathcal{N} = 4$ SYM with different ranks of the gauge group, N and $N - k$, along a co-dimension one surface. The difference in the rank of the gauge group on the two sides of the defect is accounted for by Nahm pole boundary conditions on certain components of the classical fields, i.e.

$$\phi_i^{\text{cl}} = \frac{1}{x_3} t_i \oplus 0_{(N-k)\times(N-k)}\,, \qquad x_3 > 0, \qquad i = 1, 2, 3, \tag{1}$$

where the t_i's generate a k-dimensional irreducible representation of SU(2), i.e.

$$[t_i, t_j] = \epsilon_{ijk} t_k. \tag{2}$$

À priori, we do not impose any boundary conditions on the fields $\phi_{4,5,6}$ at the defect. In the holographically dual string theory picture the defect corresponds to a single probe $D5$-brane on which k of the N D3-branes of the standard AdS/CFT construction end. The configuration is summarized in Fig. 1. It constitutes a 1/2 BPS set-up preserving the residual OSp(4|4) symmetry. For a more detailed description we refer to [12–14].

Being a defect CFT the Nahm pole version of $\mathcal{N} = 4$ SYM theory is endowed with additional non-trivial conformal data as compared to a standard CFT. These include one-point functions, two-point functions of operators with unequal conformal dimensions, and correlation functions involving bulk and boundary fields. In the

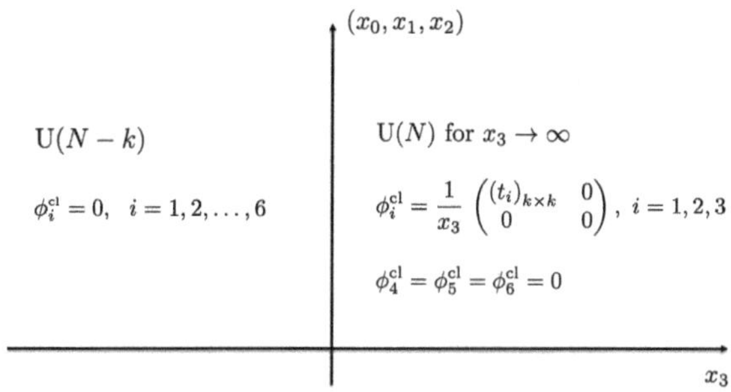

Fig. 1 The Nahm pole defect in $\mathcal{N} = 4$ SYM

present discussion we will concentrate on one-point functions which are bound to take the form

$$\langle \mathcal{O}_\Delta \rangle = \frac{C}{|x_3|^\Delta}, \tag{3}$$

with Δ being the conformal dimension. The single trace conformal operators built entirely from scalars are at one-loop order in one-to-one correspondence with the eigenstates, $|\mathbf{u}\rangle$, of an integrable SO(6) spin chain [15]

$$\mathcal{O} = \Psi^{I_1 \ldots I_L} \operatorname{tr} \phi_{I_1} \ldots \phi_{I_L} \sim |\mathbf{u}\rangle \tag{4}$$

The one-point function of such an operator is at leading order obtained simply by replacing each quantum field with its classical value. This process can be implemented by computing the overlap of the Bethe eigenstate with a matrix product state defined as follows

$$\mathrm{MPS}_k^{I_1 I_2 \ldots I_L} = \operatorname{tr} t_{I_1} t_{I_2} \ldots t_{I_L}, \tag{5}$$

with the understanding that $t_4 = t_5 = t_6 = 0$. More precisely, up to a field theoretical pre-factor the overlap C can be expressed as [3, 16]

$$C_k(\mathbf{u}) = \frac{\langle \mathrm{MPS}_k | \mathbf{u} \rangle}{\langle \mathbf{u} | \mathbf{u} \rangle^{\frac{1}{2}}}, \qquad k \geqslant 2. \tag{6}$$

In the special case $k = 1$ one can still construct a defect CFT in such a way that the configuration becomes 1/2 BPS by appropriately imposing Dirichlet respectively Neuman boundary conditions on the field components outside the $(N-k) \times (N-k)$ block for $x_3 \to 0_+$ [17]. In this case the matrix product state must be replaced by a valence bond state (i.e. a two-site product state), more precisely [18]

$$\mathrm{VBS}^{I_1 I_2 \ldots I_L} = K^{I_1 I_2} \ldots K^{I_{L-1} I_L}, \qquad K^{IJ} = 2 \sum_{I,J=1}^{3} \delta^{IJ} - 2 \sum_{I,J=4}^{6} \delta^{IJ}. \tag{7}$$

Using the language and tools of integrability the overlap between the Bethe eigenstate of the integrable $SO(6)$ spin chain and this valence bond state was determined to be [19]

$$\frac{\langle \mathrm{VBS}\,|\mathbf{u}\rangle}{\langle \mathbf{u}\,|\mathbf{u}\rangle^{\frac{1}{2}}} = 2^{-L}\sqrt{\frac{Q_1(0)Q_2(0)Q_3(0)}{Q_1\left(\frac{i}{2}\right)Q_2\left(\frac{i}{2}\right)Q_3\left(\frac{i}{2}\right)}\mathrm{Sdet}G}\,. \tag{8}$$

Here and in the later overlap formulas $Q_a(x)$ is the Baxter polynomial involving the Bethe roots associated with the a'th node of the Dynkin diagram chosen to represent the underlying symmetry algebra, and SdetG is the superdeterminant of the so-called Gaudin matrix which encodes the norm of the Bethe eigenstate. We notice, in particular, that the factor SdetG contains no reference to the defect. In Fig. 2 we show the $SO(6)$ Dynkin diagram and our convention for the numbering of the nodes. Furthermore, we have included an appendix with a brief introduction to the language of integrability.

The overlap with the matrix product state can likewise be found in a closed form [20].

$$C_k^{SO(6)} = \sqrt{\frac{Q_2(0)Q_2(\frac{i}{2})Q_2(\frac{ik}{2})Q_2(\frac{ik}{2})}{Q_3(0)Q_3(\frac{i}{2})Q_1(0)Q_1(\frac{i}{2})}}\cdot \mathbb{T}_{k-1}(0)\cdot\sqrt{\mathrm{Sdet}G}, \tag{9}$$

where

$$\mathbb{T}_n(x) = \sum_{a=-\frac{n}{2}}^{\frac{n}{2}}(x+ia)^L\frac{Q_3(x+ia)Q_1(x+ia)}{Q_2(x+i(a+\frac{1}{2}))Q_2(x+i(a-\frac{1}{2}))}. \tag{10}$$

Starting from the overlap with the valence bond state in the SO(6) case and requiring that the overlap is covariant under fermionic dualities one can deduce the dependence of the leading order overlap on Baxter Q-functions for the full defect $\mathcal{N}=4$ SYM theory for any of the possible Dynkin diagrams of its underlying symmetry algebra $\mathfrak{psu}(2,2|4)$ [21]. Here we give the formula for the overlap in the grading depicted in Fig. 3:

$$\frac{\langle \mathrm{VBS}\,|\mathbf{u}\rangle}{\langle \mathbf{u}\,|\mathbf{u}\rangle^{\frac{1}{2}}} = 2^{-L}\sqrt{\frac{Q_1(0)Q_3(0)Q_4(0)Q_5(0)Q_7(0)}{Q_2(0)Q_2(\frac{i}{2})Q_4(\frac{i}{2})Q_6(0)Q_6(\frac{i}{2})}S\det G} \tag{11}$$

In the so-called alternating grading, it is possible to derive an expression for the the overlap with the valence bond state valid to any loop order in the asymptotic

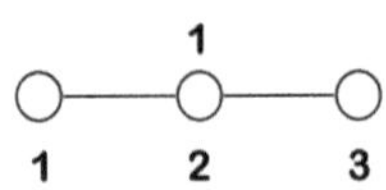

Fig. 2 The Dynkin diagram of $SO(6)$. Above the nodes are the weights of the relevant spin representation while the numbers below the nodes are the labels of the Q-functions

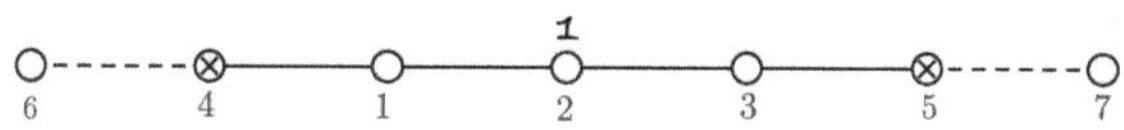

Fig. 3 One of the Dynkin diagrams of $PSU(2,2|4)$ and its associated overlap formula in Eq. (11)

limit [22–24] (i.e. excluding wrapping corrections [25]). The derivation is based on applying bootstrap arguments to the boundary reflection matrix of either the spin chain [22, 24] or the string sigma model involved [23]. It would be interesting to extend the analysis to including wrapping corrections e.g. by the thermodynamical Bethe ansatz approach. From the string theory perspective, it can be argued that the model is integrable in the classical limit [26].

3 The Nahm Pole Defect in ABJM Theory

The understanding of supersymmetric boundary conditions is much less developed for ABJM theory than for $\mathcal{N}=4$ SYM. Possible Nahm pole boundary conditions for the four complex scalar fields of ABJM theory were identified in [4] and can be used to define a domain wall version of ABJM theory in analogy with the construction for $\mathcal{N}=4$ SYM. The resulting defect set-up is illustrated in Fig. 4. It conserves half of the supersymmetries with the remaining symmetry group being $OSp(3|2)\times OSp(3|2)\times U(1)$. Two out of the four complex fields have a non-trivial Nahm pole configuration for $x_2>0$ where we notice that the x_2 scaling is now exactly as needed to produce a dCFT in three dimensions and where the two rectangular matrices S_i are given by

$$S^1_{ij}=\delta_{i,j-1}\sqrt{i},\qquad S^2_{ij}=\delta_{ij}\sqrt{q-i}\,,\qquad i=1,\ldots,q-1\qquad j=1,\ldots,q. \tag{12}$$

In the dual string theory description the defect corresponds to a single probe D4-brane on which q of the N D2-branes of the ABJM construction end. The matrices S^i are related to $SU(2)$ representation matrices in the following way: Defining the composite classical field

Fig. 4 The Nahm pole defect in ABJM theory

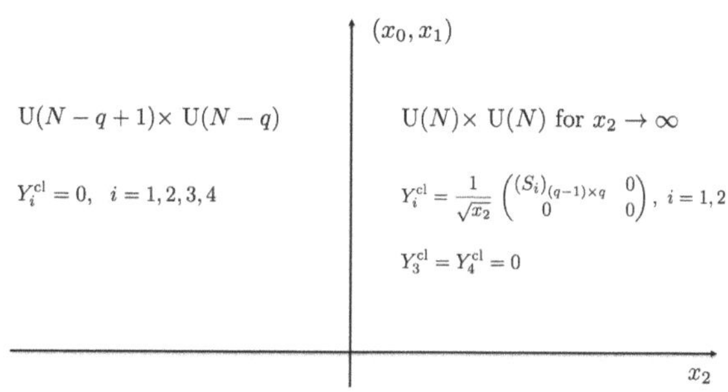

$$\Phi^{\alpha}{}_{\beta} = Y^{\alpha} Y^{\dagger}_{\beta} \equiv \Phi^{i} \sigma^{\alpha}_{i\,\beta} + \Phi \delta^{\alpha}{}_{\beta}, \tag{13}$$

one can show that [5]

$$\Phi^{i} = \frac{t^{i}}{x_2}, \qquad \Phi = \frac{q\mathbb{1}}{2x_2}. \tag{14}$$

where the t_i fulfill the $SU(2)$ commutation relations (2).

The single trace conformal operators of ABJM theory built entirely from scalars take the following form and are in one-to-one correspondence with the eigenstates, $|\mathbf{u}\rangle$, of an alternating $SU(4)$ spin chain [27]

$$\mathcal{O} = \Psi_{A_1 \dots A_{2L-1}}^{A_2 \dots A_{2L}} \operatorname{tr} Y^{A_1} Y^{\dagger}_{A_2} \dots Y^{A_{2L-1}} Y^{\dagger}_{A_{2L}}. \tag{15}$$

To compute the one-point functions of these operators in the Nahm pole background (for $x_2 > 0$) one needs to evaluate the overlap of the wavefunction Ψ with the wave function describing an appropriate matrix product state [5], i.e.

$$\mathrm{MPS_q}{}^{A_1 \dots A_{2L-1}}_{A_2 \dots A_{2L}} = \operatorname{tr} S^{A_1} S^{\dagger}_{A_2} \dots S^{A_{2L-1}} S^{\dagger}_{A_{2L}}. \tag{16}$$

We notice that for the special case $q = 2$ the matrix product state degenerates and becomes a valence bond state

$$\mathrm{VBS}\,{}^{\alpha_1 \dots \alpha_{2L-1}}_{\alpha_2 \dots \alpha_{2L}} = \mathrm{MPS_2}\,{}^{\alpha_1 \dots \alpha_{2L-1}}_{\alpha_2 \dots \alpha_{2L}} = \delta^{\alpha_1}_{\alpha_2} \dots \delta^{\alpha_{2L-1}}_{\alpha_{2L}}. \tag{17}$$

The overlap between the valence bond state and the eigenstates of the alternating $SU(4)$ spin chain was found in [28]:

$$\frac{\langle \mathrm{VBS}\,|\mathbf{u}\rangle}{\langle \mathbf{u}\,|\mathbf{u}\rangle^{\frac{1}{2}}} = 2^{-L}\, Q_2(i) \sqrt{\frac{\mathrm{Sdet}\, G}{Q_2(0) Q_2\left(\frac{i}{2}\right)}}. \tag{18}$$

where the relevant Dynkin diagram and associated spin representation labels can be seen in Fig. 5. By an iterative argument including various assumptions on factorization this overlap formula can be generalized to higher values of q [29]:

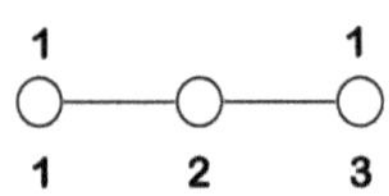

Fig. 5 The Dynkin diagram of $SU(4)$ with the associated overlap formula given in Eq. (18). Above the nodes are the weights of the relevant spin representation while the numbers below the nodes are the labels of the Q-functions

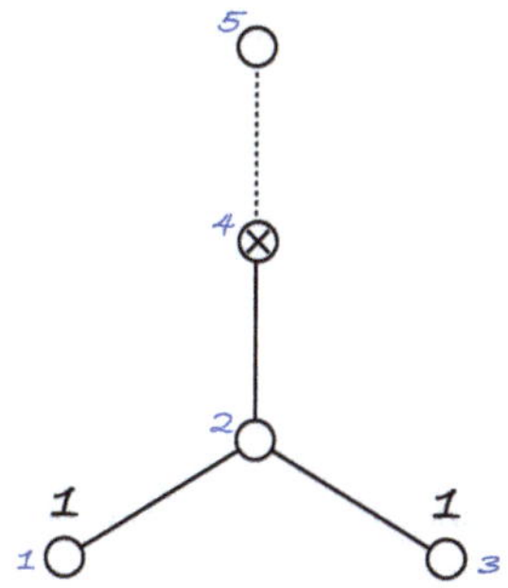

Fig. 6 One of the possible Dynkin diagrams of Osp(6|4) with the associated overlap formula given in Eq. (20)

$$\frac{\langle \mathrm{MPS}_q | \mathbf{u}\rangle}{\sqrt{\langle \mathbf{u}|\mathbf{u}\rangle}} = \sum_{k=1}^{q-1} \frac{Q_1(\frac{i}{2})Q_1(i(q-\frac{1}{2}))}{Q_1(i(k-\frac{1}{2}))Q_1(i(k+\frac{1}{2}))} \times \left(\frac{k}{2}\right)^L Q_2(ik)\sqrt{\frac{\mathrm{Sdet}\, G}{Q_2(0)Q_2(i/2)}}. \tag{19}$$

Invoking again invariance of the formula (18) under fermionic duality transformations one can deduce the form of the overlap with the valence bond state for any of the five possible Dynkin diagrams for the algebra $Osp(6|4)$ [5]. Here we give the formula for the diagram which has the $SU(4)$ Dynkin diagram above as a sub-diagram (Fig. 6):

$$\frac{\langle \mathrm{VBS}\,|\mathbf{u}\rangle}{\langle \mathbf{u}\,|\mathbf{u}\rangle^{\frac{1}{2}}} = 2^{-L}\, Q_2(i)\sqrt{\frac{Q_4(0)Q_5(0)}{Q_2(0)\, Q_2\left(\frac{i}{2}\right) Q_5\left(\frac{i}{2}\right)}}\ \mathrm{Sdet}\, G\,. \tag{20}$$

Unlike what is the case for $\mathcal{N} = 4$ SYM the overlap formula has not yet been extended to higher loop orders. The dual string theory can be argued to be integrable in the classical limit [30].

4 't Hooft Loops

The 't Hooft loop is a line defect of co-dimension three. Line defects describe point particles moving along predefined trajectories while interacting with the quantized gauge fields. Perhaps the best known example of a line operator is the Wilson loop whose expectation value measures the response of the vacuum to an external color charge. Here we consider 't Hooft loops [31] which correspond to point-like magnetic monopoles. In the $\mathcal{N} = 4$ SYM theory Wilson and 't Hooft loops are interchanged by S-duality, an exact quantum counterpart to the electric-magnetic symmetry of electromagnetism [32].

The 't Hooft loop is a disorder operator defined by singular boundary conditions on the field. Those are, for the straight line in the time direction:

$$F_{ij} \overset{r\to 0}{\simeq} q\varepsilon_{ijk}\frac{x^k}{r^3},$$
$$\phi_I \overset{r\to 0}{\simeq} q\,\frac{n_I}{r}, \tag{21}$$

where n_I is a unit six-vector defining the R-symmetry orientation of the monopole, and $q = \mathrm{diag}(q_1, \ldots, q_N)$ is the diagonal embedding of the monopole charges in the $U(N)$ gauge group. For simplicity we consider the case of just one non-zero charge $q_1 \equiv q$ and all other $q_i = 0$. Only the 11 element of the background field is then non-trivial (cf. Fig. 1). The $U(N)$ gauge symmetry gets broken to $U(N-1) \times U(1)$, with the off-diagonal components ϕ_{1i} and $\phi_{i1} = \phi^*_{1i}$ charged under the $U(1)$ subgroup where the monopole resides.

The path integral computed with these boundary conditions defines the expectation value and correlators of the 't Hooft loop. The boundary conditions are consistent with the equations of motion, so (21) can be regarded as the classical background field, just like the Nahm solution for the domain wall. In a highly supersymmertic gauge theory such as $\mathcal{N} = 4$ SYM, quantum corrections to the external field should vanish to all orders. The scalar coupling of the 't Hooft loop is essential for that [6, 33]: with this choice the monopole preserves half of the supersymmetry. Together with the residual conformal symmetries of the line the unbroken supercharges form the $\mathfrak{osp}(4^*|4)$ superalgebra [34].

The AdS dual of the 't Hooft line is a D1-brane in the bulk of $AdS_5 \times S^5$. The boundary conditions of strings ending on such a D-brane preserve integrability [35], and we expect that the 't Hooft line defines an integrable defect CFT.

The magnetic charges q_i are quantized in half-integer units. This is the famous Dirac quantization condition [36], arising from consistency of charged fields in the monopole background. Charge quantization is an important point that deserves an extended discussion.

One way to see why the monopole charge cannot be arbitrary (not the simplest one but with important implications) is to require rotational symmetry. Invariance under rotations seems an obvious property of the monopole background (21) but subtle quantum effects may in fact break it unless certain quantization conditions are imposed. Rotations are generated by the angular momentum operator, which for the monopole was found in [37]:

$$L_i = -i\varepsilon_{ijk}x_j D_k - q\,\frac{x_i}{r}\,. \tag{22}$$

This operator satisfies the $\mathfrak{su}(2)$ commutation relations $[L_i, L_j] = i\varepsilon_{ijk}L_k$ and commutes with the Hamiltonian[1]:

$$K = -D_\mu^2 + \Phi_I^2 = -\partial_t^2 - D_i^2 + \frac{q^2}{r^2}, \tag{23}$$

where $D_i = \partial_i + iA_i$ is the covariant derivative in the field of the monopole. The scalar coupling comes from the $[\phi_I, \phi_J]^2$ term in the SYM Lagrangian, and combines nicely with the centrifugal energy [37] $[\ell(\ell+1) - q^2]/r$. The q^2 term cancels out giving rise to the conventional radial Hamiltonian of a free particle, as if the monopole were not present. The only influence of the background field on the radial motion is through the allowed range of ℓ's.

The mere existence of the angular momentum does not in itself guarantee rotational symmetry, L_i must have a representation as operators in the Hilbert space. The $\mathfrak{su}(2)$ multiplets are then constructed as usual from the highest weight by raising and lowering with $L_\pm = L_1 \pm iL_2$:

$$L_3 Y_{\ell\ell} = \ell Y_{\ell\ell}, \qquad L_+ Y_{\ell\ell} = 0, \qquad Y_{\ell m} \propto L_-^{\ell - m} Y_{\ell\ell}, \tag{24}$$

thus defining the full set of spherical functions. This is not different from textbook Quantum Mechanics. The tricky point is defining L_i, for which we need to choose a gauge. One (slightly unconventional) choice is[2]

$$A = -q \cos\theta \, d\varphi. \tag{25}$$

The components of the angular momentum are then given by

$$L_3 = -i\partial_\varphi, \qquad L_\pm = \mathrm{e}^{\pm i\varphi}\left(\pm\partial_\theta + i\cot\theta\,\partial_\varphi - \frac{q}{\sin\theta}\right). \tag{26}$$

It is now straightforward to solve the highest-weight conditions. By simple integration we get:

$$Y_{\ell\ell} = \mathrm{e}^{i\ell\varphi} \sin^l\theta \tan^q\frac{\theta}{2}. \tag{27}$$

The key point is that this should be a regular function on S^2. The condition of regularity at $\theta = 0$ and $\theta = \pi$ (such that $Y_{\ell\ell} \in L^\infty(S^2)$) requires that

---

[1] The Hamiltonian is the Klein-Gordon operator of a charged scalar in the monopole background. For vectors [38, 39] and fermions [40], **L** does not commute with the Hamiltonian by itself. One needs to include spin, since only the total angular momentum is conserved.

[2] Then $F = dA$ is the volume form on S^2 as required by (21). In this gauge, the Dirac string pierces through the monopole in both directions. Normally one eliminates half of the string by a gauge transformation $A \to A \pm q\,d\varphi$, which makes the gauge potential regular on the upper/lower hemisphere. Our choice is more singular but also more symmetric, and has an advantage of leaving L_3 unaffected thus removing some unphysical phases in the Green's functions.

$$\ell \pm q \in \mathbb{Z}_+, \tag{28}$$

which means that $\ell = q, q+1, \ldots$ and both $2q \in \mathbb{Z}_+$ and $2l \in \mathbb{Z}_+$. The lower harmonics are obtained by repeated application of L_-, and can be expressed through the Jacobi polynomials [41]. It is important to notice that the monopole leads to fractionalization of the angular momentum: for the minimal magnetic charge $q = 1/2$ the orbital angular momentum is quantized in *half-integers*: $\ell = 1/2, 3/2, \ldots$

The monopole harmonics $Y_{\ell m}(\theta, \varphi)$ retain many properties of the usual spherical functions [42, 43], and can be used to construct the eigenbasis of the Klein-Gordon operator (23) and its cousins. The scalar propagator, for example, admits the following spectral representation:

$$D(x, x') = \frac{1}{8\pi^2 r r'} \left(\frac{1+\eta}{2}\right)^q \sum_{\ell=q}^{\infty} (2\ell+1) Q_\ell(\xi) P_{\ell-q}^{(0,2q)}(\eta), \tag{29}$$

where $P_n^{(\alpha,\beta)}(\eta)$ are the Jacobi polynomials normalized as $P_n^{(0,\beta)}(1) = 1$ and $Q_\ell(\xi)$ is the Legendre function of the second kind. The variables ξ and η are the standard conformal cross-ratios of two points and a line [44]:

$$\xi = \frac{(t-t')^2 + r^2 + r'^2}{2rr'},$$
$$\eta = \frac{\mathbf{x} \cdot \mathbf{x}'}{rr'}. \tag{30}$$

Expansion in spherical harmonics can be alternatively viewed as a Kaluza-Klein reduction on the sphere, with the radial coordinate and time combining into AdS_2:

$$ds^2 = r^2 \left(\frac{dt^2 + dr^2}{r^2} + d\Omega_{S^2}^2\right). \tag{31}$$

The cross-ratios ξ and η are simply related to the geodesic distances on AdS_2 and S^2, respectively. The function $Q_\ell(\xi)$ appearing in the spectral decomposition can be identified with the AdS_2 propagator of a minimally coupled scalar of mass $m^2 = \ell(\ell+1)$.[3] By the usual AdS/CFT logic each KK mode corresponds to a boundary operator of dimension $\Delta = \ell + 1$. In the present context these are the defect operators, loosely equivalent to field insertions in the 't Hooft line.

[3] The generic AdS_d propagator is a hypergeometric function [45]. The latter is equivalent to the Legendre function when specified to $d = 2$.

4.1 Protected Operators

The propagators in the monopole background are necessary to develop perturbation theory for the 't Hooft loop. Its complexity is comparable to perturbation theory in AdS/CFT, the propagators are just the same. The simplest among them are the "anomalous" propagators that vanish in the absence of the monopole due to R-symmetry constraints. Consider, as an example, a complex scalar

$$Z = (n_I + i n'_I)\phi_I, \tag{32}$$

where n_I is the vector from the definition of the 't Hooft loop and n'_I is a unit six-vector orthogonal to it: $n \cdot n' = 0$. The longitudinal and transverse scalars enter with the opposite signs in the $\langle ZZ \rangle$ correlator, and tend to cancel one another. The cancellation is complete in vacuo, when the scalars have the same propagator. In the monopole background the propagators are different,[4] but drastic simplifications still occur: as shown in [7] for $q = 1/2$ and in the collinear limit $\eta = 1$, the sum over modes telescopes leaving a simple finite residue:

$$\left\langle Z_{1i}(x) Z_{j1}(x') \right\rangle = \delta_{ij} \frac{Q_{-1/2}(\xi) - Q_{1/2}(\xi)}{32\pi^2 r r'} = \delta_{ij} \frac{E\left(\frac{1+\xi}{2}\right)}{16\pi^2 r r'}. \tag{33}$$

Here i and j are color indices of the unbroken $U(N-1)$, and $E(x)$ is the complete elliptic integral of the second kind. Importantly, the anomalous propagator has a finite OPE limit $\xi \to 1$:

$$\left\langle Z_{1i}(x) Z_{j1}(x) \right\rangle = \delta_{ij} \frac{1}{16\pi^2 r^2}. \tag{34}$$

We can use this result to calculate the expectation values of the chiral primary operators:

$$\mathcal{O}_{\text{CPO}} = \operatorname{tr} Z^L. \tag{35}$$

At the leading order, just substituting the classical solution (21) gives $(2r)^{-L}$. The one-loop correction is given by the $Z - Z$ propagator evaluated at coincident points. From (34) we get:

$$\langle \mathcal{O}_{\text{CPO}} \rangle = (2r)^{-L} \left(1 + \frac{\lambda L}{4\pi^2} + \ldots \right). \tag{36}$$

A comment on combinatorics is in order here. Only Z_{1j} fields have a non-trivial two-point function and only Z_{11} has a tree-level expectation value. We may connect any pair of Zs by a propagator, creating a loop with the color index 1 outside and

[4] The transverse scalar $n' \cdot \phi$ literally has the propagator (29). The story is little bit more complicated for the longitudinal scalar $n \cdot \phi$ which mixes with the gauge fields [7]. The spin-ℓ term in its spectral decomposition contains three terms with different AdS masses: Q_ℓ, $Q_{\ell-1}$ and $Q_{\ell+1}$ (see [7] for more details).

$i \neq 1$ inside, but because $\langle Z_{ii} \rangle = 0$ no fields should be left inside, meaning that only nearest neighbors can interact. That gives a combinatorial factor of L.

The loop corrections can be actually computed to all orders using localization of the path integral, worked out for the 't Hooft loop in [46]. We can also start (following [7]) with the known expectation value of a CPO in the presence of the Wilson line [47, 48] and then apply S-duality. The S-duality inverts the coupling: $\lambda \to 16\pi^2 N^2/\lambda$ and interchanges Wilson and 't Hooft loops. That gives, in the planar approximation [7][5]:

$$\langle \mathcal{O}_{\mathrm{CPO}} \rangle = (4r)^{-L} \left[\left(\sqrt{1 + \frac{\lambda}{\pi^2}} + 1 \right)^L - \left(\sqrt{1 + \frac{\lambda}{\pi^2}} - 1 \right)^L \right]. \tag{37}$$

Expanding in λ we recover (36).

4.2 $SO(6)$ Sector

As for non-protected operators, their one-point functions are way more complicated but amenable to integrability methods quite similar to those for the domain wall defects. We shall discuss a few particular cases that illustrate the general approach (yet to be developed), restricting to the tree-level one-point functions obtained by replacing constituent fields of the operator by the classical background (21). The result can be always represented as an overlap of the spin-chain wavefunction, e.g. Ψ from (4), with a fixed boundary state. If we are lucky and this boundary state has special integrability properties, its overlaps are given by a determinant expression similar to the ones encountered in Sects. 2 and 3.

For the scalar operators in the $SO(6)$ sector, of generic form (4), the tree-level expectation value is the overlap with a tensor product state:

$$C^{SO(6)}_{\mathrm{tHooft}}(\mathbf{u}) = 2^{-L} \frac{\langle \mathrm{TPS} \,|\mathbf{u}\rangle}{\langle \mathbf{u}\,|\mathbf{u}\rangle^{\frac{1}{2}}}, \qquad \mathrm{TPS}^{I_1 \ldots I_L} = n^{I_1} \ldots n^{I_L}. \tag{38}$$

In other words, the one-point function measures a particular component of the wavefunction, for instance $\Psi_{1\ldots1}$. This state is integrable and its overlaps admit the following determinant representation [49]:

$$\frac{\langle \mathrm{TPS} \,|\mathbf{u}\rangle}{\langle \mathbf{u}\,|\mathbf{u}\rangle^{\frac{1}{2}}} = 2^{-\frac{L}{2}} \sqrt{\frac{Q_2(0) Q_2\left(\frac{i}{2}\right)}{Q_1(0) Q_1\left(\frac{i}{2}\right) Q_3(0) Q_3\left(\frac{i}{2}\right)}}\ \mathrm{Sdet}\, G\,. \tag{39}$$

The existence of a determinant formula signals integrability of the boundary state thus confirming expectations that the 't Hooft line defines an integrable dCFT.

[5] Our normalization conventions differ from those in [7] where the operators are rescaled to have a unit two-point functions far away from the defect.

4.3 $SL(2)$ Sector

Operators with spin can acquire expectation values only for defects of co-dimension greater than one. To illustrate this new feature we consider operators obtained from the chiral primary tr Z^L by dressing it with the light-cone derivatives, $D \equiv D_0 + D_3$ (in Minkowski signature):

$$\mathcal{O} = \sum_{\{n_\ell\}} \frac{1}{n_1! \dots n_L!} \Psi_{n_1 \dots n_L} \operatorname{tr} D^{n_1} Z \dots D^{n_L} Z. \tag{40}$$

An operator with $S = n_1 + \dots + n_L$ derivatives has spin S, twist L and bare dimension $L + S$.

It is often convenient to consider the generating function instead of the local operators. For a single field the generating function is

$$U Z(x + \xi v) U^{-1} = \sum_{n=0}^{\infty} \frac{\xi^n}{n!} D^n Z(x), \tag{41}$$

where $v^\mu = (1, 0, 0, 1)$ and U is the light-like Wilson line:

$$U = \mathrm{P}\exp\left(\int_0^\xi d\eta \, v^\mu A_\mu(x + \eta v) \right). \tag{42}$$

The generating function for generic operators is a light-ray Wilson line with L insertions of Z. The dilatation operator in the $\mathfrak{sl}(2)$ sector can be reformulated entirely in terms of the light-ray Wilson lines [50].

The expectation values in the monopole background at the leading order are obtained by substituting the classical background (21) into the operator at hand. The classical field is Abelian and time-independent, so the Wilson line cancels and the light-cone shift effectively reduces to spacial translations along $\mathbf{v} = (0, 0, 1)$. Comparing the Legendre expansion [51]:

$$\frac{1}{|\mathbf{x} + \text{¸}\mathbf{v}|} = \sum_{n=0}^{\infty} \frac{(-\xi)^n}{r^{n+1}} P_n(\cos\theta), \tag{43}$$

with (41) we can read off the boundary state:

$$|B(\theta)\rangle = \sum_{n=0}^{\infty} (-1)^n P_n(\cos\theta) \, |n\rangle \, . \tag{44}$$

This gives an overlap representation for the one-point function:

$$\langle \mathcal{O} \rangle = \frac{1}{2^L r^{L+S}} \frac{\langle \mathrm{Bsl}\,|\mathbf{u}\rangle}{\langle \mathbf{u}\,|\mathbf{u}\rangle^{\frac{1}{2}}}\,, \tag{45}$$

where $|\mathrm{Bsl}\rangle$ is again of the tensor-product form:

$$|\mathrm{Bsl}\rangle = |B\rangle \otimes \cdots \otimes |B\rangle\,. \tag{46}$$

Overlaps with the eigenstates of the dilatation operator admit a determinant representation:

$$\frac{\langle \mathrm{Bsl}\,|\mathbf{u}\rangle}{\langle \mathbf{u}\,|\mathbf{u}\rangle^{\frac{1}{2}}} = (\sin\theta)^S \sqrt{\frac{Q(0)}{Q\left(\frac{i}{2}\right)}}\ \mathrm{Sdet}\, G\,. \tag{47}$$

This formula was conjectured [7] on account of structural similarity to other integrable boundary states and was rigorously proven in [52]. Noticeably, its prefactor vanishes at $\theta = 0$, unless $S = 0$. This follows from the representation theory of $\mathfrak{sl}(2)$. The Bethe eigenstates $|\mathbf{u}\rangle$ are $\mathfrak{sl}(2)$ primaries annihilated by the superconformal generator: $K\,|\mathbf{u}\rangle = 0$. Since $P_n(1) = 1$,

$$|B(0)\rangle_\ell = \sum_{n=0}^{\infty}(-1)^n\,|n\rangle_\ell = \sum_{n=0}^{\infty}\frac{(-1)^n}{n!}P_\ell^n\,|n\rangle_\ell = \mathrm{e}^{-P_\ell}\,|0\rangle_\ell\,, \tag{48}$$

where we explicitly label the position ℓ in the tensor product. The total tensor product is thus a vacuum descendant:

$$|\mathrm{Bsl}(0)\rangle = \mathrm{e}^{-P}\,|0\rangle\,. \tag{49}$$

Since $P^\dagger = K$, we find that $\langle \mathrm{Bsl}(0)\,|\mathbf{u}\rangle = 0$ unless $\mathbf{u}$ is the trivial ground state.

4.4 Gluon Operators

Another set of operators with spin, forming a closed sector under renormalization,[6] is constructed from the (anti-)self-dual components of the field strength, which we single out with the help of the ’t Hooft symbols [53]:

$$F_i^{\pm} = \frac{1}{2}\eta^{\pm}_{i\mu\nu}F^{\mu\nu}, \qquad \eta_{i\mu\nu} = \varepsilon_{0i\mu\nu} \pm \delta_{0\mu}\delta_{i\nu} \mp \delta_{0\nu}\delta_{i\mu}. \tag{50}$$

[6] At one loop.

The linear combinations of the form

$$\mathcal{O} = \Psi^{i_1 \ldots i_L} \operatorname{tr} F^{\pm}_{i_1} \ldots F^{\pm}_{i_L} \tag{51}$$

can be interpreted as states in the quantum $SU(2)$ spin chain of spin one.[7]

The boundary state is a tensor product:

$$\mathrm{Bgl}_{i_1 \ldots i_L} = x_{i_1} \ldots x_{i_L}, \tag{52}$$

whose overlaps with the spin-chain eigenstates again admit a determinant representation [7]:

$$\frac{\langle \mathrm{Bgl}\,|\mathbf{u}\rangle}{\langle \mathbf{u}\,|\mathbf{u}\rangle^{\frac{1}{2}}} = \left(\frac{r}{\sqrt{2}}\right)^{L} (\sin\theta)^{S} \sqrt{\frac{Q(0)}{Q\left(\frac{i}{2}\right)}}\ \mathrm{Sdet}\, G\,, \tag{53}$$

where S is the $SU(2)$ spin.

5 Summary and Outlook

We have reviewed three holographic set-ups where the introduction of a defect breaks half of the supersymmetries and limits conformal symmetry to a sub-space but where nevertheless integrability is preserved. Whereas all-loop asymptotic integrability has only been explicitly demonstrated for the Nahm pole version of $\mathcal{N} = 4$ SYM theory, it is expected to hold for the domain wall version of ABJM theory and for the 't Hooft loop in $\mathcal{N} = 4$ SYM theory as well. Other Nahm pole versions of $\mathcal{N} = 4$ SYM with no supersymmetry are known to be integrable only at the leading order [19, 22] or not integrable at all [54]. For ABJM theory, no other Nahm pole solutions apart from the one presented here are known but it is an interesting open question whether monopole operators could be described in terms of integrable boundary states in the same way as the 't Hooft line in $\mathcal{N} = 4$ SYM theory. It would likewise be interesting to investigate whether there exists integrable defects of even co-dimension in any of these theories. All of the known integrable cases correspond to defects of odd co-dimension.

Finally, progress on integrability of lower dimensional versions of the AdS/CFT correspondence [55–58] opens an avenue for revealing novel types of holographic integrable defect CFTs.

Acknowledgements The authors wish to thank Manu Paranjape for the invitation to the Conference in honor of Gordon Semenoff. C.K. was supported by DFF-FNU through grant number 1026-00103B. K.Z. was supported by VR grant 2021-04578.

---

[7] The 't Hooft symbols decompose the field strength into irreducible $(\mathbf{3}, 0) \oplus (0, \mathbf{3})$ representations of $SO(4) \simeq SU_L(2) \times SU_R(2)$.

A On Spin Chain Integrability

Good conformal operators of $\mathcal{N} = 4$ SYM and of ABJM theory correspond to eigenstates of integrable spin chains with underlying symmetry algebras respectively $\mathfrak{psu}(2,2|4)$ and $\mathfrak{osp}(6|4)$ [15, 27]. The starting point of AdS/CFT integrability is the fact that spin chains with nearest neighbor interactions and with these symmetries, or more generally symmetries derivable from $\mathfrak{gl}(N|M)$ symmetry are solvable by means of the Bethe ansatz approach [59]. The choice of Dynkin diagram of the symmetry algebra (which is not unique for super Lie algebras) plays an important role in this description. An eigenstate of the spin chain in question is characterized by a set of Bethe roots $u_{a,j}$ where a is an index that numbers the nodes of the Dynkin diagram and j is an index which simply counts the number of roots associated to a particular node, i.e. $j = 1, \ldots, K_a$. Knowing the Bethe roots one can explicitly write down the wave function of the eigenstate and likewise the expression for its corresponding conformal operator. The Bethe roots have to fulfil the following equations to produce an eigenstate:

$$\left(\frac{u_{a,j} - \frac{iq_a}{2}}{u_{a,j} + \frac{iq_a}{2}}\right)^L \prod_{b,k} \frac{u_{a,j} - u_{b,k} + \frac{iM_{ab}}{2}}{u_{a,j} - u_{b,k} - \frac{iM_{ab}}{2}} \equiv \mathrm{e}^{i\chi_{a,j}} = -1,$$

where M_{ab} is the Cartan matrix of the symmetry algebra, and a, b hence run from 1 to n with n being the number of nodes of the Dynkin diagram. Furthermore, q_a consists of the labels of the representation in which the spin variable of the chain lives. A root $u_{a,j}$ belonging to a node a for which $q_a \neq 0$ is said to be momentum carrying. A convenient way of representing a Bethe eigenstate is via the associated Baxter polynomials

$$Q_a(u) = \prod_{j=1}^{K_a} \left(u - u_{a,j}\right), \tag{54}$$

with $a = 1, \ldots, n$, $j = 1, \ldots, K_a$. The norm of a Bethe eigenstate is encoded in the determinant of the so-called Gaudin matrix [60] which is defined as follows

$$G_{aj,bk} = \frac{\partial \chi_{a,j}}{\partial u_{b,k}}. \tag{55}$$

A boundary state is just a particular spin chain state, which is not a simple linear combination of eigenstates, such as e.g. the matrix product state defined by the wave function in Eq. (5). An interesting question is when a boundary state can be characterized as being integrable meaning in this connection that its overlaps with the Bethe eigenstates can be found in closed form. Generalizing the concept of an integrable boundary state from field theory [61] one can argue that an integrable spin chain boundary state must be a state which constitutes a superposition of pairs of excitations with momenta $+p$ and $-p$ [62]. This translates into the statement that the

Bethe roots at the momentum carrying nodes must come in pairs with opposite sign. This can happen in two different ways which are denoted as chiral or a-chiral [22]. In the chiral case the roots at a single momentum carrying node are paired with roots from the same node. In the a-chiral case roots from one node are paired with roots from another node which appears in a symmetric way in the Dynkin diagram. The SO(6) overlaps calculated on the basis of the Dynkin diagram in Fig. 2 are chiral whereas the SU(4) overlaps calculated on the basis of the Dynkin diagram in Fig. 5 are a-chiral. In both the chiral and the a-chiral cases the collection of roots has a Z_2 symmetry which can be used to define the super determinant of the Gaudin matrix, SdetG, [18], which appears in the various overlap formulas. Special care is needed to define the Gaudin super determinant if there are Bethe roots at zero or at $\pm i/2$. Likewise, in overlap formulas singular roots should be excluded from the Baxter polynomials [63–65].

References

1. N. Beisert, C. Ahn, L.F. Alday, Z. Bajnok, J.M. Drummond, et al., Review of AdS/CFT Integrability: an overview. Lett. Math. Phys. **99**, 3–32 (2012). arXiv:1012.3982, https://doi.org/10.1007/s11005-011-0529-2
2. D. Gaiotto, E. Witten, Supersymmetric boundary conditions in N = 4 super Yang-Mills theory. J. Statist. Phys. **135**, 789–855 (2009). arXiv:0804.2902, https://doi.org/10.1007/s10955-009-9687-3
3. M. de Leeuw, C. Kristjansen, K. Zarembo, One-point functions in defect CFT and integrability. JHEP **08**, 098 (2015). arXiv:1506.06958, https://doi.org/10.1007/JHEP08(2015)098
4. S. Terashima, On M5-branes in $\mathcal{N} = 6$ membrane action. JHEP **08**, 080 (2008). arXiv:0807.0197, https://doi.org/10.1088/1126-6708/2008/08/080
5. C. Kristjansen, D.-L. Vu, K. Zarembo, Integrable domain walls in ABJM theory. JHEP **02**, 070 (2022). arXiv:2112.10438, https://doi.org/10.1007/JHEP02(2022)070
6. A. Kapustin, Wilson-'t Hooft operators in four-dimensional gauge theories and S-duality. Phys. Rev. D **74**, 025005 (2006). arXiv:hep-th/0501015, https://doi.org/10.1103/PhysRevD.74.025005
7. C. Kristjansen, K. Zarembo, 't Hooft loops and integrability. JHEP **08**, 184 (2023). arXiv:2305.03649, https://doi.org/10.1007/JHEP08(2023)184
8. D. Gaiotto, E. Witten, Janus configurations, Chern-Simons couplings, and the theta-angle in N = 4 super Yang-Mills theory. JHEP **06**, 097 (2010). arXiv:0804.2907, https://doi.org/10.1007/JHEP06(2010)097
9. W. Nahm, A simple formalism for the BPS monopole. Phys. Lett. B **90**, 413 (1980). https://doi.org/10.1016/0370-2693(80)90961-2
10. D.-E. Diaconescu, D-branes, monopoles and Nahm equations. Nucl. Phys. **B503**, 220–238 (1997). arXiv:hep-th/9608163, https://doi.org/10.1016/S0550-3213(97)00438-0
11. N.R. Constable, R.C. Myers, O. Tafjord, The Noncommutative bion core. Phys. Rev. D **61**, 106009 (2000). arXiv:hep-th/9911136, https://doi.org/10.1103/PhysRevD.61.106009
12. M. de Leeuw, A.C. Ipsen, C. Kristjansen, M. Wilhelm, Introduction to integrability and one-point functions in $\mathcal{N} = 4$ supersymmetric Yang-Mills theory and its defect cousin. Les Houches Lect. Notes **106**, 352 (2019). arXiv:1708.02525, https://doi.org/10.1093/oso/9780198828150.003.0008
13. M. de Leeuw, One-point functions in AdS/dCFT. J. Phys. A **53**(28), 283001 (2020). arXiv:1908.03444, https://doi.org/10.1088/1751-8121/ab15fb

14. G. Linardopoulos, Solving holographic defects. PoS CORFU2019 (2020) 141. arXiv:2005.02117, https://doi.org/10.22323/1.376.0141
15. J.A. Minahan, K. Zarembo, The bethe-ansatz for $\mathcal{N} = 4$ super yang-mills. JHEP **0303**, 013 (2003). arXiv:hep-th/0212208
16. I. Buhl-Mortensen, M. de Leeuw, C. Kristjansen, K. Zarembo, One-point functions in AdS/dCFT from matrix product states. JHEP **02**, 052 (2016). arXiv:1512.02532, https://doi.org/10.1007/JHEP02(2016)052
17. M. de Leeuw, A.C. Ipsen, C. Kristjansen, K.E. Vardinghus, M. Wilhelm, Two-point functions in AdS/dCFT and the boundary conformal bootstrap equations. JHEP **08**, 020 (2017). arXiv:1705.03898, https://doi.org/10.1007/JHEP08(2017)020
18. C. Kristjansen, D. Müller, K. Zarembo, Integrable boundary states in D3–D5 dCFT: beyond scalars. JHEP **08**, 103 (2020). arXiv:2005.01392, https://doi.org/10.1007/JHEP08(2020)103
19. M. De Leeuw, T. Gombor, C. Kristjansen, G. Linardopoulos, B. Pozsgay, Spin chain overlaps and the twisted Yangian. JHEP **01**, 176 (2020). arXiv:1912.09338, https://doi.org/10.1007/JHEP01(2020)176
20. M. De Leeuw, C. Kristjansen, G. Linardopoulos, Scalar one-point functions and matrix product states of AdS/dCFT. Phys. Lett. B **781**, 238–243 (2018). arXiv:1802.01598, https://doi.org/10.1016/j.physletb.2018.03.083
21. C. Kristjansen, D. Müller, K. Zarembo, Overlaps and fermionic dualities for integrable super spin chains. JHEP **03**, 100 (2021). arXiv:2011.12192, https://doi.org/10.1007/JHEP03(2021)100
22. T. Gombor, Z. Bajnok, Boundary states, overlaps, nesting and bootstrapping AdS/dCFT. JHEP **10**, 123 (2020). arXiv:2004.11329, https://doi.org/10.1007/JHEP10(2020)123
23. S. Komatsu, Y. Wang, Non-perturbative defect one-point functions in planar $\mathcal{N} = 4$ super-Yang-Mills. Nucl. Phys. B **958**, 115120 (2020). arXiv:2004.09514, https://doi.org/10.1016/j.nuclphysb.2020.115120
24. T. Gombor, Z. Bajnok, Boundary state bootstrap and asymptotic overlaps in AdS/dCFT. JHEP **03**, 222 (2021). arXiv:2006.16151, https://doi.org/10.1007/JHEP03(2021)222
25. J. Ambjorn, R.A. Janik, C. Kristjansen, Wrapping interactions and a new source of corrections to the spin-chain/string duality. Nucl. Phys. **B736**, 288–301 (2006). arXiv:hep-th/0510171, https://doi.org/10.1016/j.nuclphysb.2005.12.007
26. G. Linardopoulos, K. Zarembo, String integrability of defect CFT and dynamical reflection matrices. JHEP **05**, 203 (2021). arXiv:2102.12381, https://doi.org/10.1007/JHEP05(2021)203
27. J.A. Minahan, K. Zarembo, The Bethe ansatz for superconformal Chern-Simons. JHEP **09**, 040 (2008). arXiv:0806.3951, https://doi.org/10.1088/1126-6708/2008/09/040
28. T. Gombor, On exact overlaps for gl(N) symmetric spin chains. Nucl. Phys. B **983**, 115909 (2022). arXiv:2110.07960, https://doi.org/10.1016/j.nuclphysb.2022.115909
29. T. Gombor, C. Kristjansen, Overlaps for matrix product states of arbitrary bond dimension in ABJM theory. Phys. Lett. B **834**, 137428 (2022). arXiv:2207.06866, https://doi.org/10.1016/j.physletb.2022.137428
30. G. Linardopoulos, String integrability of the ABJM defect. JHEP **06**, 033 (2022). arXiv:2202.06824, https://doi.org/10.1007/JHEP06(2022)033
31. G. 't Hooft, On the phase transition towards permanent quark Confinement. Nucl. Phys. B138, 1 (1978). https://doi.org/10.1016/0550-3213(78)90153-0
32. C. Montonen, D.I. Olive, Magnetic monopoles as gauge particles? Phys. Lett. B **72**, 117 (1977). https://doi.org/10.1016/0370-2693(77)90076-4
33. J. Gomis, T. Okuda, D. Trancanelli, Quantum 't Hooft operators and S-duality in N = 4 super Yang-Mills. Adv. Theor. Math. Phys. **13**, 1941–1981 (2009). arXiv:0904.4486
34. P. Liendo, C. Meneghelli, Bootstrap equations for $\mathcal{N}$ = 4 SYM with defects. JHEP **01**, 122 (2017). arXiv:1608.05126, https://doi.org/10.1007/JHEP01(2017)122
35. A. Dekel, Y. Oz, Integrability of Green-Schwarz sigma models with boundaries. JHEP **1108**, 004 (2011). arXiv:1106.3446, https://doi.org/10.1007/JHEP08(2011)004
36. P.A.M. Dirac, Quantised singularities in the electromagnetic field. Proc. Roy. Soc. Lond. A **133**(821), 60–72 (1931). https://doi.org/10.1098/rspa.1931.0130

37. M. Fierz, Zur Theorie magnetisch geladener Teilchen. Helvetica Physica Acta **17**, 27 (1944)
38. H.A. Olsen, P. Osland, T.T. Wu, On the existence of bound states for a massive spin 1 particle and a magnetic monopole. Phys. Rev. D **42**, 665–689 (1990). https://doi.org/10.1103/PhysRevD.42.665
39. E.J. Weinberg, Monopole vector spherical harmonics. Phys. Rev. D 49 (1994) 1086–1092. arXiv:hep-th/9308054, https://doi.org/10.1103/PhysRevD.49.1086
40. Y. Kazama, C.N. Yang, A.S. Goldhaber, Scattering of a Dirac particle with charge Ze by a fixed magnetic monopole. Phys. Rev. D **15**, 2287–2299 (1977). https://doi.org/10.1103/PhysRevD.15.2287
41. I. Tamm, Die verallgemeinerten Kugelfunktionen und die Wellenfunktionen eines Elektrons im Felde eines Magnetpoles. Z. Phys. **71**(3–4), 141–150 (1931). https://doi.org/10.1007/BF01341701
42. T.T. Wu, C.N. Yang, Dirac's monopole without strings: classical Lagrangian theory. Phys. Rev. D **14**, 437–445 (1976). https://doi.org/10.1103/PhysRevD.14.437
43. T.T. Wu, C.N. Yang, Some properties of monopole harmonics. Phys. Rev. D **16**, 1018–1021 (1977). https://doi.org/10.1103/PhysRevD.16.1018
44. E.I. Buchbinder, A.A. Tseytlin, Correlation function of circular Wilson loop with two local operators and conformal invariance. Phys. Rev. **D87**(2), 026006 (2013). arXiv:1208.5138, https://doi.org/10.1103/PhysRevD.87.026006
45. E. D'Hoker, D.Z. Freedman, Supersymmetric gauge theories and the AdS / CFT correspondence, in *Theoretical Advanced Study Institute in Elementary Particle Physics (TASI 2001): Strings, Branes and EXTRA Dimensions* (2002), pp. 3–158. arXiv:hep-th/0201253
46. J. Gomis, T. Okuda, V. Pestun, Exact results for 't Hooft loops in Gauge theories on S^4. JHEP **05**, 141 (2012). arXiv:1105.2568, https://doi.org/10.1007/JHEP05(2012)141
47. G.W. Semenoff, K. Zarembo, More exact predictions of SUSYM for string theory. Nucl. Phys. **B616**, 34–46 (2001). arXiv:hep-th/0106015, https://doi.org/10.1016/S0550-3213(01)00455-2
48. K. Okuyama, G.W. Semenoff, Wilson loops in $\mathcal{N} = 4$ SYM and fermion droplets. JHEP **06**, 057 (2006). arXiv:hep-th/0604209, https://doi.org/10.1088/1126-6708/2006/06/057
49. M. De Leeuw, T. Gombor, C. Kristjansen, G. Linardopoulos, B. Pozsgay, Spin chain overlaps and the twisted Yangian. JHEP **01**, 176 (2020). arXiv:1912.09338, https://doi.org/10.1007/JHEP01(2020)176
50. A.V. Belitsky, A.S. Gorsky, G.P. Korchemsky, Gauge/string duality for QCD conformal operators. Nucl. Phys. B **667**, 3–54 (2003). arXiv:hep-th/0304028, https://doi.org/10.1016/S0550-3213(03)00542-X
51. A. Legendre, Recherches sur l'attraction des sphéroïdes homogènes. Mem. Math. Phys. Acad. Sci. Paris **10**, 411 (1785)
52. T. Gombor, Exact overlaps for all integrable two-site boundary states of $\mathfrak{gl}(N)$ symmetric spin chains (11 2023). arXiv:2311.04870
53. G. 't Hooft, Computation of the quantum effects due to a four-dimensional pseudoparticle. Phys. Rev. **D14**, 3432–3450 (1976), [Erratum: Phys. Rev. **D18**, 2199 (1978)]. https://doi.org/10.1103/PhysRevD.18.2199.3, https://doi.org/10.1103/PhysRevD.14.3432
54. M. de Leeuw, C. Kristjansen, K.E. Vardinghus, A non-integrable quench from AdS/dCFT. Phys. Lett. B **798**, 134940 (2019). arXiv:1906.10714, https://doi.org/10.1016/j.physletb.2019.134940
55. A. Babichenko, B. Stefanski, K. Zarembo, Integrability and the AdS_3/CFT_2 correspondence. JHEP **03**, 058 (2010). arXiv:0912.1723, https://doi.org/10.1007/JHEP03(2010)058
56. A. Sfondrini, Towards integrability for $\mathrm{AdS_3/CFT_2}$. J. Phys. **A48**(2), 023001 (2015). arXiv:1406.2971, https://doi.org/10.1088/1751-8113/48/2/023001
57. D. Sorokin, A. Tseytlin, L. Wulff, K. Zarembo, Superstrings in AdS(2)xS(2)xT(6). J. Phys. A **A44**, 275401 (2011). arXiv:1104.1793, https://doi.org/10.1088/1751-8113/44/27/275401
58. B. Hoare, A. Pittelli, A. Torrielli, Integrable S-matrices, massive and massless modes and the $\mathrm{AdS_2} * \mathrm{S2}$ superstring. JHEP **11**, 051 (2014). arXiv:1407.0303, https://doi.org/10.1007/JHEP11(2014)051

59. H. Saleur, The continuum limit of sl(N/K) integrable superspin chains. Nucl. Phys. B **578**, 552–576 (2000). arXiv:solv-int/9905007, https://doi.org/10.1016/S0550-3213(00)00002-X
60. M. Gaudin, *The Bethe Wavefunction* (Cambridge University Press, Cambridge, 2014)
61. S. Ghoshal, A.B. Zamolodchikov, Boundary S matrix and boundary state in two-dimensional integrable quantum field theory. Int. J. Mod. Phys. **A9**, 3841–3886 (1994), [Erratum: Int. J. Mod. Phys. **A9**, 4353 (1994)]. arXiv:hep-th/9306002, https://doi.org/10.1142/S0217751X94001552
62. L. Piroli, B. Pozsgay, E. Vernier, What is an integrable quench?, Nucl. Phys. B **925**, 362–402 (2017). arXiv:1709.04796, https://doi.org/10.1016/j.nuclphysb.2017.10.012
63. M. Brockmann, J. De Nardis, B. Wouters, J.-S. Caux, Néel-XXZ state overlaps: odd particle numbers and Lieb-Liniger scaling limit. J. Phys. A: Math. Theor. **47**, 345003 (2014). arXiv:1403.7469
64. M. de Leeuw, C. Kristjansen, S. Mori, AdS/dCFT one-point functions of the SU(3) sector. Phys. Lett. B **763**, 197–202 (2016). arXiv:1607.03123, https://doi.org/10.1016/j.physletb.2016.10.044
65. C. Kristjansen, D. Müller, K. Zarembo, Duality relations for overlaps of integrable boundary states in AdS/dCFT. JHEP **09**, 004 (2021). arXiv:2106.08116, https://doi.org/10.1007/JHEP09(2021)004

Elliptic Calogero-Sutherland Model and Conformal Field Theory

Edwin Langmann

Abstract In a project with Gordon Semenoff on 1+1 dimensional QCD many years ago (when he was my postdoc advisor), we stumbled over a method to solve Calogero-Moser-Sutherland models using gauge theories. Since then, these models have reappeared in different forms in many of my research projects. In this contribution, I describe a recent such project where a second quantization of the elliptic Calogero-Sutherland model led us to a new soliton equation and a non-relativistic variant of the Coleman correspondence. (Work with Bjorn Berntson and Jonatan Lenells.)

Keywords Quantum field theory · Soliton equations · Integrable systems

1 Introduction

I was postdoc during 1991–1994 at UBC in Vancouver with Gordon Semenoff as my advisor. This was a wonderful time in my life, and it was very important for my career. I feel lucky that I got this opportunity to work with Gordon during a critical phase of my career—I don't think I would be in science any more without him. Gordon gave me self-confidence and encouraged me to work on what interested me (I remember him saying that people working on what interests them most do better work). His office door was always open: when I needed feedback or advise, I could just drop by, any time. He listened, always, and then said something helpful (today I appreciate more how difficult it is to say the right thing—somehow, Gordon always did; moreover, I was one of 10 students and postdocs working with Gordon at the time, and he had time for all of us—I still don't know how he could do this). I learnt a lot from Gordon. I remember my struggling with writing papers, or my being uncomfortable with a way of doing quantum field theory which I was not familiar with (everybody else around seemed to be fluent at this)—Gordon helped me to improve, kindly, efficiently. Some things Gordon said still influence me; e.g., I

E. Langmann (✉)
Department of Physics, KTH Royal Institute of Technology, Stockholm, Sweden
e-mail: langmann@kth.se

M. Paranjape et al. (eds.), *Proceedings of Gravity, Strings and Fields*,
CRM Series in Mathematical Physics,
https://doi.org/10.1007/978-3-031-91266-5_7

remember him saying one should always try to have one paper under review—I did not always live up to this over the years, but I have tried.

Thank you, Gordon!

One of my projects at UBC with Gordon was quantum gauge theories in 1+1 dimensions and, in particular, the non-Abelian version of the Schwinger model (or QCD_{1+1} with massless quarks, as we called it). I hoped to find an exact solution of this model; I banged my head against this for a while, and then decided it was better to stay away from trying to solve models exactly (well, it did not exactly turn out like this in the long run). But one thing we could do was to work out what it meant to do gauge fixing for a non-Abelian Yang-Mills gauge theory on spacetime which is a cylinder [1]. What we wrote in the paper was, as I realized shortly afterwards, a way to relate gauge theories on a cylinder to integrable systems of Calogero-Moser-Sutherland (CMS) type. Looking back, this was very important for me: it was the first time when I stumbled over CMS-type system and, somehow, CMS-type systems have remained with me every since: a lot of my work over the years is related to these systems in one way or another.

In this paper, I describe a recent project where a CMS-type system features in a prominent way (this was together with Bjorn Berntson and Jonatan Lenells [2, 3]). Actually, in this story, there are three different players: (i) a quantum CMS model, (ii) a Conformal Field Theory (CFT), (iii) a soliton equation. In a special case which was known before, (i) is the (trigonometric) Calogero-Sutherland model, (ii) is a conventional chiral CFT, and (iii) is the so-called Benjamin-Ono equation; see e.g. [4–8]. Our contribution was a generalization of this to the elliptic case. We were lucky that, in this case, the soliton equation we found was not known before, and what we learnt from this allowed us to find and solve a few other previously unknown soliton equations [9]. In this project, several different strands in my research over the years got together in a way that surprised me.

My plan is as follows. In Sect. 2 I describe the story in a nutshell (like I would when meeting Gordon again after a while and trying to give an answer to his question "What have you been up to recently?" that would interest him). In Sect. 3 I give some background (if some things I write in Sect. 2 are not clear, they become hopefully clearer when reading this section). I end with some final remarks in Sect. 4.

2 The Story in a Nutshell

2.1 The eCS Model

The elliptic Caloger-Sutherland (eCS) model describes a quantum mechanical system of an arbitrary number, N, of identical particles moving on a circle of circumference $2\ell > 0$ and interacting with with two-body interactions given by the Weierstrass elliptic $\wp$-function with half-periods $(\ell, \mathrm{i}\delta)$:

$$\wp_1(x) = \sum_{n\in\mathbb{Z}} \frac{(\pi/2\ell)^2}{\sin^2(\pi(x-2ni\delta)/2\ell)} \tag{1}$$

where $\delta > 0$ [10]; see [11] for review. In units $m = \hbar = 1$, this model can can be defined by the Hamiltonian

$$H_{N;g}(\boldsymbol{x}) = -\frac{1}{2}\sum_{j=1}^{N} \frac{\partial^2}{\partial x_j^2} + g(g-1)\sum_{1\le j<k\le N} \wp_1(x_j - x_k) + c_{N;g} \tag{2}$$

where $x_j \in [-\ell, \ell)$ are the particle positions, $g \geq 1$ is a coupling parameter, and $c_{N;g}$ a known constant[1] (I add this constant to simplify some formulas further below). Note that, in the limit where the circle becomes the real line, the Weierstrass $\wp$-function reduces to $1/\sinh^2$:

$$\lim_{\ell\to\infty} \wp_1(x) = \frac{(\pi/2\delta)^2}{\sinh^2(\pi x/2\delta)}.$$

Thus, $\wp_1(x)$ can be regarded as 2ℓ-periodized version of this $1/\sinh^2$ decaying exponentially like $(\pi/\delta)^2\exp(-\pi|x|/\delta)$ at large distances $|x|$: the parameter δ controls the exponential decay of the two-body interaction potential at distances $|x|$ such that $\delta \ll |x| \ll \ell$.

2.2 2nd Quantization of eCS Model

In the limit where the interactions becomes long-range, i.e.,

$$\lim_{\delta\to\infty} \wp_1(x) = \frac{(\pi/2\ell)^2}{\sin^2(\pi x/2\ell)} = \sum_{n\in\mathbb{Z}} \frac{1}{(x-2n\ell)^2},$$

the eCS model reduces to the Sutherland model exactly solved a long time ago; in particular, the eigenfunctions of the Sutherland model are given by the Jack polynomials which are well-understood.

In 2000, I became interested in the problem to construct the elliptic generalizations of the Jack polynomials providing the eigenfunctions of the eCS model. I thus constructed a CFT that allowed me to obtain a mathematical tool for this purpose [12] (this was based on previous results in the trigonometric case I partly obtained in collaboration with Alan Carey). In this work, I constructed vertex operators $\phi_\nu(x)$ and a CFT operator $\mathcal{H}_\nu$ such that the following relations holds true,

$$[\mathcal{H}_\nu, \phi_\nu(x_1)\cdots\phi_\nu(x_N)]|0\rangle = H_{N,\nu^2}(\boldsymbol{x})\phi_\nu(x_1)\cdots\phi_\nu(x_N)|0\rangle \tag{3}$$

[1] $c_{N;g} = Ng^2(\pi/2\ell)^2\big(1/3 - (1/8)\sum_{n=1}^{\infty} np^n/(1-p^n)\big)/2$ with $p = \exp(-2\pi\delta/\ell)$.

where $|0\rangle$ is the vacuum state in the CFT. I called this *2nd quantization of the eCS model* since a single CFT operator, $\mathcal{H}_\nu$, accounts for the eCS Hamiltonians for coupling parameter $g = \nu^2$ for arbitrary particle numbers N. At that time, I regarded this CFT as useful device without interest in physics. In particular, while I knew the formula for the CFT operator $\mathcal{H}_\nu$, I did not understand its full significance: I was satisfied with it giving me the means to construct elliptic generalizations of the Jack polynomials.

2.3 Relation to Solitons

In 2020 (prompted by a questions to me by Junichi Shiraishi), we realized that this CFT providing a 2nd quantization of the eCS model was part of a non-chiral CFT with left- and right moving degrees of freedom combined in a symmetric way [2]: this CFT is generated by two massless boson field $\varphi_r(x)$, $r = \pm$, such that $\rho_r(x) \equiv \partial_x \varphi_r(x)$ satisfy the commutator relations

$$[\rho_r(x), \rho_{r'}(x')] = -2\pi \mathrm{i} \delta_{r,r'} \partial_x \delta(x - x') \tag{4}$$

for $r, r' = \pm$ and $x, x' \in [-\ell, \ell)$. Moreover, using these operators, the CFT operator $\mathcal{H}_\nu$ above can be written as

$$\mathcal{H}_\nu = \frac{1}{4\pi} \int_{-\ell}^{\ell} \sum_{r=\pm} {}^{\times}_{\times} \frac{\nu}{3} \rho_r^3 + \frac{\nu^2 - 1}{2} \Big(\rho_r T \rho_{r,x} + \rho_{-r} \tilde{T} \rho_{r,x} \Big) {}^{\times}_{\times} \, \mathrm{d}x \tag{5}$$

where $\rho_{r,x}(x) = \partial_x \rho_r$ and ${}^{\times}_{\times} \cdots {}^{\times}_{\times}$ denotes boson normal ordering, with T and $\tilde{T}$ integral operators given by

$$\begin{aligned} (Tf)(x) &= \frac{1}{\pi} \fint_{-\ell}^{\ell} \zeta_1(x' - x) f(x') \, \mathrm{d}x', \\ (\tilde{T} f)(x) &= \frac{1}{\pi} \int_{-\ell}^{\ell} \zeta_1(x' - x + \mathrm{i}\delta) f(x') \, \mathrm{d}x', \end{aligned} \tag{6}$$

with

$$\zeta_1(x) = \lim_{M \to \infty} \sum_{m=-M}^{M} \left(\frac{\pi}{2\delta} \right) \cot \left(\frac{\pi}{2\delta} (x - 2\mathrm{i} m \delta) \right) \tag{7}$$

the Weierstrass zeta functions with half-periods $(\ell, \mathrm{i}\delta)$ closely related to the Weierstrass $\wp$-function in (1), $\wp_1(x) = -\partial_x \zeta_1(x)$.

When looking at (5), it is natural to ask if this operator can be regarded as Hamiltonian generating an interesting dynamics for the boson fields. It is straightforward to compute the Heisenberg time evolution equations

$$\partial_t \rho_r = \mathrm{i}[\mathcal{H}_\nu, \rho_r],$$

and, by scaling: $u \equiv \nu\rho_+/2$ and $v \equiv \nu\rho_-/2$, we obtained the following equations,

$$\begin{aligned} u_t + 2 \,{}^{\times}_{\times} u u_x {}^{\times}_{\times} + \tfrac{1}{2}(g-1)\big(T u_{xx} + \tilde{T} v_{xx}\big) &= 0, \\ v_t - 2 \,{}^{\times}_{\times} v v_x {}^{\times}_{\times} - \tfrac{1}{2}(g-1)\big(T v_{xx} + \tilde{T} u_{xx}\big) &= 0. \end{aligned} \tag{8}$$

By taking a classical limit (which amounts to interpreting u and v as functions and dropping the normal ordering) and setting $\frac{1}{2}(g-1) = 1$ (which can be done without loss of generality on the classical level), we found that this leads to a soliton equation which we called *non-chiral intermediate long-wave (ILW) equation* [2]; see Sect. 3.1 for a motivation of this name. In subsequent work, we could solve this equation in nearly the same strong sense as the KdV equation is solved; see [3] for references.

Thus, the same object $\mathcal{H}_\nu$ providing a 2nd quantization of the eCS model defines, at the same time, a quantum version of a soliton equation.

2.4 Generalized eCS Models

The eCS model is only the tip of an iceberg: as I now describe, we found that $\mathcal{H}_\nu$ is the 2nd quantization of a generalization of the eCS model which can describe arbitrary numbers of four different types of particles.

For fixed ν, we have two vertex operators in the CFT, namely

$$\phi_{\nu,r}(x) = {}^{\times}_{\times} \exp\big(-\mathrm{i}r\nu\partial_x^{-1}\rho_r(x)\big) {}^{\times}_{\times} \quad (r = \pm); \tag{9}$$

the vertex operators $\phi_\nu(x)$ in (3) correspond to $\phi_{\nu,+}(x)$. Since $\mathcal{H}_\nu$ in (5) depends on $\rho_+(x)$ and $\rho_-(x)$ in a symmetric way, (3) implies that the similar equation holds true with $\phi_\nu(x_j) \equiv \phi_{\nu,+}(x_j)$ replaced by $\phi_{\nu,-}(x_j)$. Moreover, by (5), $\mathcal{H}_\nu = -\nu^2\mathcal{H}_{-1/\nu}$. Thus, if we use vertex operators $\phi_{-1/\nu,r}(x_j)$ instead of $\phi_{\nu,r}(x_j)$ in (3) for $r = \pm$, then $\mathcal{H}_\nu$ provides a second quantization of the operators $-\nu^2 H_{N,1/\nu^2}(\boldsymbol{x})$: by symmetry, we see that the CFT accommodates four different 2nd quantizations of the eCS model!

It is natural to ask if one can obtain an interesting generalization by using all four types of vertex operators, $\phi_{\nu,r}(x_j)$ and $\phi_{-1/\nu,r}(x_j)$ for $r = \pm$, at the same time. We found that this is the case: using the abbreviation

$$\phi_\nu^N(\boldsymbol{x}) \equiv \phi_\nu(x_1)\cdots\phi_\nu(x_N) \quad (N \in \mathbb{Z}_{\geq 0}) \tag{10}$$

(with $\phi_\nu^0 \equiv I$ (identity)), the generalization of (3) we found can be written as

$$\begin{aligned} &\big[\mathcal{H}_\nu, \phi_{\nu,+}^{N_1}(\boldsymbol{x})\phi_{-1/\nu,+}^{M_1}(\tilde{\boldsymbol{x}})\phi_{\nu,-}^{N_2}(\boldsymbol{y})\phi_{-1/\nu,-}^{M_2}(\tilde{\boldsymbol{y}})\big]|0\rangle \\ &= H_{N_1,M_1,N_2,M_2;\nu^2}(\boldsymbol{x},\tilde{\boldsymbol{x}},\boldsymbol{y},\tilde{\boldsymbol{y}})\phi_{\nu,+}^{N_1}(\boldsymbol{x})\phi_{-1/\nu,+}^{M_1}(\tilde{\boldsymbol{x}})\phi_{\nu,-}^{N_2}(\boldsymbol{y})\phi_{-1/\nu,-}^{M_2}(\tilde{\boldsymbol{y}})|0\rangle \end{aligned} \tag{11}$$

with the differential operator

$$\begin{aligned}H_{N_1,M_1,N_2,M_2;g}(\boldsymbol{x},\tilde{\boldsymbol{x}},\boldsymbol{y},\tilde{\boldsymbol{y}}) = \\ H_{N_1;g}(\boldsymbol{x}) - gH_{M_1;1/g}(\tilde{\boldsymbol{x}}) + H_{N_2;g}(\boldsymbol{y}) - gH_{M_2;1/g}(\tilde{\boldsymbol{y}}) \\ + g(g-1)\tilde{V}_{N_1,N_2}(\boldsymbol{x},\boldsymbol{y}) + (1-1/g)\tilde{V}_{M_1,M_2}(\tilde{\boldsymbol{x}},\tilde{\boldsymbol{y}}) \\ + (1-g)V_{N_1,M_1}(\boldsymbol{x},\tilde{\boldsymbol{x}}) + (1-g)V_{N_2,M_2}(\boldsymbol{y},\tilde{\boldsymbol{y}}) \\ + (1-g)\tilde{V}_{N_1,M_2}(\boldsymbol{x},\tilde{\boldsymbol{y}}) + (1-g)\tilde{V}_{M_1,N_2}(\tilde{\boldsymbol{x}},\boldsymbol{y})\end{aligned} \tag{12}$$

where

$$V_{N,M}(\boldsymbol{x},\boldsymbol{y}) \equiv \sum_{j=1}^{N}\sum_{k=1}^{M} \wp_1(x_j - y_j), \quad \tilde{V}_{N,M}(\boldsymbol{x},\boldsymbol{y}) \equiv \sum_{j=1}^{N}\sum_{k=1}^{M} \wp_1(x_j - y_j + \mathrm{i}\delta). \tag{13}$$

From a quantum field theory point of view, this generalization is natural: to generate the whole Fock space of the model, one needs all states

$$\phi_{\nu,+}^{N_1}(\boldsymbol{x})\phi_{-1/\nu,+}^{M_1}(\tilde{\boldsymbol{x}})\phi_{\nu,-}^{N_2}(\boldsymbol{y})\phi_{-1/\nu,-}^{M_2}(\tilde{\boldsymbol{y}})|0\rangle.$$

Thus, in this CFT, we have sectors with arbitrary numbers (N_1, M_1, N_2, M_2) of four kinds of particles (I say *particles* here and below for simplicity but, as discussed further below, it is perhaps better to think of them as solitons); the eCS Hamiltonian corresponds to the special sectors where only one kinds of particle is present and, in general sectors, one needs this generalized eCS Hamiltonian $H_{N_1,M_1,N_2,M_2;g}(\boldsymbol{x},\tilde{\boldsymbol{x}},\boldsymbol{y},\tilde{\boldsymbol{y}})$.

The physics interpretation of this generalized eCS model is somewhat unconventional in that the model describes particles which partly have negative mass. To be more specific: Note that

$$\lim_{\ell\to\infty} \wp_1(x+\mathrm{i}\delta) = -\frac{(\pi/2\delta)^2}{\cosh^2(\pi x/2\delta)},$$

i.e., $\wp_1(x+\mathrm{i}\delta)$ is the 2ℓ-periodic version of this weakly attractive interaction. The naive physics interpretation of this generalized eCS model is that it describes four different kinds of particles characterized by two quantum numbers (m,r) with $m = 1, -1/g$ and $r = \pm$. Interactions between a (m,r)-particle and a (m',r')-particle is via the singular repulsive interaction potential $\wp(x)$ if $r = r'$ and via the non-singular attractive interaction potential $\wp(x+\mathrm{i}\delta)$ if $r = -r'$; moreover, the coupling constant is given by $g(m+m')(gmm'-1)/2$. Note that the parameter m appears like a particle mass and, for this reason, particles with $m = -1/g$ have negative mass! Thus, the model with the Hamiltonian $H_{N_1,M_1,N_2,M_2;g}(\boldsymbol{x},\tilde{\boldsymbol{x}},\boldsymbol{y},\tilde{\boldsymbol{y}})$ does not have a conventional quantum mechanical interpretation. However, recent results suggest that there is a *generalized* quantum mechanical interpretation: one can define a scalar product promoting eigenfunctions of $H_{N_1,M_1,N_2,M_2;g}(\boldsymbol{x},\tilde{\boldsymbol{x}},\boldsymbol{y},\tilde{\boldsymbol{y}})$ to an orthogonal Hilbert space basis (I believe this since we were able to prove this in the trigonometric limit where the pertinent eigenfunctions are known [13]). I thus believe the strange features of this model are no problem: they simply reflect that the Hamiltonian $H_{N_1,M_1,N_2,M_2;\nu^2}(\boldsymbol{x}, \tilde{\boldsymbol{x}}, \boldsymbol{y}, \tilde{\boldsymbol{y}})$ provides a description of solitons (rather than particles) in an interacting quantum field theory preserving soliton numbers; the latter is a characteristic feature of an integrable quantum field theory. That such a description is possible is non-trivial—it allows to reduce the solution of a quantum field theory to the solution of quantum mechanics-like models in the different sectors. As explained in Sect. 3.3, the idea that this is possible is due to Ruijsenaars in a particular integrable quantum field theory.

Anyway, (10) suggests that $H_{N_1,M_1,N_2,M_2;g}(\boldsymbol{x}, \tilde{\boldsymbol{x}}, \boldsymbol{y}, \tilde{\boldsymbol{y}})$ is a natural generalization of the eCS models: it would be interesting to construct the eigenfunctions of this model and prove their orthogonality. I should mention that the special case $(N_1, M_1, N_2, M_2) = (N, M, 0, 0)$ of $H_{N_1,M_1,N_2,M_2;g}(\boldsymbol{x}, \tilde{\boldsymbol{x}}, \boldsymbol{y}, \tilde{\boldsymbol{y}})$ is known as *deformed eCS model* in the mathematics literature; see [14] and references therein. Moreover, if the integrability of the deformed eCS models is known, the integrability of the general model $H_{N_1,M_1,N_2,M_2;g}(\boldsymbol{x}, \tilde{\boldsymbol{x}}, \boldsymbol{y}, \tilde{\boldsymbol{y}})$ follows by a trick due to Calogero; see [3]. Out contribution was to show that this generalized model appears naturally in a quantum field theory; this adds motivation to construct its eigenfunctions (motivation is needed since this is hard work).

2.5 Interpretation as Interacting Fermion Model

We wrote $\mathcal{H}_\nu$ in (5) as an interacting boson model. As I now explain, by using the boson-fermion correspondence, one can also write $\mathcal{H}_\nu$ as interacting fermion Hamiltonian.

The Fock space of the CFT I am discussing can be interpreted as Fock space of massless boson fields or, equivalently, using the boson-fermion correspondence, as Fock space of Dirac fermions (the fermion operators are vertex operators for $\nu = \pm 1$). Moreover, using the boson-fermion correspondence, the operator $\mathcal{H}_\nu$ can be written as interacting fermion Hamiltonian:

$$
\begin{aligned}
\mathcal{H}_\nu^{\mathrm{F}} = \frac{\nu}{2}\int_{-\ell}^{\ell} \sum_{r=\pm} :\psi_r^\dagger(x)(-\partial_x^2)\psi_r(x): \mathrm{d}x \\
+ \frac{1}{2}(\nu^2 - 1)G^2 \int_{-\ell}^{\ell} \fint_{-\ell}^{\ell} \sum_{r=\pm} J_r(x)\wp_1(x - x')J_r(x')\,\mathrm{d}x\,\mathrm{d}x' \\
+ \frac{1}{2}(\nu^2 - 1)G^2 \int_{-\ell}^{\ell} \int_{-\ell}^{\ell} \sum_{r=\pm} J_r(x)\wp_1(x - x' + \mathrm{i}\delta)J_{-r}(x')\,\mathrm{d}x\,\mathrm{d}x',
\end{aligned}
\tag{14}
$$

with $\psi_r^{(\dagger)}(x)$ fermion operators satisfying the usual canonical anti-commutator relations, $J_r(x) = :\psi_r^\dagger(x)\psi_r(x):$ fermion densities, G a known constant,[2] and the colons indicating normal ordering; see [3] for further details.

As elaborated in Sect. 3.3, this suggests to us that this CFT is a non-relativistic limit of the massive Thirring model which, by a famous proposal due to Coleman [15], is equivalent to quantum sine-Gordon theory.

It is interesting to note that (14) looks very similar to a naive 2nd quantization of the generalized eCS model as a non-relativistic quantum many-body Hamiltonian describing arbitrary numbers of two kinds of particles; this generalized eCS model is given by

$$H = -\frac{\nu}{2}\sum_{j=1}^{N_1}\frac{\partial^2}{\partial x_j^2} - \frac{\nu}{2}\sum_{j=1}^{N_2}\frac{\partial^2}{\partial y_j^2} + (\nu^2-1)G^2\sum_{1\le j<k\le N_1}\wp_1(x_j-x_k)$$
$$+ (\nu^2-1)G^2\sum_{1\le j<k\le N_2}\wp_1(y_j-y_k) + (\nu^2-1)G^2\sum_{j=1}^{N_1}\sum_{k=1}^{N_2}\wp_1(x_j-y_k+\mathrm{i}\delta),$$

which is similar to the special case $M_1 = M_2 = 0$ of the generalized eCS Hamiltonian in (12) (it is the same except that the mass is $1/\nu$ instead of 1 and that the coupling constant is $(\nu^2-1)G^2$ instead of $g(g-1)$). To avoid misunderstanding, I stress that the present construction is different: it is on a fermion Fock space with a filled Dirac sea $|0\rangle$ such that the conventional Dirac Hamiltonian is positive definite; see [3] for further details.

3 Background

To set the results described in Sect. 2 in perspective, I discuss some background to my story. My discussion is informal, and I try to keep the references to a minimum—the interested reader can find more references in [3].

3.1 Soliton Equations

Soliton equations is about integrable non-linear partial differential equations or differential-integral equations; see e.g. [16] for an introductory textbook and [17] for a monograph.

The KdV equation is probably the most famous example. It can be written as partial differential equation

[2] $G = \prod_{m=1}^{\infty}(1-p^m)$ with $p = \exp(-2\pi\delta/\ell)$.

$$u_t + 2uu_x + \frac{\delta}{3}u_{xxx} = 0$$

for a function $u = u(x, t)$ of position $x \in \mathbb{R}$ and time $t \geq 0$, with $\delta > 0$ a parameter. The KdV equation is one example in a family of *chiral soliton equations* of the form

$$u_t + 2uu_x + \mathrm{i}\omega(-\mathrm{i}\partial_x)u = 0 \tag{15}$$

where $\omega(k)$ is a dispersion relation which is odd, $\omega(-k) = -\omega(k)$; the KdV corresponds to $\omega(k) = -k^3\delta/2$. These equations are chiral in the sense that the solitons described by them can only move in one direction: from left to right. Another well-known such soliton equation is the *Benjamin-Ono (BO) equation* where $\omega(k) = \mathrm{sign}(k)k^2$; since the sign function corresponds to the Hilbert transform

$$(Hf)(x) = \frac{1}{\pi} \fint_{\mathbb{R}} \frac{1}{x' - x} f(x')\mathrm{d}x'$$

in position space, the BO equation can be written as

$$u_t + 2uu_x + Hu_{xx} = 0; \tag{16}$$

it thus is a non-linear integro-differential equation.

Both the KdV and the BO equation appear in classical hydrodynamics describing non-linear waves in shallow and deep channels, and there is a soliton equation interpolating between these two known as *intermediate long-wave (ILW) equation*; the ILW is usually written as

$$u_t + 2uu_x + \frac{1}{\delta}u_x + Tu_{xx} = 0 \tag{17}$$

with the following integral transform generalizing the Hilbert transform,

$$(Tf)(x) = \frac{1}{2\delta} \fint_{\mathbb{R}} \coth\left(\frac{\pi}{2\delta}(x' - x)\right) f(x')\mathrm{d}x'; \tag{18}$$

see [18] and references therein. One can show that the ILW equation is of the form as in (15) with the dispersion relation $k^2 \coth(k\delta) - k/\delta$ interpolating between $-k^3\delta/3$ for small δ and $k^2\mathrm{sign}(k)$ for $\delta \to \infty$.[3]

All solitons described above come in two variants: (i) the real-line version described above, (ii) the 2ℓ-periodic version where we restrict to solutions satisfying $u(x, t) = u(x + 2\ell, t)$, i.e., space is changed from $\mathbb{R}$ to a circle of circumference 2ℓ. To change from (i) to (ii), the dispersion relation $\omega(k)$ remains the same by the wave numbers k become quantized, $k \in (\pi/\ell)\mathbb{Z}$. On the level of integral operators, it is written as the $\mathbb{R}$-version but replacing the Hilbert transform by its 2ℓ-periodic version given by

[3] The limit $\delta \to 0$ is slighly subtle since it requires a re-scaling.

$$(Hf)(x) = \frac{1}{2\ell} \int_{-\ell}^{\ell} \cot\left(\frac{\pi}{2\ell}(x' - x)\right) f(x')\mathrm{d}x'.$$

In the same way, one can get a 2ℓ-periodic variant of the ILW equation by replacing the integral operator T above by the one in (6).

3.2 Soliton-CMS Correspondence

There is a famous relation between the KdV equation and the (classical) Calogero model describing identical particles moving on the real line and interacting with a two-body potential $V(x) = 1/x^2$, where x is the inter-particle distance [19]. This model can be defined by the Hamiltonian

$$H = \sum_{j=1}^{N} \frac{1}{2} p_j^2 + g^2 \sum_{1 \leq j < k \leq N} V(x_j - x_k) \tag{19}$$

where $g^2 > 0$ is the coupling constant (we use units where the particle mass $m = 1$).

The Calogero model is the simplest example in a family of integrable models known as (classical) Calogero-Moser-Sutherland (CMS) model. As is well-known, these models come in four kinds: (I) rational, (II) trigonometric, (III) hyperbolic, (IV) elliptic, and these cases are distinguished by the kind of function appearing in the two-body potential:

$$V(x) = \begin{cases} 1/x^2 & \text{(rational case I)} \\ (\pi/2\ell)^2/\sin^2(\pi x/2\ell) & \text{(trigonometric case II)} \\ (\pi/2\delta)^2/\sinh^2(\pi x/2\delta) & \text{(hyperbolic case III)} \\ \wp_1(x) & \text{(elliptic case IV)} \end{cases} \tag{20}$$

with $\wp_1(x)$ defined in (1); see [11] for further details.

It is interesting to note that the integral operators arising for the chiral soliton equations discussed in Sect. 3.1 can be written as

$$\frac{1}{\pi} \fint \alpha(x' - x) f(x')\mathrm{d}x'$$

with

$$\alpha(x) = \begin{cases} 1/x & \text{(BO on } \mathbb{R}) \\ (\pi/2\ell)\cot(\pi x/2\ell) & (2\,\ell - \text{ periodic BO}) \\ (\pi/2\delta)\coth(\pi x/2\delta) & \text{(ILW on } \mathbb{R}) \\ \zeta_1(x) & (2\,\ell - \text{periodic ILW}). \end{cases} \tag{21}$$

Note that the functions V in (20) and α in (21) are related in a simple way: $V(x) = -\alpha'(x)$. Moreover, it is known that the two versions of the BO equation can be solved by a pole ansatz with the dynamics of the poles given by the (classical) rational CMS model in the real-line case and the trigonometric CMS model in the 2ℓ-periodic case [20]. This suggested that the hyperbolic and elliptic CMS models would be related to the ILW equation on $\mathbb{R}$ and the 2ℓ-periodic ILW equation, respectively [8]. It thus was surprising to us that we found a different equation in the hyperbolic and elliptic cases:

$$\begin{aligned} u_t + 2uu_x + Tu_{xx} + \tilde{T}v_{xx} &= 0, \\ v_t - 2vv_x - Tu_{xx} - \tilde{T}v_{xx} &= 0. \end{aligned} \tag{22}$$

Note that, in the limit $\delta \to \infty$, $T \to H$ and $\tilde{T} \to 0$, the non-chiral ILW equation reduces to two decoupled BO equations related by a parity transformation: the v-equation is obtained by the u-equation by the transformation $v(x,t) = u(-x,t)$. (I should mention that this results does not rule out the possibility that there still is a relation between the periodic ILW equation and the elliptic CMS model, as suggested in [8].)

We called (22) non-chiral ILW equation since it clearly is a relative of the ILW equation (17). However, different from the ILW equation, it is invariant under the parity transformation $(u(x,t), v(x,t)) \to (v(-x,t), u(-x,t))$ and, as suggested by this, its solitons can move in both directions—this is one reason for why we called it non-chiral ILW equation. Another reason is that, while the BO equation is related to a chiral CFT, the non-chiral ILW equation is related to a non-chiral CFT.

A basic insight we got from this is the following: there are many known soliton equations where the Hilbert transform H appears. The Hilbert transform has the property that it squares to minus the identity: $H^2 = -I$; the integral operator T in (6) does not have this property. However, using both integral operators T and $\tilde{T}$ in (6), one can construct a matrix integral operator

$$\mathcal{T} = \begin{pmatrix} T & \tilde{T} \\ -\tilde{T} & -T \end{pmatrix}$$

that satisfies $\mathcal{T}^2 = -I$, and one can write the non-chiral ILW equation naturally using the operator; see [2]. I mention this insight since it allowed us to guess other soliton equations which previously were unknown [9, 21].

What I discussed in this section is a close relation between classical soliton equations and classical CMS models. My story in Sect. 2 contains a quantum version of this relation.

3.3 Coleman Correspondence

In the story in Sect. 2, there is a quantum field theory which can be written either as interacting boson model or as interacting fermion model. When written as a boson model, it corresponds to a quantum version of a soliton equation. Moreover, this quantum field theory is related to CMS models and generalizations thereof.

There is a previously known story which is similar: as proposed by Coleman, the massive Thirring model is equivalent to quantum sine-Gordon theory [15]; moreover, as proposed by Ruijsenaars, there is a relativistic generalization of the hyperbolic CMS model related to quantum sine Gordon theory [22, 23]. We believe that the story in Sect. 2 is actually the non-relativistic limit of the Coleman correspondence and Ruijsenaars' proposal.

More specifically, the sine-Gordon equation is a famous soliton equation and, for this reason, the massive Thirring model/quantum sine Gordon theory is similar to the CFT described in Sect. 2. Moreover, Ruijsenaars proposed the models know known under his name as a means to construct special functions allowing to construct and exactly solve the massive Thirring model/quantum sine-Gordon theory [22]. It was challenging to develop the mathematics needed to construct these functions but, by now, these functions are well understood; see e.g. [24]. I believe that our results in [3] can guide towards realizing Ruijsenaars' vision [23]. In the trigonometric non-relativistic case, we were able to realize Ruijsenaars' vision in [25].

4 Conclusions

The relation between CFT, CMS-type models and soliton equations is useful to find new integrable systems: this relation suggests that, to any CMS-type system, there should be a corresponding soliton equation. When going through the list of CMS models, we found a few cases where the corresponding soliton equation was not known and, sure enough, we were able to find these missing soliton equations; see e.g. [9]. I believe that it will be possible to find several other soliton equations related to the relativistic generalizations of the CMS models. Moreover, in recent works with collaborators mentioned in the Acknowledgements, we were able to generalize some of the results reported here to the relativistic (Ruijsenaars) case; however, much remains to be done.

Until recent, I thought I have been working on various different unrelated topics during my career: seemingly unrelated problems in condensed matter physics, particle physics, mathematical physics and mathematics. It now seems that these topics all hang together: this is the story I tried to tell in this paper.

Acknowledgements I am grateful to Bjorn Berntson, Martin Hallnäs, Jonatan Lenells, Masatoshi Noumi, Junichi Shiraishi and Hjalmar Rosengren for collaborations and discussions that were important for my understanding. I would like to thank Manu Paranjape for organizing this meeting and for his patience.

References

1. E. Langmann, G.W. Semenoff, Gauge theories on a cylinder. Phys. Lett. B. **296**(1–2), 117–120 (1992)
2. B.K. Berntson, E. Langmann, J. Lenells, Nonchiral intermediate long-wave equation and interedge effects in narrow quantum Hall systems. Phys. Rev. B. **102**(15), 155308 (2020)
3. B.K. Berntson, E. Langmann, J. Lenells, Conformal field theory, solitons, and elliptic Calogero–Sutherland models. Commun. Math. Phys. **406**(2), 33 (2025)
4. A. Jevicki, Non-perturbative collective field theory. Nucl. Phys. B. **376**(1), 75–98 (1992)
5. H. Awata, Y. Matsuo, S. Odake, J. Shiraishi, Excited states of the Calogero-Sutherland model and singular vectors of the W_N algebra. Nuclear Phys. B. **449**(1–2), 347–374 (1995)
6. A.P. Polychronakos, Waves and solitons in the continuum limit of the Calogero-Sutherland model. Phys. Rev. Lett. **74**(26), 5153 (1995)
7. A.G. Abanov, P.B. Wiegmann, Quantum hydrodynamics, the quantum Benjamin-Ono equation, and the Calogero model. Phys. Rev. Lett. **95**(7), 076402 (2005)
8. A.G. Abanov, E. Bettelheim, P. Wiegmann, Integrable hydrodynamics of Calogero-Sutherland model: bidirectional Benjamin-Ono equation. J. Phys. A: Math. Theor. **42**(13), 135201 (2009)
9. B.K. Berntson, E. Langmann, J. Lenells, Spin generalizations of the Benjamin-Ono equation. Lett. Math. Phys. **112**(3), 50 (2022)
10. M.A. Olshanetsky, A.M. Perelomov, Quantum completely integrable systems connected with semi-simple Lie algebras. Lett. Math. Phys. **2**, 7–13 (1977)
11. M.A. Olshanetsky, A.M. Perelomov, Quantum integrable systems related to Lie algebras. Phys. Rep. **94**(6), 313–404 (1983)
12. E. Langmann, Anyons and the elliptic Calogero-Sutherland model. Lett. Math. Phys. **54**(4), 279–289 (2000)
13. F. Atai, M. Hallnäs, E. Langmann, Orthogonality of super-Jack polynomials and a Hilbert space interpretation of deformed Calogero-Moser-Sutherland operators. Bull. London Math. Soc. **51**(2), 353–370 (2019)
14. A.N. Sergeev, A.P. Veselov, Generalised discriminants, deformed Calogero-Moser-Sutherland operators and super-Jack polynomials. Adv. Math. **192**(2), 341–375 (2005)
15. S. Coleman, Quantum sine-Gordon equation as the massive Thirring model. Phys. Rev. D. **11**(8), 2088 (1975)
16. P.G. Drazin, R.S. Johnson, *Solitons: An Introduction*, vol. 2 (Cambridge University Press; 1989)
17. M.J. Ablowitz, P.A. Clarkson, *Solitons, Nonlinear Evolution Equations and Inverse Scattering*, vol. 149 (Cambridge University Press, 1991)
18. Y. Kodama, J. Satsuma, M.J. Ablowitz, Nonlinear intermediate long-wave equation: analysis and method of solution. Phys. Rev. Lett. **46**, 687 (1981)
19. H. Airault, H.P. McKean, J. Moser, Rational and elliptic solutions of the Korteweg-de Vries equation and a related many-body problem. Commun. Pure Appl. Math. **30**(1), 95–148 (1977)
20. H.H. Chen, Y.C. Lee, N.R. Pereira, Algebraic internal wave solitons and the integrable Calogero-Moser-Sutherland N-body problem. Phys. Fluids **22**(1), 187–188 (1979)
21. B.K. Berntson, R. Klabbers, E. Langmann, The non-chiral intermediate Heisenberg ferromagnet equation. J. High Energy Phys. **2022**(3), 1–54 (2022)
22. S.N.M. Ruijsenaars, Complete integrability of relativistic Calogero-Moser systems and elliptic function identities. Commun. Math. Phys. **110**(2), 191–213 (1987)
23. S.N.M. Ruijsenaars, Sine-Gordon solitons vs. relativistic Calogero–Moser particles, in *Integrable Structures of Exactly Solvable Two-Dimensional Models of Quantum Field Theory* (Springer, 2001), pp. 273–292
24. M. Hallnäs, S. Ruijsenaars, Joint eigenfunctions for the relativistic Calogero–Moser Hamiltonians of hyperbolic type. III. Factorized asymptotics. Int. Math. Res. Not. **2021**(6), 4679–4708 (2021)
25. F. Atai, E. Langmann, Deformed Calogero-Sutherland model and fractional quantum Hall effect. J. Math. Phys. **58**(1), 011902 (2017)

Anomalies in Hadronic *B* Decays

David London

Abstract In this talk, I describe a global fit to $B \to PP$ decays, where $B = \{B^0, B^+, B_s^0\}$ and the pseudoscalar $P = \{\pi, K\}$, under the assumption of flavour SU(3) symmetry [SU(3)$_F$]. It is found that the individual fits to $\Delta S = 0$ or $\Delta S = 1$ decays are good, but the combined fit is very poor: there is a 3.6σ disagreement with the standard model. (This is quite a bit larger than the anomaly in $b \to s\ell^+\ell^-$ transitions.) This discrepancy can be removed by adding SU(3)$_F$-breaking effects, but 1000% SU(3)$_F$ breaking is required, considerably more than the $\sim$ 20% breaking of $f_K/f_\pi - 1$. These results are rigorous, group-theoretically – no theoretical assumptions have been made. But when one adds a single assumption motivated by QCD factorization, the discrepancy grows to 4.4σ. These are the anomalies in hadronic *B* decays. Although one cannot yet claim that new physics is present, it is clear that something very unexpected is going on.

Keywords Anomalies in hadronic *B* decays · New physics

1 Beyond the Standard Model

The development of the standard model (SM) in particle physics is one of the great triumphs in all of physics. The SM has made a great many predictions, almost all of which have been verified, including the existence of the Higgs boson. There is no question that, just like Newton's laws, the SM is correct.

However, also like Newton's laws, the SM is not complete. There are a number of reasons that lead to this conclusion. The "philosophical" reasons include the large number of arbitrary parameters in the SM, the many unanswered questions, the fact that not all forces are unified, etc. There are also practical reasons, and these are

D. London (✉)
Physique des Particules, Université de Montréal, Montréal, QC, Canada
e-mail: london@lps.umontreal.ca

M. Paranjape et al. (eds.), *Proceedings of Gravity, Strings and Fields*,
CRM Series in Mathematical Physics,
https://doi.org/10.1007/978-3-031-91266-5_8

probably more important: the SM has no explanation for certain physics observations such as neutrino masses, dark matter, and the matter-antimatter asymmetry in the universe.

We therefore conclude that there must be physics beyond the SM. Most research in particle physics today, both experimental and theoretical, focuses on the search for this *new physics*.

Is this really true? Gordon Semenoff is best known for his work on various aspects of field theory, such as theories in $2+1$ dimensions, super-Yang-Mills theories, finite-temperature field theory, etc. One can hardly say that the focus of this research is new physics. On the other hand, a search through INSPIRE reveals that, since 2014, Gordon also has nine publications with the MoEDAL experiment (in addition to his work on field theory). What is this? According to its CERN website [1], "The MoEDAL-MAPP experiment is designed to search for hypothetical exotic particles that would point to new physics beyond the Standard Model of particle physics." So even Gordon Semenoff is participating in the search for new physics.

2 Searching for New Physics (Experiment)

Experiments look for new physics (NP) in two distinct ways. In direct searches, one aims to produce new particles in high-energy collisions, and then to detect their presence via their decay products. If something is seen, we will not only have found NP, we will also know what it is. On the other hand, in indirect searches, one uses the fact that the new particles may contribute virtually to certain observables in some rare low-energy processes. The signal of this would then be a discrepancy between the measured value of the observable and its SM prediction. Here, if a disagreement is found, we can infer that NP is present, but we will have no idea what it is.

The LHC has been performing direct searches for NP for many years. Unfortunately, to date no new particles have been observed. This suggests that the NP, whatever it is, must be heavy. In this case, the most promising way of finding NP may well be via indirect signals.

3 Describing New Physics (Theory)

In general, there are two ways of describing NP. Traditionally, this was done via model building. Here a NP model is constructed by extending the SM, either by adding new symmetries (i.e., new gauge bosons), and/or by adding new fermions or scalars. Given a model, one can compute the effects of these new particles on various observables. Note that one can also consider "simplified models," in which one considers the addition of one or more NP particles, without building a complete model.

The modern approach to describing NP uses effective field theories (EFTs). The canonical example of an EFT is the Fermi theory. Here, the weak interactions are described as a direct interaction among four fermions. Later this was recognized as the low-energy limit of an interaction mediated by the exchange of a W boson. That is, the Fermi interaction is obtained when the W is "integrated out."

This can be generalized: the low-energy effective field theory (LEFT) is obtained when all heavy SM particles (W, Z, t, H) are integrated out (see, e.g., Ref. [2]). The LEFT Hamiltonian obeys the unbroken $SU(3)_C \times U(1)_{em}$ symmetry, and is applicable below the weak scale. It contains operators of all dimensions, in which the coefficients of these operators obey a power series in $1/M_W$.

This idea can also be applied to NP. We know the NP is heavy, so that we can integrate it out, generating the SM effective field theory (SMEFT) (see, e.g., Refs. [3, 4]). The SMEFT Hamiltonian obeys the unbroken $SU(3)_C \times SU(2)_L \times U(1)_Y$ symmetry, and is applicable below the NP scale [O(TeV)] and above the weak scale. Once again, there are operators of all dimensions; their coefficients obey a power series in $1/\Lambda_{\rm NP}$. (Most of the dimension-4 terms correspond to the SM.)

EFTs are energy-dependent, i.e., the Hamiltonians change as one goes from one energy scale to another. This evolution is described by the renormalization-group equations. This means that (i) the values of the coefficients vary with energy, and (ii) the operators mix as we change energy scales. (This second point is very important!)

The bottom line is that all this must be taken into account when looking for a (high-energy) NP explanation of a low-energy discrepancy with the SM.

4 Semileptonic $b \to s\ell^+\ell^-$ Anomalies

For over ten years, the values of a number of observables in processes involving $b \to s\mu^+\mu^-$ transitions have shown significant disagreements with the predictions of the SM. Examples of such processes include $B \to K^{(*)}\mu^+\mu^-$, $B_s^0 \to \phi\mu^+\mu^-$, etc. In addition, the ratios

$$R_{K^{(*)}} \equiv \frac{\mathcal{B}(\bar{B} \to K^{(*)}\mu^+\mu^-)}{\mathcal{B}(\bar{B} \to K^{(*)}e^+e^-)} \tag{1}$$

were measured, and for a number of years, their values differed significantly from 1 (the SM prediction). The most compelling explanation was that NP affects $b \to s\mu^+\mu^-$, but not $b \to se^+e^-$ [5].

In December, 2022, LHCb announced that they now find $R_{K^{(*)}} \simeq 1$, in agreement with the SM [6, 7]. This implies that the NP contributes equally to $b \to s\mu^+\mu^-$ and $b \to se^+e^-$ [8–10]; this is referred to as "lepton-flavour-universal NP".

It is possible to construct models to explain this. For example, a Z' boson could contribute to $b \to s\ell^+\ell^-$ at tree level. The Z' must have an FCNC coupling to $\bar{s}b$ and must couple equally to e^+e^- and $\mu^+\mu^-$.

On the other hand, it is more informative to examine the problem from a model-independent (EFT) point of view. Here's what we know. The $b \to s\ell^+\ell^-$ anomalies are observed at the scale m_b. The fits [8–10] show that the preferred solution requires equal new contributions to the LEFT operators $\bar{s}\gamma^\mu P_L b\bar{\mu}\gamma_\mu\mu$ and $\bar{s}\gamma^\mu P_L b\bar{e}\gamma_\mu e$. We therefore have the following question. These LEFT operators are independent; their coefficients are a-priori unrelated. Can we find SMEFT operators that generate equal coefficients for these operators when the theory is run down from $\Lambda_{\rm NP}$ to m_b using the renormalization group?

Yes! Here is one example: Suppose there is a $b \to s\tau^+\tau^-$ SMEFT operator at scale $\Lambda_{\rm NP}$. When we run the theory down to m_b, the LEFT operators $\bar{s}\gamma^\mu P_L b\bar{\mu}\gamma_\mu\mu$ and $\bar{s}\gamma^\mu P_L b\bar{e}\gamma_\mu e$ are generated at one loop through γ exchange, see Fig. 1. And since $Q^e_{em} = Q^\mu_{em}$, their coefficients are equal!

I note in passing that there are other SMEFT operators that generate LEFT $\bar{s}\gamma^\mu P_L b\bar{\ell}\gamma_\mu\ell$ operators in this way; this type of operator mixing is very common. Also, SMEFT doesn't say how these operators are produced – that's for the model builders to do. For example, the $b \to s\tau^+\tau^-$ SMEFT operator could be produced via a Z' or a leptoquark (LQ) exchange. But models that contain a Z' or a LQ may also have other new particles. And SMEFT operators other than $b \to s\tau^+\tau^-$ may be generated when all NP particles are integrated out, in which case the fit must be repeated. This shows how the model-building and EFT approaches are complementary: both must be considered when searching for a NP explanation of a low-energy anomaly.

5 Hadronic *B* Anomalies

Returning to Fig. 1, note that the γ can decay to particles other than just e^+e^- and $\mu^+\mu^-$. In particular, it can decay to $\bar{u}u$ and $\bar{d}d$. That is, on very general grounds, one also expects NP contributions to $b \to s\bar{u}u$ and $b \to s\bar{d}d$. In other words, we can expect to see NP effects in $\Delta S = 1$ hadronic B decays!

In fact, there are several $\Delta S = 1$ hadronic B decays that exhibit deviations from the SM predictions, but they have not attracted nearly as much attention as the $b \to s\ell^+\ell^-$ anomalies. They include:

1. The $B \to \pi K$ puzzle has been around for about 20 years [11]. The amplitudes for the four decays $B^+ \to \pi^0 K^+$, $B^+ \to \pi^+ K^0$, $B^0 \to \pi^- K^+$, and $B^0 \to \pi^0 K^0$ form an isospin quadrilateral. Nine observables have been measured in these

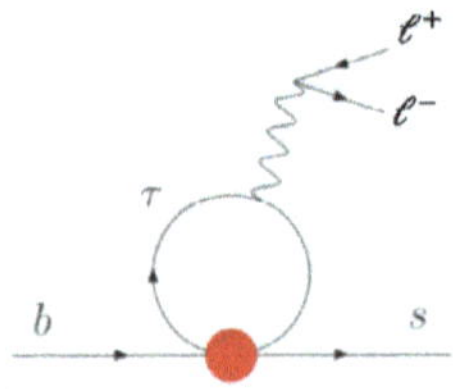

Fig. 1 Generation of the operators $\bar{s}\gamma^\mu P_L b\bar{\mu}\gamma_\mu\mu$ and $\bar{s}\gamma^\mu P_L b\bar{e}\gamma_\mu e$ with equal coefficients

decays, and a combined analysis shows that these results are not entirely consistent with one another. There is a $\sim 2\sigma$ discrepancy with the SM.

2. The U-spin puzzle [12]: U-spin symmetry ($d \leftrightarrow s$) relates the six decays $B^0_{d,s} \to P^{\pm}P'^{\mp}$, where P and P' are each π or K. 12 observables have been measured in these decays; once again, a combined analysis reveals a certain inconsistency. One can explain the data by allowing for U-spin-breaking effects, but we find that 100% U-spin breaking is required, much larger than the ~25–30% expected in SM. This results hints at NP in $B^0_s \to K^+K^-$.
3. Three puzzles involving $B^0_s \to K^0\bar{K}^0$ [13]: the decay $B^0_s \to K^0\bar{K}^0$ is related to $B^0_s \to K^+K^-$ by isospin, to $B^0_d \to K^0\bar{K}^0$ by U spin, and to $B^+ \to \pi^+K^0$ by virtue of having the same quark-level decay, $\bar{b} \to \bar{s}d\bar{d}$. In all three cases, there is a disagreement of ~2–3σ in the branching ratios of the pair of decays.

The point here is that we have five different sets of B decays that exhibit a significant discrepancy with the SM. All are $\Delta S = 1$ decays, i.e., they involve the $b \to s$ transition, just like the semileptonic $b \to s\ell^+\ell^-$ anomalies. This is very intriguing.

This talk was given on July 27, 2023. At the time, I noted that members of my group were examining these hadronic B anomalies, and I presented some preliminary results. Since then, the analysis has been completed and the paper written up in Ref. [14]. Below I summarize the analysis and its results.

All of the above processes are $B \to PP$ decays, where $B = \{B^0, B^+, B^0_s\}$, and the pseudoscalar $P = \{\pi, K\}$. These decays are all related to one another by flavour SU(3) symmetry [SU(3)$_F$]. One can therefore express all amplitudes in terms of SU(3)$_F$ reduced matrix elements (RMEs). By performing a global fit to all the $B \to PP$ observables under the assumption of SU(3)$_F$, these puzzles can be combined, and one can quantify just how well (or poorly) the SM explains the data.

It is straightforward to show that there are a total of seven independent RMEs. Defining $\lambda^{(q)}_{q'} \equiv V^*_{q'b}V_{q'q}$, $q = d, s$, $q' = u, c, t$, two RMEs are proportional to $\lambda^{(q)}_u$, two are proportional to $\lambda^{(q)}_t$, and three RMEs involve both $\lambda^{(q)}_u$ and $\lambda^{(q)}_t$. If SU(3)$_F$ is unbroken, these RMEs are the same for $\Delta S = 0$ ($q = d$) and $\Delta S = 1$ ($q = s$) decays. However, they can be different if SU(3)$_F$-breaking effects are allowed.

Many years ago, it was shown that an equivalent description of these SU(3)$_F$ amplitudes is provided by topological amplitudes (quark diagrams) [15, 16]. There are eight topologies. The tree (T), color-suppressed tree (C), annihilation (A), and W-exchange (E) diagrams are associated with $\lambda^{(q)}_u$. The electroweak penguin (P_{EW}) and color-suppressed electroweak penguin (P^C_{EW}) are associated with $\lambda^{(q)}_t$. The penguin (P) comes in two types—P_{uc} and P_{tc} are associated with $\lambda^{(q)}_u$ and $\lambda^{(q)}_t$, respectively—and similarly for the penguin-annihilation (PA) diagram. These ten diagrams appear in seven linear combinations in the amplitudes; these are related to the seven RMEs.

One advantage of diagrams over RMEs is that it is easy to write amplitudes in terms of diagrams—no (tricky) SU(3)$_F$ Clebsch-Gordan coefficients need to be computed. Another advantage is that one can estimate the relative sizes of different diagrams. For example, it has been argued that E, A and PA can be neglected compared to the other diagrams because they involve an interaction with the spectator quark [15, 16].

But this is also problematic: results that use dynamical assumptions such as this are not rigorous group-theoretically. In our analysis, we use diagrams, but we make no assumptions about their sizes. These are fixed only by the data.

There are eight $\Delta S = 0$ $B \to PP$ decays. These are parametrized by 7 (effective) diagrams, corresponding to 13 unknown theoretical parameters (7 magnitudes and 6 relative strong phases). But 15 observables have been measured. These include CP-averaged branching ratios, direct CP asymmetries, and indirect CP asymmetries. With more observables than unknowns, a fit can be done with only $\Delta S = 0$ decays. We find $\chi^2_{\rm min}/{\rm d.o.f.} = 0.35/2$, for a p-value of 0.84. This is an excellent fit.

There are also eight $\Delta S = 1$ $B \to PP$ decays, again parametrized by 7 (effective) diagrams. For these decays, there are again 15 observables, so a fit can be performed. We find $\chi^2_{\rm min}/{\rm d.o.f.} = 1.8/2$, for a p-value of 0.4. This is a very good fit.

But here's the kicker. Under SU(3)$_F$, $\Delta S = 0$ and $\Delta S = 1$ decays are the same. We can then perform a combined fit, including all the data. But when such a fit is performed, we find that $\chi^2_{\rm min}/{\rm d.o.f.} = 43.8/17$, for a p-value of 3.6×10^{-4}. This corresponds to a discrepancy with the SM at the level of 3.6σ.

Once again, I stress that no dynamical assumptions have been made regarding the diagrams. This result is completely rigorous from a group-theoretical point of view.

Note that the 3.6σ is not the "pull," which has become popular in the analysis of the $b \to s\ell^+\ell^-$ anomalies. The pull, which often reaches 4–5σ or even larger, indicates how much better than the SM a particular NP model describes the data. But it says nothing about how well the SM explains the data. In fact, even when the measured values of $R_{K^{(*)}}$ differed from 1, the overall discrepancy with the SM was only around 2.3σ. That is, even though the analyses of the $b \to s\mu^+\mu^-$ anomalies found NP scenarios with pulls of $\sim 6\sigma$, in actual fact, the SM did not do such a poor job of explaining the data. So the hadronic B anomalies are quite a bit more problematic.

Of course, the analysis has assumed unbroken SU(3)$_F$ symmetry. However, we know that SU(3)$_F$ is, in fact, broken. For example, $f_K/f_\pi - 1 = \sim 20\%$; this is typically taken to be the expected size of SU(3)$_F$ breaking in the SM. Could the addition of such SU(3)$_F$-breaking effects remove the discrepancy?

Fortunately, the fit contains enough information to address this question. Above, we found an excellent fit when one considers only $\Delta S = 0$ decays and a very-good fit for $\Delta S = 1$ decays. In both cases, the best-fit values of the (effective) diagrams can be extracted. These are shown in Table 1. Denoting the $\Delta S = 0$ and $\Delta S = 1$ diagrams by D and D', respectively, in the SU(3)$_F$ limit, $D_i = D'_i$. However, we see in the Table that one typically finds $|D'_i/D_i| = O(10)$, which corresponds to 1000% SU(3)$_F$ breaking! That is, if the SM really does explain the data, then the breaking of the flavour SU(3)$_F$ symmetry must be at an unexpectedly large level.

But there's more. Until now, no theoretical assumptions about the diagrams have been made; the results are rigourous, from a group-theoretical point of view. Now, two of the diagrams are T and C, respectively the colour-allowed and colour-suppressed tree diagrams. Naively we expect $|C/T| = 1/3$, simply by counting colors. In QCD factorization, this ratio is computed for $B \to \pi K$ decays ($\Delta S = 1$), and $|C/T| \simeq 0.2$ is found [17–21]. To a good approximation, $|\widetilde{C}/\widetilde{T}| \simeq |C/T|$. If we fix $|\widetilde{C}^{(\prime)}/\widetilde{T}^{(\prime)}|$ to

Table 1 Best-fit values of the magnitudes of the effective diagrams for the $\Delta S = 0$ and $\Delta S = 1$ fits, as well as their ratios. For definitions of the diagrams, see Ref. [14]

Fit $\Delta S = 0$	$\lvert\widetilde{T}\rvert$	$\lvert\widetilde{C}\rvert$	$\lvert\widetilde{P}_{uc}\rvert$	$\lvert\widetilde{A}\rvert$
	4.0 ± 0.5	6.6 ± 0.7	3 ± 4	6 ± 5
	$\lvert\widetilde{PA}_{uc}\rvert$	$\lvert P_{tc}\rvert$	$\lvert PA_{tc}\rvert$	
	0.7 ± 0.8	0.8 ± 0.4	0.2 ± 0.4	
Fit $\Delta S = 1$	$\lvert\widetilde{T}'\rvert$	$\lvert\widetilde{C}'\rvert$	$\lvert\widetilde{P}'_{uc}\rvert$	$\lvert\widetilde{A}'\rvert$
	48 ± 14	41 ± 14	48 ± 15	81 ± 28
	$\lvert\widetilde{PA}'_{uc}\rvert$	$\lvert P'_{tc}\rvert$	$\lvert PA'_{tc}\rvert$	
	7 ± 4	0.78 ± 0.16	0.24 ± 0.04	
	$\lvert\widetilde{T}'/\widetilde{T}\rvert$	$\lvert\widetilde{C}'/\widetilde{C}\rvert$	$\lvert\widetilde{P}'_{uc}/\widetilde{P}_{uc}\rvert$	$\lvert\widetilde{A}'/\widetilde{A}\rvert$
	12 ± 4	6.6 ± 2.2	16 ± 22	14 ± 13
	$\lvert\widetilde{PA}'_{uc}/\widetilde{PA}_{uc}\rvert$	$\lvert P'_{tc}/P_{tc}\rvert$	$\lvert PA'_{tc}/PA_{tc}\rvert$	
	10 ± 13	0.97 ± 0.52	1.3 ± 2.7	

0.2 and redo the fits, we now find that, when the $\Delta S = 0$ and $\Delta S = 1$ decays are combined (i.e., perfect $SU(3)_F$ symmetry is assumed), the best fit has $\chi^2_{\min}/\text{d.o.f.} = 55.8/18$, for a p-value of 9.4×10^{-6}. The discrepancy with the SM has grown significantly: it is now 4.4σ!

These are the anomalies in hadronic *B* decays. It is far too early to claim that we are seeing signs of new physics, but it is clear that something very unexpected is going on. This needs to be investigated further.

Acknowledgements This work was financially supported in part by NSERC of Canada.

References

1. The link is https://home.cern/science/experiments/moedal-mapp
2. E.E. Jenkins, A.V. Manohar, P. Stoffer, Low-energy effective field theory below the electroweak scale: operators and matching. JHEP **03**, 016 (2018). https://doi.org/10.1007/JHEP03(2018)016. [arXiv:1709.04486 [hep-ph]]
3. W. Buchmüller, D. Wyler, Effective Lagrangian analysis of new interactions and flavor conservation. Nucl. Phys. B **268**, 621 (1986). https://doi.org/10.1016/0550-3213(86)90262-2
4. For a review, see I. Brivio, M. Trott, The standard model as an effective field theory. Phys. Rept. **793**, 1 (2019). https://doi.org/10.1016/j.physrep.2018.11.002. [arXiv:1706.08945 [hep-ph]]
5. For a review as of 2021, see D. London, J. Matias, *B* flavour anomalies: 2021 theoretical status report. Ann. Rev. Nucl. Part. Sci. **72**, 37–68 (2022). [arXiv:2110.13270 [hep-ph]]
6. R. Aaij et al. [LHCb], Test of lepton universality in $b \to s\ell^+\ell^-$ decays. Phys. Rev. Lett. **131**(5), 051803 (2023). [arXiv:2212.09152 [hep-ex]]
7. R. Aaij et al. [LHCb], Measurement of lepton universality parameters in $B^+ \to K^+\ell^+\ell^-$ and $B^0 \to K^{*0}\ell^+\ell^-$ decays. Phys. Rev. D **108**(3), 032002 (2023). [arXiv:2212.09153 [hep-ex]]

8. A. Greljo, J. Salko, A. Smolkovič, P. Stangl, Rare b decays meet high-mass Drell-Yan. JHEP **05**, 087 (2023). [arXiv:2212.10497 [hep-ph]]
9. M. Algueró, A. Biswas, B. Capdevila, S. Descotes-Genon, J. Matias, M. Novoa-Brunet, To (b)e or not to (b)e: no electrons at LHCb. Eur. Phys. J. C **83**(7), 648 (2023). [arXiv:2304.07330 [hep-ph]]
10. T. Hurth, F. Mahmoudi, S. Neshatpour, B anomalies in the post $R_{K^{(*)}}$ era. Phys. Rev. D **108**(11), 115037 (2023). [arXiv:2310.05585 [hep-ph]]
11. See N.B. Beaudry, A. Datta, D. London, A. Rashed, J.S. Roux, The $B \to \pi K$ puzzle revisited. JHEP **01**, 074 (2018). [arXiv:1709.07142 [hep-ph]], and references therein
12. B. Bhattacharya, S. Kumbhakar, D. London, N. Payot, U-spin puzzle in B decays. Phys. Rev. D **107**(1), L011505 (2023). [arXiv:2211.06994 [hep-ph]]
13. Y. Amhis, Y. Grossman, Y. Nir, The branching fraction of $B_s^0 \to K^0\overline{K}^0$: three puzzles. JHEP **02**, 113 (2023). [arXiv:2212.03874 [hep-ph]]
14. R. Berthiaume, B. Bhattacharya, R. Boumris, A. Jean, S. Kumbhakar and D. London, Anomalies in Hadronic B Decays. [arXiv:2311.18011 [hep-ph]]
15. M. Gronau, O.F. Hernandez, D. London, J.L. Rosner, Decays of B mesons to two light pseudoscalars. Phys. Rev. D **50**, 4529 (1994). [arXiv:hep-ph/9404283 [hep-ph]]
16. M. Gronau, O.F. Hernandez, D. London, J.L. Rosner, Electroweak penguins and two-body B decays. Phys. Rev. D **52**, 6374 (1995). [arXiv:hep-ph/9504327 [hep-ph]]
17. M. Beneke, G. Buchalla, M. Neubert, C.T. Sachrajda, QCD factorization in B —> pi K, pi pi decays and extraction of Wolfenstein parameters. Nucl. Phys. B **606**, 245 (2001). [arXiv:hep-ph/0104110 [hep-ph]]
18. G. Bell, NNLO vertex corrections in charmless hadronic B decays: imaginary part. Nucl. Phys. B **795**, 1 (2008). [arXiv:0705.3127 [hep-ph]]
19. G. Bell, NNLO vertex corrections in charmless hadronic B decays: real part. Nucl. Phys. B **822**, 172 (2009). [arXiv:0902.1915 [hep-ph]]
20. M. Beneke, T. Huber, X.Q. Li, NNLO vertex corrections to non-leptonic B decays: tree amplitudes. Nucl. Phys. B **832**, 109 (2010). [arXiv:0911.3655 [hep-ph]]
21. G. Bell, M. Beneke, T. Huber, X.Q. Li, Two-loop current–current operator contribution to the non-leptonic QCD penguin amplitude. Phys. Lett. B **750**, 348 (2015). [arXiv:1507.03700 [hep-ph]]

Edge States and Their Cousins in an Open SSH Model

Alexei Bissonnette, Nicolas Delnour, Andrew Mckenna, Hichem Eleuch, Michael Hilke, and Richard MacKenzie

Abstract The Su-Schrieffer-Heeger model is perhaps the simplest model with topological edge stages, states that are an essential property of materials known as topological insulators. In this talk, I will briefly describe the SSH model and will discuss edge states in a finite isolated SSH chain and also in a chain with leads attached to the ends. The effect of the leads on the details of the states, and in particular on the energy spectrum, will be discussed. As the SSH-lead coupling constant varies, a transition akin to a topological phase transition is observed. More specifically, the spectrum of the SSH chain changes from nontrivial to trivial (or vice versa), the most prominent sign being the disappearance (or appearance) of edge states.

1 Introduction

In the 1970s, polyacetylene was something of a puzzle in polymer physics. The molecule consists of a chain of carbon atoms, each of which is bonded to a hydrogen atom, with one free electron per (CH) unit. The chain was known to dimerize, in

A. Bissonnette · R. MacKenzie (✉)
Département de physique, Université de Montréal, Complexe des Sciences, Montréal, QC, Canada
e-mail: richard.mackenzie@umontreal.ca

A. Bissonnette
e-mail: alexei.bissonnette@outlook.com

N. Delnour · A. Mckenna · M. Hilke
Department of Physics, McGill University, Montréal, QC, Canada
e-mail: nicolas.delnour@mail.mcgill.ca

A. Mckenna
e-mail: andrew.mckenna@mail.mcgill.ca

M. Hilke
e-mail: hilke@physics.mcgill.ca

H. Eleuch
Department of Applied Physics and Astronomy, University of Sharjah, Sharjah, UAE
e-mail: heleuch@fulbrightmail.org

M. Paranjape et al. (eds.), *Proceedings of Gravity, Strings and Fields*,
CRM Series in Mathematical Physics,
https://doi.org/10.1007/978-3-031-91266-5_9

accordance with the Peierls instability [10, 11], developing a gap in the process. This should have resulted in an insulator, yet polyacetylene is a conducting polymer.

Su, Schrieffer and Heeger [13] proposed a model wherein the conductivity of polyacetylene is ascribed to the presence of solitons interpolating between the two equivalent ground state configurations. SSH convincingly demonstrated that solitons, being charged and mobile, are the charge carriers in polyacetylene.

A glance at the citation history of the Su-Schrieffer-Heeger paper is quite revealing. The paper received about a hundred citations per year up until around 2010, when the citation rate climbed dramatically, with over 500 citations already in the first seven months of 2023 according to Google Scholar.

This renewed interest in the SSH model is largely due to an explosion of interest in topological insulators beginning in the mid-2000s. Although much of the underlying physics of the subject had been known for many years, the theoretical study of topological edge states in the spin quantum Hall system by Kane and Mele [8] fueled much of the interest in the subject. This interest spilled over into the SSH model, being arguably the simplest model with topological edge states. Thus, the SSH model serves not only as a toy model of topological insulators, but it is also very interesting in its own right since it describes real materials such as polyacetylene.

The relevant features of topological insulators for this work are: a gapped bulk energy spectrum describing an insulator; a gapless edge state spectrum describing a conducting state; a symmetry that protects the topological order. These features barely scratch the surface of the subject of topological insulators; for more details, see, for instance, [7].

In this talk, I will quickly review some of the relevant properties of the SSH model. I will then consider a finite SSH chain which can support edge states. I will then discuss an open SSH chain formed when leads are attached to the chain. The main focus will be an examination of the effect of the leads on the energy spectrum. We will see that something resembling a topological phase transition takes place as the coupling between the SSH chain and the leads grows. Finally, I will discuss how the work described here relates to earlier work.

2 SSH Basics: Infinite Chain

At a fundamental level, a reasonable starting point for the study of polyacetylene would be a model of electrons and phonons, and indeed this is the approach taken in the SSH paper. Here, we study a simpler model, taking the underlying atomic positions as predetermined by the underlying physics, namely an electron hopping along a dimerized chain of atoms. Dimerization causes the hopping parameter to alternate between two fixed values t_1 and t_2, which are presumed to be positive. The Hamiltonian (in matrix form and in a pictorial notation) is

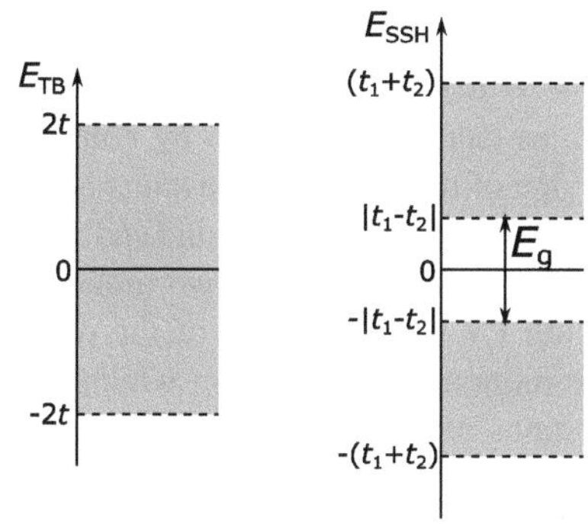

Fig. 1 Energy spectrum of the usual tight-binding model and of the SSH model. Dimerization introduces a gap of width $E_g = 2\Delta = 2|t_1 - t_2|$

$$H_{\text{SSH}} = \begin{pmatrix} \ddots & \ddots & & & & & \\ \ddots & 0 & t_1 & & & & \\ & t_1 & 0 & t_2 & & & \\ & & t_2 & 0 & t_1 & & \\ & & & t_1 & 0 & t_2 & \\ & & & & t_2 & 0 & \ddots \\ & & & & & \ddots & \ddots \end{pmatrix}$$
$$= \bullet\bullet\bullet \circ \overset{t_1}{-} \circ \overset{t_2}{-} \circ \overset{t_1}{-} \circ \overset{t_2}{-} \circ \overset{t_1}{-} \circ \bullet\bullet\bullet \tag{1}$$

The energy spectrum is well known (see, for instance, [1]):

$$E_{\text{SSH}}(k)^2 = t_1^2 + t_2^2 + 2t_2t_2\cos(2k). \tag{2}$$

For the special case $t_1 = t_2 = t$, the spectrum reduces to the standard result for a tight-binding Hamiltonian:

$$E_{\text{TB}}(q) = 2t\cos(q). \tag{3}$$

These spectra are displayed in Fig. 1.

3 SSH Basics: Finite Chain

A finite SSH chain can be either even or odd in length. An even-length chain starts and finishes with the same coupling constant (for example, $t_1 - t_2 - \cdots - t_2 - t_1$)

whereas an odd-length chain starts and finishes with opposite coupling constants (for example, $t_1 - t_2 - \cdots - t_1 - t_2$). The first and last couplings have an important impact on the spectrum, as can easily be seen by considering the limit of a fully dimerized chain (where one of the coupling constants is set to zero).

Consider a chain of even length N beginning and ending with t_1. If $t_1 = 0$ (Fig. 2a), the chain divides into $(N-2)/2$ dimers and two monomers. The Hamiltonian (1) block-diagonalizes, having $(N-2)/2$ copies of the dimer Hamiltonian $H_{\text{dimer}} = t_2\sigma_1$ and 2 copies of the monomer Hamiltonian which is simply the number zero. Thus, there is a pair of states on each dimer with energies $\pm t_2$ as well as a pair of monomer states of energy zero. The energy spectrum consists of $(N-2)/2$ states of energy $+t_2$, $(N-2)/2$ states of energy $-t_2$, and two states of energy zero.

If $t_2 = 0$ (Fig. 2b), the chain divides into $N/2$ dimers. The same reasoning leads us to conclude that the energy spectrum consists of $N/2$ states of energy $+t_2$ and $N/2$ states of energy $-t_2$. Notably absent are the zero-energy states seen in the opposite limit.

We can study the interpolation of the spectrum between these limits by fixing $t_1 + t_2 = 1$ and varying t_1 from 0 to 1 (Fig. 2c). The main states of interest are the edge states (dotted lines). These have $E = 0$ and are strictly localized on the first and last sites of the chain for $t_1 = 0$. As t_1 is turned on, the edge states persist, with exponentially small but nonzero energies, becoming less and less localized; eventually their decay length goes to infinity and they join the continuum. Example bulk and edge states are displayed in Fig. 2d.

What are the observable consequences of edge states in the SSH model? In higher dimensional topological insulators, edge states give rise to surface currents. This is of course not possible in a one-dimensional model, for which the "surface" consists of two points. However, the edge states have an important effect on transmission through the SSH chain. To see this, we must consider an SSH chain connected to leads which can exchange electrons with the chain. Once the leads are attached, the chain itself becomes an open quantum system.

4 Open SSH Chain

Consider the system depicted in the top panel of Fig. 3, consisting of two semi-infinite leads connected to the ends of the SSH chain. The leads are described by a (gapless) tight-binding model; we choose the energy scale such that the hopping parameter in the leads is $t = 1$. The coupling between the SSH chain and the leads is t_L.

The isolated leads have the same energy spectrum as an infinite lead (left panel of Fig. 1); the SSH chain in isolation has a spectrum consisting of a number of states lying within the SSH bands (right panel of Fig. 1) and, possibly, two edge states lying within the SSH gap (Fig. 2c). When connected, there will still be two states of every energy in the lead band $-1 < E < 1$; there may also be states outside the lead band. Such states would decay exponentially in the leads, but they could be bulk or edge

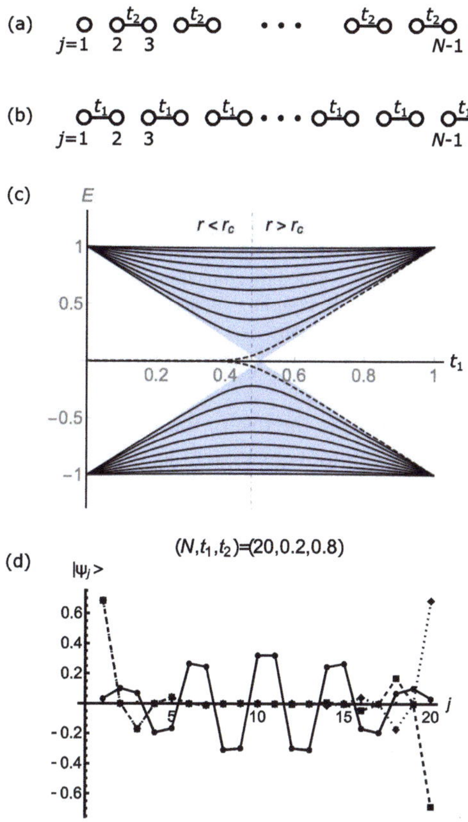

Fig. 2 Fully dimerized SSH chain for **a** $t_1 = 0$ and **b** $t_2 = 0$. **c** Energy spectrum for a chain of length $N = 20$ as a function of t_1, keeping $t_1 + t_2 = 1$. The shaded region represents the bands in the thermodynamic limit $N \to \infty$. Note that the gap disappears for $t_1 = t_2 = 0.5$. The spectra of the two limiting cases are seen on the left and right edges. The dotted state is an edge state for $r \equiv t_1/t_2 < r_c \equiv N/(N+2)$. The state gets more and more delocalized (its decay length increases) until $r = r_c$, at which point the decay length diverges and the state becomes a bulk state. See [1, 2, 15] for details. **d** Example bulk (solid) and edge (dotted, dashed) states. (*Note* (c) from [2] (copyright by M. Bédard, 2022)

states from the point of view of the SSH chain, depending on whether the energy is within or outside the SSH bands, respectively.

Note that there are three possible scenarios as far as the bands are concerned: the SSH bands can lie entirely within the lead band, they can lie partially within the lead band (this is the scenario depicted in Fig. 1), or they can lie outside the lead band.

It is interesting to consider the energy spectrum as a function of the SSH-lead coupling t_L. For definiteness, we consider the third scenario, where the SSH bands are *outside* the lead band. The limit $t_L = 0$ is well understood (but not particularly interesting). More interesting is the situation if t_L is large (much larger than any of the other hopping amplitudes). Then, the sites linked by t_L are so strongly coupled

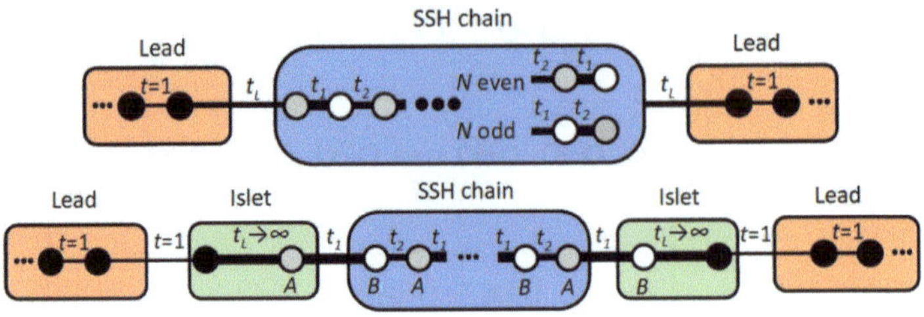

Fig. 3 Top panel: Open SSH system formed by connecting semi-infinite leads to the ends of the SSH chain. Bottom panel: in the strong coupling limit (t_L large), the pairs of sites linked by t_L decouple from their neighbours, forming islets

that they essentially decouple from the rest of the system as $t_L \to \infty$ to form *islets*, resulting in an effective five-component system as depicted in the bottom panel of Fig. 3 [3].

Much like the dimers discussed earlier, an isolated islet is described by the Hamiltonian $H_{\text{islet}} = t_L \sigma_1$ and has eigenenergies $\pm t_L$; the corresponding eigenstates are perfectly localized on the islets. With the coupling of the islets to the rest of the system, the states and energies are of course modified. The pair of states of energy t_L combine to form spatially even and odd states with a tiny energy splitting between them, and similarly for the pair of states of energy $-t_L$. The states are no longer strictly confined to the islets; rather, they decay exponentially as one moves into the leads or SSH chain. Such states are known in the literature as *Tamm states* [6, 12, 14].

Note that the SSH chain essentially loses its first and last sites as t_L increases, as can be seen by comparing the two depictions of the system in Fig. 3. To see the importance of this observation, suppose that $t_1 > t_2$. Then for small t_L the eigenstates of the SSH chain will be only mildly affected by the leads. In particular, there will be $N/2$ states in each of the positive and negative SSH bands.

For large t_L, with the formation of the islets the SSH chain is effectively reduced to length $N - 2$, starting and finishing with the *smaller* coupling constant t_2. Thus, there are $(N - 4)/2$ states within each SSH band as well as two edge states near zero energy.

In addition, for large t_L there are two pairs of Tamm states, one pair with energies near $+t_L$ and one pair with energies near $-t_L$. The part of the spectrum outside the lead band and with positive energy ($E > 2$) is depicted in Fig. 4.

The alert reader might be puzzled by the fact that for small t_L I mentioned a total of N states ($N/2$ within each SSH band) whereas for large t_L I mentioned a total of $N + 2$ states ($(N - 4)/2$ within each SSH band, two edge states near zero energy and four Tamm states). It appears that two states have been created out of thin air. Not quite: one must remember that there is also a continuum of states, so, while states cannot be created out of thin air, they can be created out of the continuum. Clearly, one must be careful about the counting of states.

Note that in the above scenario the edge states that form as t_L gets large lie within the lead band, so in terms of the energy spectrum their presence goes unnoticed.

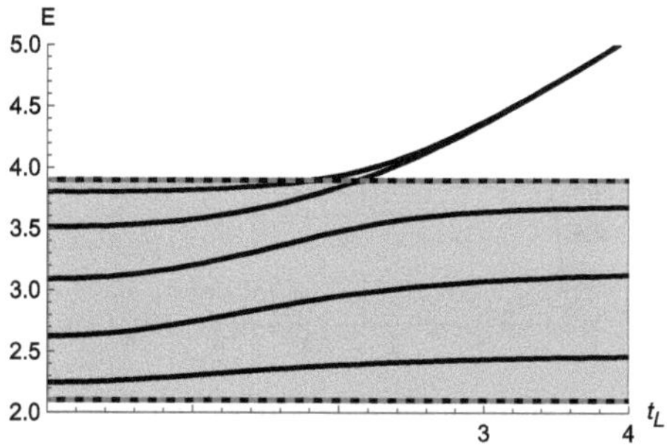

Fig. 4 Energy spectrum for $(N, t_1, t_2) = (10, 3, 0.9)$ as a function of t_L. Only $E > 2$ is shown because the lead continuum occupies $-2 < E < 2$ and furthermore the spectrum is symmetric under $E \to -E$

However, they have an important effect on various properties of the system such as the local density of states and, as we will see below, the transport properties of the SSH chain.

The newly-formed edge states that appear as t_L increases come as something of a surprise since the isolated SSH chain does not have edge states. Furthermore, unlike normal edge states whose largest amplitude is on the first and last sites of the SSH chain, the newly-formed edge states are localized on the *second and second-last* sites of the SSH chain. For this reason, we refer to these states as *phase-inverted edge states*, or as *PIE states* [3].

The important conclusion of this section is that the SSH system undergoes something akin to a topological phase transition as t_L passes from zero to infinity: if at $t_\mathrm{L} = 0$ the chain is topological trivial/nontrivial, then as $t_\mathrm{L} \to \infty$ it becomes topologically nontrivial/trivial.

5 Transmission

We can study transmission across the SSH chain for energies within the lead band. This is most interesting if the SSH bands lie partly or wholly within the lead band, so for this section we take $(t_1, t_2) = (0.8, 1.2)$. The transmission coefficient can be determined by elementary methods; transmission at $E = 0$ is particularly easy and instructive [9]. The result is

$$T = \frac{4t_L^4 t_1^2 r^{N-2}}{(t_L^4 + t_1^2 r^{N-2})^2}, \tag{4}$$

where $r = t_1/t_2$. The transmission coefficient can be written in terms of the combination of parameters $x = t_L^4/(t_1^2 r^{N-2})$, giving

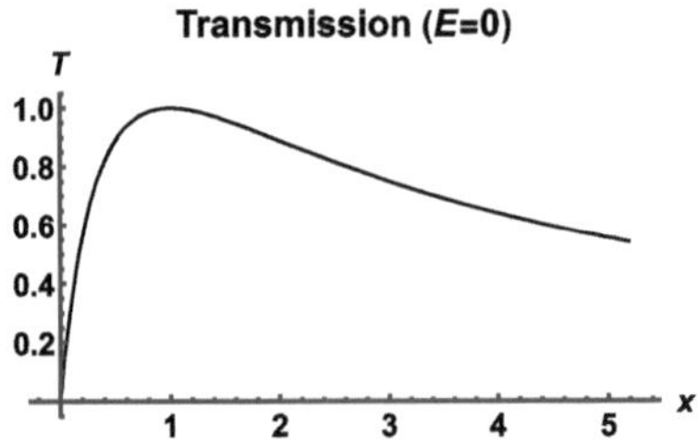

Fig. 5 Zero-energy transmission coefficient T as a function of the parameter $x = t_L^4/(t_1^2 r^{N-2})$, where $r = t_1/t_2$

$$T = \frac{4x}{(x+1)^2}. \tag{5}$$

This is depicted in Fig. 5.

Note that $T(x = 1) = 1$. One way to achieve $x = 1$ is to choose $t_1 = t_2 = t_L = 1$. This must certainly result in perfect transmission since the system is then an uninterrupted tight-binding model with no inhomogeneities to scatter off. But other combinations of hopping parameters correspond to $x = 1$ as well (for instance, $t_1 = t_2 = {t_\mathrm{L}}^2$), resulting in a nontrivial system with perfect transmission.

For other values of E the transmission can still be determined by elementary methods, but a powerful approach is to determine the effective Green's function for the SSH chain. This approach also provides a great deal of additional information about the system (local density of states, etc.), although we will not be concerned with that here. The effective Green's function and the effective Hamiltonian are defined as follows. We note first that the Hamiltonian can be written in terms of diagonal blocks which represent the subsystems and nondiagonal blocks which represent how the subsystems interact:

$$H = \left(\begin{array}{c|c|c} (H_\mathrm{L}) & t_\mathrm{L} & \\ \hline t_\mathrm{L} & (H_\mathrm{SSH}) & t_\mathrm{L} \\ \hline & t_\mathrm{L} & (H_\mathrm{R}) \end{array}\right) \tag{6}$$

where (H_x) represents a matrix that fills the corresponding block. The full Green's function is defined via $G = (E - H)^{-1}$, and the effective Green's function G_eff for the SSH chain is that part of G that corresponds to the SSH block; that is to say, G_eff and the effective Hamiltonian H_eff are defined via

$$G = \left(\begin{array}{c|c|c} (x) & (x) & (x) \\ \hline (x) & (G_{\text{eff}}) & (x) \\ \hline (x) & (x) & (x) \end{array}\right), \qquad G_{\text{eff}} = (E - H_{\text{eff}})^{-1}. \tag{7}$$

Here the matrix is written in block form and the entries (x), while calculable, are not needed. A detailed calculation [4, 5] results in the following effective Hamiltonian written in pictorial notation:

$$H_{\text{eff}} =$$

(8)

where Σ is the self-energy describing the influence of the leads on the SSH chain:

$$\Sigma = \frac{t_L^2}{2}\left(E - \text{sign}(E+2)\sqrt{E^2-4}\right) \tag{9}$$

The effective Hamiltonian has two unusual properties. First, it is energy-dependent, reflecting the fact that the influence on the leads depends on the energy. Second, it is not necessarily Hermitian. In particular, for energies within the lead band a particle in the SSH chain can escape into the leads, so from the point of view of the SSH chain probability is not conserved. This is reflected in the fact that Σ is complex for $-2 < E < 2$.

It turns out that the transmission coefficient can be written in terms of the $(N, 1)$ coefficient of the effective Green's function [4]:

$$T = 4\,(\text{Im}\,\Sigma)^2 |\,(G_{\text{eff}})_{N1}\,|^2. \tag{10}$$

The analytical expression is not particularly transparent. More useful is a colour plot of the transmission coefficient as a function of E and t_{L} for fixed (N, t_1, t_2) (see Fig. 6). The bright bands correspond to strong transmission. Looking at $E = 0$, we see that the transmission peaks at certain value of t_{L} and then drops off, consistent with Fig. 5. More importantly, we see very clearly the behaviour noted earlier: the system transitions from a topologically nontrivial SSH chain for small t_{L} (note the strong transmission at energies corresponding to the energies of the full SSH chain, and in particular at the edge state energies) to the topologically trivial reduced SSH chain (with no edge states) for large t_{L}. While the presence or absence of edge states is clearly visible in the transmission coefficient, the properties of the edge states (for instance, their width) are not readily apparent. A technique for determining such properties was discussed in [5].

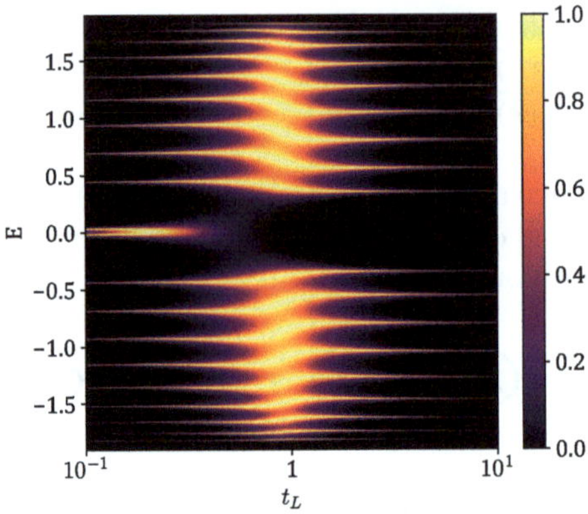

Fig. 6 Transmission across the SSH chain as a function of t_L and E for $(N, t_1, t_2) = (20, 0.8, 1.2)$

6 Conclusion

To conclude, we see that the SSH chain is a simple model with many of the properties of topological insulators. When coupled to leads, there is something akin to a topological phase transition as the coupling varies. For very small coupling the system is in a well-defined topological phase; for very large coupling it is in the opposite phase; in the intermediate region, the system is in a transition regime. These observations are clearly visible in the energy spectrum and in the transmission coefficient across the SSH chain.

Acknowledgements We thank Maude Bédard for the use of several figures from her MSc thesis in the talk, one of which appears in this paper. This work was supported in part by the Natural Science and Engineering Research Council of Canada (#RGPIN-2020-05094) and by the Fonds de Recherche Nature et Technologies du Québec via the INTRIQ strategic cluster grant (#2018-RS-201919).

References

1. J.K. Asbóth, L. Oroszlány, A. Pályi, *A Short Course on Topological Insulators* (Springer International Publishing, 2016)
2. M. Bédard, MSc thesis (2022). https://doi.org/10.1866/27986
3. A. Bissonnette, N. Delnour, A. Mckenna, H. Eleuch, M. Hilke, R. MacKenzie, (2023). arXiv:2306.17761
4. S. Datta, *Quantum Transport: Atom to Transistor* (Cambridge University Press, 2005)
5. N. Delnour, A. Bissonnette, H. Eleuch, R. MacKenzie, M. Hilke, Phys. Lett. A **466**, 128716 (2023)
6. T. Goto, A.V. Dorofeenko, A.M. Merzlikin, A.V. Baryshev, A.P. Vinogradov, M. Inoue, A.A. Lisyansky, A.B. Granovsky, Phys. Rev. Lett. **101**, 113902 (2008)

7. M.Z. Hasan, C.L. Kane, Rev. Mod. Phys. **82**, 3045–3067 (2010)
8. C.L. Kane, E.J. Mele, Phys. Rev. Lett. **95**, 226801 (2005)
9. B. Ostahie, A. Aldea, Phys. Lett. A **387**, 127030 (2021)
10. R. Peierls, Annalen der Physik **396**, 121–148 (1930)
11. R. Peierls, *More Surprises in Theoretical Physics* (Princeton University Press, 1991)
12. D. Pile, Nat. Photon. **7**, 932 (2013)
13. W.P. Su, J.R. Schrieffer, A.J. Heeger, Phys. Rev. Lett. **42**, 1698–1701 (1979)
14. I. Tamm, Phys. Z. Sowjetunion **1**, 733–735 (1932)
15. M. Zaimi, C. Boudreault, N. Baspin, N. Delnour, H. Eleuch, R. MacKenzie, M. Hilke, Phys. Lett. A **388**, 127035 (2020)

Strings from Nambu-Goto to Polyakov and Back

Yuri Makeenko

Abstract I discuss the recent progress in bypassing the KPZ barrier for the existence of nonperturbative bosonic strings in $1 < d < 25$. I consider string anomalies which emerge from higher terms of the DeWitt-Seeley expansion as $\varepsilon \times \varepsilon^{-1}$ with ε being a UV cutoff. I show they give a nonvanishing contribution to the central charge and to the string susceptibility, telling the Nambu-Goto and Polyakov strings apart. I describe an exact solution of the emerging four-derivative two-dimensional conformal theory and its relation to minimal models. Talk at "Gravity, Strings and Fields: A Conference in Honour of Gordon Semenoff", Montreal July 24–28, 2023.

1 Brief Introduction

This Talk is about the problem of the existence of nonperturbative bosonic strings in $d = 4$ space-time dimensions inherited from 1980's. The quantity of interest is the string susceptibility index $\gamma_{\rm str}$ defined through the number of surfaces of large area A by

$$\left\langle \delta\left(\int \sqrt{g} - A\right)\right\rangle \propto A^{\gamma_{\rm str}-3}\, {\rm e}^{CA}. \tag{1}$$

For closed Polyakov's string it was calculated by Knizhnik-Polyakov-Zamolodchikov [1] (KPZ) and David [2], Distler-Kawai [3] (DDK) using the technique of conformal field theory (CFT). The result

$$\gamma_{\rm str} = 2 + (h-1)\frac{12 + (d_+ + d_-)/2 - d + \sqrt{(d_+ - d)(d_- - d)}}{12} \quad \text{(genus } h\text{)} \tag{2}$$

with $d_+ = 25$, $d_- = 1$ was real only if $d \le 1$ or $d \ge 25$ but not in $d = 4$ (the KPZ barrier).

Y. Makeenko (✉)
NRC "Kurchatov Institute" – ITEP, Moscow, Russia
e-mail: makeenko@itep.ru

M. Paranjape et al. (eds.), *Proceedings of Gravity, Strings and Fields*,
CRM Series in Mathematical Physics,
https://doi.org/10.1007/978-3-031-91266-5_10

I argue in this Talk that a possible solution of the problem may be that the barrier is shifted to $d_- > 4$ for the Nambu-Goto string. Such a situation is realized for the four-derivative CFT I shall momentarily describe.

2 From the Nambu-Goto String to Four-Derivative CFT

2.1 Nambu-Goto Versus Polyakov Strings

The action of the Polyakov string is quadratic in the target-space coordinate X^μ and has an independent metric tensor g_{ab}:

$$\mathcal{S} = \frac{K_0}{2} \int \mathrm{d}^2\omega \sqrt{g} g^{ab} \partial_a X \cdot \partial_b X, \qquad K_0 = \frac{1}{2\pi\alpha'}. \tag{3}$$

The action of the Nambu-Goto string can also be written likewise by introducing the (imaginary) Lagrange multiplier λ^{ab}:

$$K_0 \int \mathrm{d}^2\omega \sqrt{\det \partial_a X \cdot \partial_b X} = K_0 \int \mathrm{d}^2\omega \sqrt{g} + \frac{K_0}{2} \int \mathrm{d}^2\omega\, \lambda^{ab} \left(\partial_a X \cdot \partial_b X - g_{ab} \right). \tag{4}$$

The two string formulations are equivalent classically when $g^{\mathrm{cl}}_{ab} = \partial_a X \cdot \partial_b X$ coincides with the induced metrics (and $\lambda^{ab}_{\mathrm{cl}} = \sqrt{g} g^{ab}$). I shall return soon to the general argument from the book [4] about the equivalence.

Let us split $X^\mu = X^\mu_{\mathrm{cl}} + X^\mu_{\mathrm{q}}$ and do the Gaussian path integral over the quantum part X^μ_{q}:

$$S_{\mathrm{ind}} = K_0 \int \mathrm{d}^2\omega \sqrt{g} + \frac{K_0}{2} \int \mathrm{d}^2\omega\, \lambda^{ab} \left(\partial_a X_{\mathrm{cl}} \cdot \partial_b X_{\mathrm{cl}} - g_{ab} \right) + \frac{d}{2} \mathrm{tr} \log \mathcal{O}, \tag{5}$$

where the operator $\mathcal{O} = -\frac{1}{\sqrt{g}} \partial_a \lambda^{ab} \partial_b$ reproduces the Laplacian Δ for $\lambda^{ab}_{\mathrm{cl}}$. An additional ghost determinant emerges as usual in the conformal gauge. The action (5) is called induced (or emergent). For smooth fields it coincides with the effective action.

Two-dimensional determinants diverge and are to be regularized. For Schwinger's proper-time regularization the integrals over τ are simply cut from below at $a^2 = 1/4\pi\Lambda^2$. Alternatively, one can use Pauli-Villars' regularization when

$$\mathrm{tr} \log \mathcal{O}\Big|_{\mathrm{reg}} = -\int_0^\infty \frac{\mathrm{d}\tau}{\tau} \mathrm{tr}\, \mathrm{e}^{-\tau\mathcal{O}} \left(1 - \mathrm{e}^{-\tau M^2} \right)^2, \qquad \Lambda^2 = \frac{M^2}{2\pi} \log 2 \tag{6}$$

is convergent at finite regulator mass M and divergent as $M \to \infty$.

For Pauli-Villars' regularization Feynman's diagrammatic technique applies and the det's can be exactly computed for certain metrics by the Gel'fand-Yaglom technique to compare with the DeWitt-Seeley expansion of $\langle \omega | \,\mathrm{e}^{-\tau\mathcal{O}} | \omega \rangle$ which starts with the term $1/\tau$ in two dimensions. For $\tau \sim \Lambda^{-2}$ the higher terms are suppressed as powers of R/Λ^2 for smooth fields but revive if they are not smooth. Higher-derivative terms are regularization-dependent but related by simple formulas [5] for the two regularizations.

2.2 Emergent Action $\mathcal{S}[\varphi, \lambda]$

Integrating out X_q^μ we get (a part of) the emergent action

$$\frac{d}{2} \operatorname{tr} \ln \left[-\frac{1}{\rho} \partial_a \lambda^{ab} \partial_b \right]_{\text{reg}} = \sum_n \frac{1}{n}$$

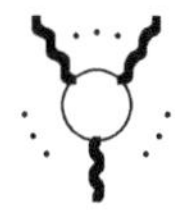

(7)

where the wavy lines correspond to fluctuations $\delta\lambda^{ab}$ or δg_{ab} about the ground state. In the conformal gauge

$$g_{ab} = \hat{g}_{ab}\, \mathrm{e}^{\varphi} \tag{8}$$

with $\hat{g}_{ab}$ being the background metric tensor we find the following contribution from X^μ and its Pauli-Villars' regulators to the emergent action [6]:

$$\begin{aligned} \mathcal{S}_X[\varphi, \lambda] = \frac{d}{2} \int \Bigg[&-\frac{\sqrt{\hat{g}}\, \mathrm{e}^{\varphi} \Lambda^2}{\sqrt{\det \lambda^{ab}}} + \frac{1}{48\pi} \left(\sqrt{\hat{g}} \varphi \hat{\Delta} \varphi + \lambda^{ab} \hat{g}_{ab} \hat{\Delta} \varphi + 2\lambda^{ab} \nabla_a \partial_b \varphi \right) \\ &+ \frac{\sqrt{\hat{g}}\, \mathrm{e}^{-\varphi}}{160\pi M^2} (\hat{\Delta}\varphi)^2 \Bigg] + \mathcal{O}(M^{-4}), \qquad \Lambda^2 = \frac{M^2}{2\pi} \log 2 \end{aligned} \tag{9}$$

with ∇_a being the covariant derivative for the metric tensor g_{ab} given by Eq. (8) and $\hat{\Delta} = \mathrm{e}^{\varphi} \Delta$ is the Laplacian for the metrics tensor $\hat{g}_{ab}$. The ghost contribution to the emergent action is as usual.

Expanding to the quadratic order in $\delta\lambda^{ab}$, we see that its mass squared $\sim\Lambda^2$, so $\delta\lambda^{ab}$ is expected to decouple [4]. Rather surprisingly, its private life which occurs at the distances $\sim \Lambda^{-1}$ is nevertheless observable [7] like anomalies in quantum field theory (QFT). I shall now start describing this issue.

2.3 Four-Derivative Liouville Action $\mathcal{S}[\varphi]$

Path-integrating out $\delta\lambda^{ab}$ in (9) and integrating the emergent action by parts, we obtain modulo boundary terms the four-derivative action [7]

$$\mathcal{S}[\varphi] = \frac{1}{16\pi b_0^2} \int \sqrt{g} \left[g^{ab} \partial_a \varphi \partial_b \varphi + 2\mu_0^2 + \varepsilon R \left(R + G g^{ab} \partial_a \varphi \partial_b \varphi \right) \right], \quad b_0^2 = \frac{6}{26-d} \tag{10}$$

where $\varphi = -\Delta^{-1} R$ becomes a local field in the conformal gauge (8) and $\varepsilon \propto M^{-1}$ is the worldsheet cutoff. Only the two four-derivative terms shown in Eq. (10) are independent. The first additional term R^2 appears already for Polyakov's string from the DeWitt-Seeley expansion of the heat kernel but the second one with $G \neq 0$ is specific to the Nambu-Goto string. We have dropped terms of higher orders in ε (to be discussed below).

Of course the higher-derivative terms vanish classically for smooth metrics with $\varepsilon R \ll 1$, reproducing the usual Liouville action. However, the quartic derivative provides both an ultraviolet (UV) cutoff ε^{-1} and also an interaction whose coupling constant is ε. We thus encounter uncertainties like powers of $\varepsilon \times \varepsilon^{-1}$ so the higher-derivative terms revive quantumly. In other words the smallness of ε is compensated by a change of the metric (the shift of φ) what is specific to the theory with diffeomorphism invariance.

The described procedure looks like an appearance of anomalies in QFT. We may expect for this reason that possible yet higher-derivative terms will not change the results. Such a universality at $G = 0$ was argued in [7] and proven in [6].

3 Conformal Invariance of Four-Derivative Liouville Action

3.1 "Improved" Energy-Momentum Tensor

To apply the methods of CFT to the four-derivative Liouville theory (10), we first derive its energy-momentum tensor T_{ab}. Naively, one simply substitutes the decomposition (8) into the action (10) and takes the variation with respect to $\hat{g}_{ab}$. But this would give only a part of T_{ab} associated with the minimal coupling of φ to the background gravity $\hat{g}_{ab}$. The correct procedure is to use

$$\sqrt{g} R = \sqrt{\hat{g}} \left(q \hat{R} - \hat{\Delta} \varphi \right) \qquad (q = 1 \quad \text{classically}) \tag{11}$$

under the decomposition (8). This results in an appearance of the additional terms in the action, leading to a non-minimal interaction of φ with background gravity. These additional terms are crucial for the Weyl covariance of the four-derivative action (10).

The energy-momentum tensor associated with the minimal interaction is conserved thanks to the classical equation of motion but not traceless owing to the presence of the dimensionful parameters like μ_0^2 and ε. The "improved" energy-momentum tensor [8] that involves additional terms, coming from the non-minimal interaction with background gravity, is conserved and *traceless* thanks to the classical equation of motion for φ in spite of the presence of dimensionful μ_0^2 and ε. This is a general property because in the conformal gauge (8) we have

$$T_a^a \equiv \hat{g}^{ab}\frac{\delta S[g]}{\delta \hat{g}^{ab}} = -\frac{\delta S[g]}{\delta \varphi}, \tag{12}$$

where the left-hand side represents the trace of the "improved" energy-momentum tensor while the right-hand side represents the classical equation of motion for φ. The four-derivative action (10) is thus conformal invariant for flat backgrounds.

Before explicitly writing the energy-momentum tensor let me mention the renormalization of the parameters b_0^2 in Eq. (10) and q in Eq. (11) in the quantum case: $b_0^2 \to b^2 = b_0^2 + \mathcal{O}(b_0^4)$ and $q_0 = 1 \to q = 1 + \mathcal{O}(b_0^2)$. This is like it was pointed out by DDK [2, 3] for the usual Liouville action. We can think of the renormalization as a result of path-integrating out the regulator fields. Accounting for the renormalization, we write for the T_{zz} component [9]

$$\begin{aligned} T_{zz} = \\ & -\frac{1}{4b^2}\left[(\partial\varphi)^2 - 2\varepsilon\partial\varphi\partial\Delta\varphi - G\varepsilon(\partial\varphi)^2\Delta\varphi + 4G\varepsilon\partial\varphi\partial(\mathrm{e}^{-\varphi}\partial\varphi\bar{\partial}\varphi) + G\varepsilon\partial(\partial\varphi\Delta\varphi)\right] \\ & +\frac{q}{4b^2}\left[2\partial^2(\varphi - \varepsilon\Delta\varphi) + 4G\varepsilon\partial^2(\mathrm{e}^{-\varphi}\partial\varphi\bar{\partial}\varphi) - G\varepsilon\frac{1}{\bar{\partial}}\partial^2(\bar{\partial}\varphi\Delta\varphi)\right]. \end{aligned} \tag{13}$$

Notice the appearance of the nonlocal (last) term as inherited from the nonlocality of the covariant action (10).

3.2 DDK for the Four-Derivative Liouville Action

Given T_{zz} it is possible to compute á la DDK the conformal weight and the central charge at one loop about the classical ground state. The operator products $T_{zz}(z)\,\mathrm{e}^{\varphi(0)}$ and $T_{zz}(z)T_{zz}(0)$ are given at one loop by the diagrams depicted in Fig. 1. The diagrams (a) to (d) renormalize $b_0^2 \to b^2$ in the propagator and (e) to (g) renormalize $1 \to q$ (or $\mathrm{e}^{\varphi(0)}$ as discussed below). The diagram (k) contributes $-b^2$ to the conformal weight of $\mathrm{e}^{\varphi(0)}$ which equals 1, so we have as usual

$$1 = q - b^2. \tag{14}$$

For the central charge the diagrams (a) to (d) contribute $6q^2/b^2$ as usual. The diagram (k) contributes the usual 1 but now the nonlocal term in Eq. (13) revives and gives additionally $6Gq$. There is also a logarithmically divergent term which does not appear if T_{zz} is normal-ordered. Adding the three terms we find

$$c^{(\varphi)} = \frac{6q^2}{b^2} + 1 + 6Gq \tag{15}$$

at one loop. As will be shown below, actually Eqs. (14) and (15) are *exact*.

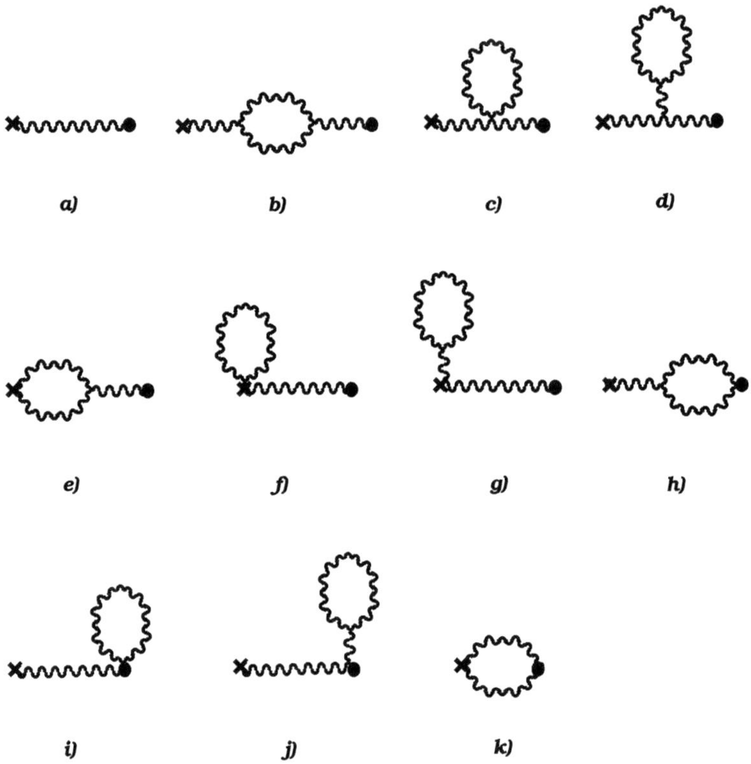

Fig. 1 One-loop diagrams contributing to the operator-product expansion of $T_{zz}(z)\,\mathrm{e}^{\varphi(0)}$ or $T_{zz}(z)T_{zz}(0)$. The points z and 0 are depicted by the cross and the dot, respectively

4 Algebraic Check of DDK

4.1 Covariant Pauli-Villars' Regularization

To describe the one-loop renormalization in QFT, we add Pauli-Villars' regulators: Grassmann Y, $\bar{Y}$ (of mass squared M^2) and normal Z (of mass squared $2M^2$). This regularizes all involved divergences including those in tadpoles. To simplify formulas I keep below only one Y which is enough to compute finite parts. The regulator action reads

$$\mathcal{S}_{\text{reg}} = \frac{1}{16\pi b_0^2}\int \sqrt{g}\left[g^{ab}\partial_a Y \partial_b Y + M^2Y^2 + \varepsilon(\Delta Y)^2 + G\varepsilon g^{ab}\partial_a Y \partial_b Y R\right]. \quad (16)$$

The regulators make a contribution to the energy-momentum tensor which is quadratic in the regulator fields and local. The total energy-momentum tensor is conserved and traceless (!) as before in spite of the mass. We thus expect conformal invariance to be maintained quantumly what can be explicitly checked by the one-loop and partially two-loop calculations.

4.2 Explicit One-Loop Results

The renormalization of b^2 comes from the usual one-loop diagrams including tadpoles:

$$\frac{1}{b^2} = \frac{1}{b_0^2} - \left(\frac{1}{6} - 4 + A + 2G \int \mathrm{d}k^2 \frac{\varepsilon}{(1+\varepsilon k^2)} - \frac{1}{2}GA\right) + \mathcal{O}(b_0^2). \tag{17}$$

Here $A(\varepsilon M^2) \sim \varepsilon M^2$ ($\varepsilon M^2 \gg 1$) is the contribution of the tadpole.

The analogous one-loop renormalization of q/b^2 in T_{zz} reads

$$\frac{q}{b^2} = \frac{1}{b_0^2} - \frac{1}{6} + 2 - \frac{1}{2}A - \frac{1}{2}G - G \int \mathrm{d}k^2 \frac{\varepsilon}{(1+\varepsilon k^2)} + \frac{1}{4}GA \tag{18}$$

or, multiplying by $6b^2$,

$$\frac{6q^2}{b^2} = \left(\frac{q}{b^2}\right)^2 \times 6b^2 = \frac{6}{b_0^2} - 1 - 6G + \mathcal{O}(b_0^2). \tag{19}$$

This precisely confirms [7] the addition to the central charge in Eq. (15) obtained by the CFT technique.

5 The Method of Singular Products

For a general higher-derivative action $S[\varphi]$ one sees tremendous cancellations when computing the operator products $T_{zz}(z)\,\mathrm{e}^{\varphi(0)}$ or $T_{zz}(z)T_{zz}(0)$. To account for the cancellation, it is convenient to write the generator of the conformal transformation as

$$\hat{\delta}_\xi = \int_{D_1} \mathrm{d}^2 z \left(q\xi'(z) \frac{\delta}{\delta\varphi(z)} + \xi(z)\partial\varphi(z) \frac{\delta}{\delta\varphi(z)} \right), \tag{20}$$

where the domain D_1 includes the singularities of $\xi(z)$ leaving outside possible singularities of a string of functions $X(\omega_i)$ on which $\hat{\delta}_\xi$ acts. Actually, the form of $\hat{\delta}_\xi$ on the right-hand side of Eq. (20) is primary. Its advantage over the standard one based on T_{zz} is that it accounts for the cancellations in the quantum case, while there are subtleties associated with singular products.

5.1 Conformal Weight

It is easy to reproduce

$$\hat{\delta}_\xi \, \mathrm{e}^{\alpha\varphi(\omega)} \stackrel{\text{w.s.}}{=} (q\alpha - b^2\alpha^2)\xi'(\omega)\, \mathrm{e}^{\alpha\varphi(\omega)} + \xi(\omega)\partial\, \mathrm{e}^{\alpha\varphi(\omega)} \tag{21}$$

for the quadratic action by (20) via the singular products:

$$\begin{aligned}\hat{\delta}_\xi \, \mathrm{e}^{\alpha\varphi(\omega)} \stackrel{\text{w.s.}}{=} {} & q\alpha\xi'(\omega)\, \mathrm{e}^{\alpha\varphi(\omega)} + \int_{D_1} \mathrm{d}^2 z\, \alpha^2 \xi(z) \big\langle \partial\varphi(z)\varphi(\omega)\big\rangle \delta^{(2)}(z-\omega)\, \mathrm{e}^{\alpha\varphi(\omega)} \\ & + \alpha\xi(\omega)\partial\varphi(\omega)\, \mathrm{e}^{\alpha\varphi(\omega)}.\end{aligned} \tag{22}$$

where the equality is understood in the week sense, i.e. under path integrals. The most interesting is the second term on the right-hand side, where the singular product equals [10]

$$\int_{D_1} \mathrm{d}^2 z\, \xi(z) \big\langle \partial^n \varphi(z)\varphi(\omega)\big\rangle \delta^{(2)}(z-\omega) = (-1)^n \frac{2b^2}{n(n+1)} \xi^{(n)}(\omega), \tag{23}$$

reproducing (21).

It is also easy to understand that Eq. (21) remains valid also for the four-derivative action (as well as for yet higher derivative actions) if the exponential is defined as the normal product. This is because $\varphi(z)$ pairs with only one $\varphi(\omega)$ modulo the renormalization of α. Among three parameters b, q and α only two are independent, so we can set $\alpha = 1$ to be nonrenormalized. The coefficient in front of the first term on the right-hand side of Eq. (21) represents the conformal weight of $\mathrm{e}^{\alpha\varphi}$ and we arrive at Eq. (14). We have thus argued it remains unchanged for the higher-derivative action.

5.2 Central Charge

The central charge $c^{(\varphi)}$ of φ can be computed from the variation of T_{zz} under the infinitesimal conformal transformation generated by (20) as

$$\Big\langle \hat{\delta} T_{zz}(\omega) \Big\rangle = \frac{c^{(\varphi)}}{12} \xi'''(\omega) \tag{24}$$

with the normal ordering in T_{zz}.

Quadratic action: The energy-momentum tensor for the quadratic action given by (13) with $\varepsilon = 0$. transforms as

$$\left\langle \hat{\delta}_{\xi} T_{zz}^{(2)}(\omega) \right\rangle = \frac{1}{2b^2} \int \mathrm{d}^2 z \langle q^2 \xi'''(z) + \xi'(z)\partial^2\varphi(z)\varphi(\omega) + \xi(z)\partial^3\varphi(z)\varphi(\omega)\rangle$$
$$\times \delta^{(2)}(z-\omega) = \frac{\xi'''(\omega)}{2}\left(\frac{q^2}{b^2} + \frac{1}{3} - \frac{1}{6}\right) = \xi'''(\omega)\left(\frac{q^2}{2b^2} + \frac{1}{12}\right), \quad (25)$$

where we have used Eq. (23). The propagator here is *exact* as before in Eq. (22): this is why b^2 cancels. In Eq. (25) $1/12$ corresponds to the usual quantum addition 1 to the central charge. The right-hand side of Eq. (25) reproduces the DDK formula for the central charge.

Four-derivative action: For the four-derivative action Eq. (25) remains unchanged for the quadratic part of T_{zz}. The variation of the part which is $\propto \varepsilon$ reads for $G = 0$ [6]

$$\left\langle \hat{\delta} T_{zz}^{(4)}(\omega) \right\rangle_{G=0} = \frac{1}{b^2} \int \mathrm{d}^2 z \left\langle \left[2q\varepsilon\xi'''(z)\partial\bar{\partial}\varphi(z) + (4q-2)\varepsilon\xi''(z)\partial^2\bar{\partial}\varphi(z)\right.\right.$$
$$\left.\left. -6\varepsilon\xi'(z)\partial^3\bar{\partial}\varphi(z) - 4\varepsilon\xi(z)\partial^4\bar{\partial}\varphi(z)\right]\varphi(\omega)\right\rangle \delta^{(2)}(z-\omega)$$
$$= \frac{\xi'''(\omega)}{4}\left(-2\cdot 2q + (4q-2)\cdot 1 + 6\frac{2}{3} - 4\frac{1}{2}\right) = 0, \quad (26)$$

where for the singular product we have used [10]

$$\int_{D_1} \mathrm{d}^2 z\, \xi(z) \langle [-4\varepsilon\partial^{n+1}\bar{\partial}\varphi(z)\varphi(\omega)] \rangle \delta^{(2)}(z-\omega) = (-1)^n \frac{2b^2}{(n+1)} \xi^{(n)}(\omega). \quad (27)$$

The derivation of Eq. (26) is similar to Eq. (22). Only the average of the quadratic in $\varphi(\omega)$ part of δT_{zz} is nonvanishing for the normal ordered T_{zz}: one $\varphi(\omega)$ is annihilated by the variational derivative $\delta/\delta\varphi(z)$, one more $\varphi(\omega)$ pairs with $\partial\varphi(z)$. The renormalization of the quadratic in φ part of T_{zz} is taken into account by the change $b_0^2 \to b^2$ in Eq. (13).

The consequence of Eq. (26) is that the central charge of φ coincides at $G = 0$ (i.e. for the usual Liouville action plus εR^2) with the one for the quadratic action. Moreover, this universality holds for the terms ε^2 as well [6] and most likely for higher terms in the DeWitt-Seeley expansion of the heat kernel and thus for the Polyakov string.

We can repeat the computation for the part of T_{zz} in Eq. (13) which involves G. Again only the average of the quadratic in $\varphi(\omega)$ part of δT_{zz} is nonvanishing for the normal ordered T_{zz}. This makes the computations pretty much similar to those at one loop. However, the contribution of higher loops is effectively taken into account by the deviations of b^2 and q from their bare values. This is why I call it an "intelligent" one loop.

The contribution of the local part of T_{zz} to $c^{(\varphi)}$ vanishes like (26):

$$\frac{1}{b^2} Gq\varepsilon \int \mathrm{d}^2 z \left\langle\left[-\xi'''(z)\partial\bar{\partial}\varphi(z) - 2\xi''(z)\partial^2\bar{\partial}\varphi(z)\right]\varphi(\omega)\right\rangle \delta_{\varepsilon}^{(2)}(z-\omega) = 0, \quad (28)$$

while for the nonlocal term we obtain

$$\left\langle \hat{\delta}_\xi \left(-\tfrac{1}{b^2} G q \varepsilon \tfrac{1}{\bar{\partial}} \partial^2 (\bar{\partial}\varphi \, \mathrm{e}^{-\varphi} \partial\bar{\partial}\varphi) \right) \right\rangle$$
$$= -\tfrac{1}{b_0^2} G q \varepsilon \int \mathrm{d}^2 z \, \xi'''(z) \left\langle \partial\bar{\partial}\varphi(z)\varphi(\omega) \right\rangle \delta^{(2)}(z-\omega) = \tfrac{1}{2} G q \xi'''(\omega). \quad (29)$$

In both cased we used Eq. (27). In contrast to the classical analysis, now a non-vanishing finite contribution arises from the singular product. Its occurrence is like in the CFT technique where it has arisen [9] at one loop from the nonlocal (last) term in (13).

Summing (25) and (29), we obtain the central charge $c^{(\varphi)}$ of φ shown in Eq. (15). We have thus proved that it is exact [11]. The vanishing of the total central charge is represented (d comes from X^μ and -26 comes from the ghosts) by the equation

$$d - 26 + c^{(\varphi)} = 0. \quad (30)$$

6 Relation to the Minimal Models

It is easy to solve Eqs. (14), (15), (30) which modify KPZ-DDK for the four-derivative action (10), adding a parameter G. The result reads

$$b^{-2} = \frac{13 - d - 6G + \sqrt{(d_+ - d)(d_- - d)}}{12}, \quad (31)$$

$$d_\pm = 13 - 6G \pm 12\sqrt{1+G} \quad (32)$$

and $\gamma_{\rm str}$ is as in Eq. (2) with the KPZ barriers thus shifted to $d_\pm$ given by (32).

Their values depend on G which has to lie in the interval $[-1, 0]$ for the the action (10) to be stable as it follows from the identity (modulo boundary terms)

$$\int \mathrm{e}^{-\varphi} \left[(\partial\bar{\partial}\varphi)^2 - G \partial\varphi \bar{\partial}\varphi \partial\bar{\partial}\varphi \right] = \int \mathrm{e}^{-\varphi} \left[(1+G)(\partial\bar{\partial}\varphi)^2 - G \nabla\partial\varphi \bar{\nabla}\bar{\partial}\varphi \right]. \quad (33)$$

Then b^{-2} is real for $d < d_-$ which increases from 1 at $G = 0$ to 19 at $G = -1$.

As in the case of the usual Liouville theory the operators

$$V_\alpha = \mathrm{e}^{\alpha\varphi}, \qquad \alpha = \frac{1-n}{2} + \frac{1-m}{2b^2} \quad (34)$$

are the BPZ null-vectors for integer n and m. Their conformal weights $\Delta_\alpha = \alpha + (\alpha - \alpha^2)b^2$ coincide with Kac's spectrum of CFT with the central charge $c = 1 + 6(b + b^{-1})^2$ of the Virasoro algebra (not to be confused with the central charge $c^{(\varphi)}$ of φ) where b is displayed in Eq. (31).

To describe minimal models, we choose like in the usual Liouville theory

$$c = 25 + 6\frac{(p-q)^2}{pq}, \quad G = \frac{(26-d-c)q}{6(q+p)}, \quad 1 - 6\frac{(p-q)^2}{pq} \le d \le 19 - 6\frac{p}{q} \tag{35}$$

with coprime $q > p$. In that case this implied $d = 26 - c = 1 - 6\frac{(p-q)^2}{pq}$ for the central charge of matter but now d is a free parameter. From Eq. (31) we find

$$b^{-2} = \begin{cases} \dfrac{q}{p} & \text{perturbative branch} \\ -1 + \dfrac{(25-d)p}{6(q+p)} & \text{the other branch} \end{cases} \quad \text{for } d > 25 - 6\frac{(p+q)^2}{p^2}.$$

The perturbative branch is as in the usual Liouville theory, but the second branch is no longer $p \leftrightarrow q$ with it. There are no obstacles against $d = 4$ for $q = p + 1$!

7 Conclusion

The four-derivative term in the emergent action (10) for φ revives, telling the Nambu-Goto and Polyakov strings apart. For the Polyakov string the higher-derivative terms do not change the KPZ-DDK results (the universality holds). The four-derivative Liouville theory is conformal invariant in spite of the presence of the dimensionful worldsheet cutoff ε and is exactly solved by the method of singular products. The results are like KPZ-DDK except the barriers are shifted for the four-derivative action which could also be the case for the Nambu-Goto string. All that is specific to the theory with diffeomorphism invariance.

I am indebted to the Organizing Committee and especially to Manu Paranjape for the opportunity to participate in the Conference and to meet old friends. My warmest Congratulations to Gordon on his Jubileum!

As is argued in [12] the Nambu-Goto string alludes to $G = -1/3$ in $d = 4$ which from Eq. (35) is associated with the (4,3) minimal model, like the Ising model on a random lattice.

References

1. V.G. Knizhnik, A.M. Polyakov, A.B. Zamolodchikov, Fractal structure of 2D quantum gravity. Mod. Phys. Lett. A **3**, 819 (1988)
2. F. David, Conformal field theories coupled to 2D gravity in the conformal gauge. Mod. Phys. Lett. A **3**, 1651 (1988)
3. J. Distler, H. Kawai, Conformal field theory and 2D quantum gravity. Nucl. Phys. B **321**, 509 (1989)
4. A.M. Polyakov, *Gauge Fields and Strings* (Harwood Academic Publishers, 1987)

5. Y. Makeenko, Pauli-Villars' regularization of ghosts in path-integral string formulation. JHEP **05**, 085 (2023). ([arXiv:2302.01954 [hep-th]])
6. Y. Makeenko, Singular products and universality in higher-derivative conformal theory. JHEP **09**, 086 (2023). ([arXiv:2307.06295 [hep-th]])
7. Y. Makeenko, Private life of the Liouville field that causes new anomalies in the Nambu-Goto string. Nucl. Phys. B **967**, 115398 (2021). ([arXiv:2102.04753 [hep-th]])
8. S. Deser, R. Jackiw, Energy-momentum tensor improvements in two dimensions. Int. J. Mod. Phys. B **10**, 1499 (1996). ([arXiv:9510145 [hep-th]])
9. Y. Makeenko, Opus on conformal symmetry of the Nambu-Goto versus Polyakov strings. Int. J. Mod. Phys. A **38**, 2350010 (2023). ([arXiv:2204.10205 [hep-th]])
10. Y. Makeenko, Notes on higher-derivative conformal theory with nonprimary energy-momentum tensor that applies to the Nambu-Goto string. JHEP **01**, 110 (2023). ([arXiv:2212.02241 [hep-th]])
11. Y. Makeenko, Exact solution of higher-derivative conformal theory and minimal models. Phys. Lett. **845**, 138170 (2023). ([arXiv:2308.05030 [hep-th]])
12. Y. Makeenko, Nambu-Goto string as a higher-derivative Liouville theory. Phys. Rev. D **110**, L081903 (2024). ([arXiv:2407.01136 [hep-th])

Complexity Equals (Almost) Anything

Robert C. Myers and Shan-Ming Ruan

Abstract Recent investigations [1–3] have introduced an infinite class of novel gravitational observables in asymptotically anti-de Sitter (AdS) space that reside on codimension-one or -zero regions of the bulk spacetime. These observables encompass well-established holographic complexity measures such as the maximum volume of the extremal hypersurfaces and the action or spacetime volume of the Wheeler-DeWitt (WDW) patch. Furthermore, this family of observables exhibits two universal properties when applied to the thermofield double (TFD) state: they exhibit linear growth at late times and faithfully reproduce the switchback effect. This implies that any observable from this class has the potential to serve as a gravitational dual for the circuit complexity of boundary states.

Keywords AdS/CFT correspondence · Black holes · Circuit complexity

1 Why Holographic Complexity?

In recent years, there has been significant progress in understanding quantum gravity from the perspective of quantum information theory, especially in the context of holography. Quantum information theory has provided fascinating and profound insights into fundamental questions about the AdS/CFT correspondence, a powerful framework for studying the holographic duality between bulk gravitational theory and quantum field theory on the boundary. In particular, the celebrated Ryu-Takayanagi formula proposed that the entanglement entropy of a subsystem of the holographic CFT is geometrically described by the area of a certain minimal surface in bulk spacetime [4, 5]. Developments in the past decade provide strong evidence

R. C. Myers (✉)
Perimeter Institute for Theoretical Physics, Waterloo, ON, Canada
e-mail: rmyers@perimeterinstitute.ca

S.-M. Ruan
Center for Gravitational Physics and Quantum Information, Yukawa Institute for Theoretical Physics, Kyoto University, Kitashirakawa Oiwakecho, Sakyo-ku, Kyoto, Japan
e-mail: ruan.shanming@yukawa.kyoto-u.ac.jp

M. Paranjape et al. (eds.), *Proceedings of Gravity, Strings and Fields*,
CRM Series in Mathematical Physics,
https://doi.org/10.1007/978-3-031-91266-5_11

that the entanglement structure of the underlying quantum mechanical degrees of freedom plays a crucial role in determining the emergent spacetime geometry and its dynamics.

However, it has become clear that entanglement entropy alone is not sufficient to fully capture the time evolution of a quantum system. For example, it grows only for a short time until the system is thermalized. From a holographic perspective, this is reflected in the fact that holographic entanglement entropy is unable to probe the bulk of spacetime far beyond the event horizon of black holes. To overcome these limitations and to gain a more complete understanding of the underlying physical dynamics of the system, it is essential to investigate alternative quantum information measures.

A fascinating concept that has recently entered this discussion is the quantum circuit complexity [6, 7]. The latter measures how difficult it is to create a particular target state $|\Psi_{\text{T}}\rangle$ from a reference state $|\Psi_{\text{R}}\rangle$ by using a set of fundamental unitary operations or "gates". For example, a quantum circuit can be realized by applying a set of elementary gates g_i:

$$|\Psi_{\text{T}}\rangle = U_{\text{TR}}|\Psi_{\text{R}}\rangle = g_n\, g_{n-1}\, \cdots\, g_2\, g_1\, |\Psi_{\text{R}}\rangle\,. \tag{1}$$

The circuit complexity of the target state $|\Psi_{\text{T}}\rangle$ is then defined as the minimum number of gates needed to construct the particular unitary U_{TR}, i.e., the cost of the optimal circuit. Various approaches have also been developed in recent years to understand the complexity of states in quantum field theory, e.g., see [8–10]. It is important to note that the definition of circuit complexity is ambiguous because the exact value of the complexity is sensitive to the choice of the reference state, the set of gates and the costs assigned to different types of gates. We emphasize that these ambiguities are inherent to the concept itself.

A fascinating aspect of complexity is that it can continue to evolve long after the entanglement entropies of the quantum system have thermalized. In the following, we will focus on holographic complexity, i.e., the bulk dual of the circuit complexity of the boundary states. Several proposals have been made to define holographic complexity. The complexity=volume (CV) proposal [7, 11] suggests that holographic complexity is dual to the maximal volume of a certain class of spacelike hypersurfaces, i.e.,

$$\mathcal{C}_{\text{V}}(\Sigma_{\text{CFT}}) = \max_{\Sigma_{\text{CFT}}=\partial\Sigma} \left[\frac{\mathcal{V}(\Sigma)}{G_{\text{N}}\,\ell_{\text{bulk}}}\right]. \tag{2}$$

Here G_{N} is the Newton's constant of the bulk gravitational theory, and ℓ_{bulk} is an arbitrary length scale, often chosen to be the AdS radius of curvature L. The codimension-one hypersurfaces Σ are all anchored on the asymptotic boundary time slice Σ_{CFT} where the CFT state lives. Different from the CV proposal, the complexity=action (CA) proposal [12] is associated with a codimension-zero subregion in the bulk spacetime. It proposes that holographic complexity is given by the gravitational action evaluated on the Wheeler-DeWitt patch, i.e., the domain of dependence for the surfaces Σ in Eq. (2),

$$\mathcal{C}_{\mathrm{A}}(\Sigma_{\mathrm{CFT}}) = \frac{I_{\mathrm{WDW}}}{\pi \hbar}. \tag{3}$$

Given the challenge of evaluating gravitational action, the complexity=spacetime volume (CV2.0) proposal [13] suggests that the complexity is simply determined by the spacetime volume of the WDW patch, viz,

$$\mathcal{C}_{\mathrm{SV}}(\Sigma_{\mathrm{CFT}}) = \frac{V_{\mathrm{WDW}}}{G_{\mathrm{N}}\, \ell_{\mathrm{bulk}}^2}. \tag{4}$$

The natural setting where holographic complexity has been investigated is the thermofield double state, which is dual to an eternal two-sided black hole. An appealing feature of the three proposals above is that they probe the black hole interior at arbitrarily late times. As evidence that they provide the holographic dual of quantum complexity, the three proposals satisfy two universal features that are argued to hold for any definition of quantum complexity applied to a TFD state. First, at late times, the complexity should grow linearly with time, with the growth rate proportional to the mass of the dual black hole [6]. In addition, there is a universal time delay in the complexity's response to perturbations, known as the switchback effect [11].

The various holographic complexity proposals have certainly drawn attention to new types of gravitational observables in the bulk, as well as highlighting their possible role in the AdS/CFT correspondence. However, as described above, the definition of complexity is inherently ambiguous, and so one might expect that the gravitational dual for complexity should also exhibit an analogous and equally diverse set of ambiguities. This idea was realized in recent studies [1–3], where the authors propose a new infinite class of gravitational observables living on general codimension-one or -zero regions of asymptotically AdS spacetimes. It is shown that these new observables exhibit both the expected linear growth at late times and the switchback effect in shockwave geometries. Consequently, any member of this infinite class of observables serves as an equally plausible candidate for holographic complexity. This proposal has been somewhat playfully dubbed "complexity equals anything".

This chapter aims to summarize the complexity=anything proposal by exploring several simple examples. We also investigate the properties of the newly introduced gravitational observables and their applications in probing the black hole interiors in the bulk spacetime.

2 Complexity Equals Anything

The complexity=anything proposal was introduced in [1, 2]. In part, the impetus behind these studies came from the inherent ambiguities associated with the definition of circuit complexity, and revealed an infinite class of gravitational observables that can be considered plausible candidates for holographic complexity. In particular, these diffeomorphism-invariant observables collectively exhibit linear growth in the late-time regime and manifest the switchback effect with shockwave perturbations, in AdS black hole backgrounds. In this section, we give a succinct overview of the complexity=anything proposal [1, 2] with both codimension-one and codimension-zero observables.

2.1 Codimension-One Observables

Recent investigations of holographic complexity proposals have led to the exploration of novel gravitational observables anchored to time slices Σ_{CFT} on the asymptotic boundary. Building upon this discourse, Ref. [1] extended the CV proposal (2) to encompass an infinite class of codimension-one observables. An essential observation in this discussion was that the CV proposal (2) involves a twofold process. First, there is an extremization procedure which identifies a distinctive codimension-one surface, i.e., the extremal hypersurface, chosen from the myriad of possible spacelike surfaces with a fixed boundary at Σ_{CFT}. Subsequently, one evaluates a specific geometric property, namely the volume, of this selected hypersurface. Motivated by this perspective, the most general gravitational observables on codimension-one hypersurfaces can be characterized by two scalar functionals, F_1 and F_2 [1]. Following the earlier discussion, the first step is to fix a specific codimension-one hypersurface via an extremization procedure. To wit,

$$\delta_{\text{X}}\left[\int_{\Sigma} d^d\sigma\sqrt{h}\, F_2(g_{\mu\nu}; X^\mu)\right] = 0\,, \tag{5}$$

where $F_2(g_{\mu\nu}; X^\mu)$ depends on the bulk metric $g_{\mu\nu}$ and the embedding functions $X^\mu(\sigma^a)$ specify the location of Σ. The extremization is performed over all spacelike hypersurfaces which are asymptotically anchored to a fixed time slice Σ_{CFT} on the boundary theory—see the left panel of Fig. 1. This procedure selects a particular codimension-one hypersurface, denoted Σ_{F_2}, and given this hypersurface, the next step is to evaluate a geometric property on the surface to define the gravitational observable,

$$O_{F_1,\Sigma_{F_2}}(\Sigma_{\text{CFT}}) = \frac{1}{G_{\text{N}}L}\int_{\Sigma_{F_2}} d^d\sigma\,\sqrt{h}\, F_1(g_{\mu\nu}; X^\mu)\,. \tag{6}$$

Again, the scalar functional F_1 depends on the bulk metric and the embedding functions. Of course, this definition is simplified with the choice $F_1 = F_2$ and the latter reduces to the CV proposal by taking $F_1 = F_2 = 1$.

To investigate the properties of the gravitational observables defined in Eqs. (5) and (6), we turn our attention to asymptotically AdS black holes as the $(d{+}1)$-dimensional background geometry. Employing Eddington-Finkelstein coordinates, the bulk metric takes the following form

$$ds^2 = -f(r)\,dv^2 + 2\,dv\,dr + r^2\,d\Omega^2_{k,d-1}\,, \tag{7}$$

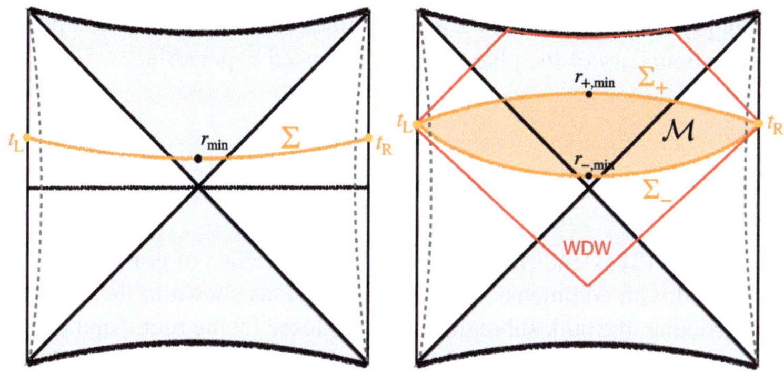

Fig. 1 The left panel shows an extremal codimension-one hypersurface $\Sigma(\tau)$, represented by the orange curve. On the right, the orange region denotes the codimension-zero subregion $\mathcal{M}$ bounded by the two surfaces $\Sigma_\pm$ as in the complexity=anything proposal. In a certain limit, the subregion can extend to the WDW patch with its null boundaries, as shown by the red lines

where $v = t - \int_r^\infty d\tilde{r}/f(\tilde{r})$ denotes the infalling time coordinate. The spatial sections of the boundary geometry are spherical, flat or negatively curved with $k = +1, 0, -1$, respectively. The full two-sided black hole geometry is dual to the TFD state entangling the CFTs on the left and right boundaries, i.e.,

$$|\psi_{\rm TFD}(t_{\rm L}, t_{\rm R})\rangle = \sum_{E_n} e^{-\beta E_n/2 - iE_n(t_{\rm L}+t_{\rm R})/2} |E_n\rangle_{\rm L} \otimes |E_n\rangle_{\rm R} \,. \tag{8}$$

Without loss of generality, we consider a symmetric setup for the boundary time slice with $t_{\rm R} = t_{\rm L} = \tau/2$, as shown in Fig. 1. The determination of the extremal surface with respect to the measure F_2 functional can be reformulated as a classical mechanics problem with involving the equation of motion for a non-relativistic particle in an effective potential determined by F_2. Importantly, it was shown in [1] that the growth rate converges to a constant at late times, namely

$$\lim_{\tau\to\infty}\left(\frac{dO_{F_1,\Sigma_{F_2}}}{d\tau}\right) \propto P_\infty \,. \tag{9}$$

Thus one recovers the desired linear growth of these observables at late times. Moreover, the growth rate P_∞ is found to be proportional to the mass of the AdS black hole (in the high-temperature limit). At the same time, the evaluation of these observables in the shockwave geometries reveals the manifestation of the switchback effect—a universal time delay in response to perturbations.

It is noteworthy that not all of these observables produce extremal surfaces, as described above. As elucidated in [3], the existence of the aforementioned in the late-time limit hinges on the presence of a local maximum of the effective potential $U(r)$ within the horizon. Otherwise, there are no locally extremal surfaces and, as described in [3], the hypersurfaces Σ_{F_2} determined by the maximization process are pushed to the boundary of the phase space of allowed hypersurfaces.

2.2 Codimension-Zero Observables

The discussion of [2] extends these ideas to an infinite class of gravitational observables associated with codimension-zero regions $\mathcal{M}$, as shown in the right panel of Fig. 1. In particular, the bulk subregion $\mathcal{M}$ is enclosed by the future and past boundaries, denoted $\Sigma_{\pm}$, both anchored to the same boundary time slice, i.e., $\partial\Sigma_{\pm} = \Sigma_{\text{CFT}}$. Hence $\partial\mathcal{M} = \Sigma_{+} \cup \Sigma_{-}$. These codimension-zero observables are also shown to exhibit the desired universal properties for holographic complexity, including the linear growth at late times and the switchback effect [2]. This construction draws inspiration from the CA and CV2.0 proposals, which evaluate different functionals on the WDW patch. In contrast, however, the complexity=anything proposal begins by identifying a specific bulk subregion through an extremization procedure, akin to the generalized approach in Eq. (5).

Therefore, the codimension-zero version of complexity=anything again follows a two-step procedure. First, we identify the extremal subregion and boundaries with three independent geometric scalar functionals $F_{2,\pm}$ and G_2 by varying the shape of the two boundaries $\Sigma_{\pm}$, i.e.,

$$\delta_{X_{\pm}}\left[\int_{\Sigma_{\pm}} d^d\sigma\sqrt{h}\, F_{2,\pm}(g_{\mu\nu}; X^{\mu}_{\pm}) + \frac{1}{L}\int_{\mathcal{M}} d^{d+1}x\sqrt{g}\, G_2(g_{\mu\nu})\right] = 0\,. \tag{10}$$

Here, $F_{2,\pm}$ are scalar functionals of the bulk metric $g_{\mu\nu}$ and the embedding functions $X^{\mu}_{\pm}(\sigma^a)$ of the corresponding boundaries $\Sigma_{\pm}$. Similarly, G_2, integrated over the codimension-zero region $\mathcal{M}$, is a scalar functional constructed from the background curvature. This procedure singles out a special codimension-zero region $\mathcal{M}_{G_2,F_{2,\pm}}$ with extremal boundaries denoted as $\Sigma_{\pm}[G_2, F_{2,\pm}]$.

Subsequently, we can evaluate a separate geometric functional on this extremal subregion. The observable is thus defined by

$$\begin{aligned} O\left[G_1, F_{1,\pm}, \mathcal{M}_{G_2,F_{2,\pm}}\right](\Sigma_{\text{CFT}}) = \frac{1}{G_N L}\int_{\Sigma_{+}[G_2,F_{2,+}]} d^d\sigma\,\sqrt{h}\, F_{1,+}(g_{\mu\nu}; X^{\mu}_{+}) \\ + \frac{1}{G_N L}\int_{\Sigma_{-}[G_2,F_{2,-}]} d^d\sigma\,\sqrt{h}\, F_{1,-}(g_{\mu\nu}; X^{\mu}_{-}) + \frac{1}{G_N L^2}\int_{\mathcal{M}_{G_2,F_{2,\pm}}} d^{d+1}x\,\sqrt{g}\, G_1(g_{\mu\nu})\,. \end{aligned} \tag{11}$$

Importantly, $F_{1,\pm}$ and G_1 are three independent scalar functionals, providing the versatility of this definition. The new complexity=anything proposal thus involves two independent bulk scalars, G_1 and G_2, and four independent boundary scalars, $F_{1,\pm}$ and $F_{2,\pm}$. Notably, this comprehensive proposal subsumes the codimension-one observables originally discussed above by setting $G_1 = G_2 = F_{1,-} = F_{2,-} = 0$, emphasizing its flexibility.

The simplest non-trivial choice for Eq. (10) is obtained by taking $F_{2,\pm}$ and G_2 all to be constants. To wit,

$$\mathcal{C}_{\text{gen}} = \frac{1}{G_N L}\left[\alpha_+ \int_{\Sigma_+} d^d\sigma \sqrt{h} + \alpha_- \int_{\Sigma_-} d^d\sigma \sqrt{h} + \frac{1}{L}\int_{\mathcal{M}} d^{d+1}x \sqrt{-g}\right], \tag{12}$$

where $\alpha_\pm$ are dimensionless (positive) constants. The result of extremizing over the shape of the boundary surfaces has an elegant geometric interpretation,

$$K_{\Sigma_+} = -\frac{1}{\alpha_+ L}, \qquad K_{\Sigma_-} = \frac{1}{\alpha_- L}. \tag{13}$$

In other words, the extremal boundaries $\Sigma_\pm$ are constant mean curvature (CMC) slices. An interesting feature of this simple result is that CMC slices can be pushed to the future and past null surfaces emanating from Σ_{CFT} in the limit $\alpha_\pm \to 0$, as indicated by the divergences $K_{\Sigma_\pm} \to \mp\infty$. That is, $\mathcal{M}$ becomes the WDW patch in this limit. So we can recover the CA and CV2.0 proposals as special cases of the complexity=anything proposal in Eqs. (10) and (11).

3 Singularity Probes

Thanks to the flexibility of the complexity=anything proposal, we are able to systematically probe the black hole interior. Furthermore, it provides gravitational observables that facilitate the understanding of the geometric properties near the spacetime singularity [3]. Of course, the latter requires that we position the extremal surface $\Sigma_\pm$ in close proximity to the singularity. However, this proximity can be achieved by considering the future CMC slice defined in Eq. (13) with $\alpha_+ \ll 1$, as shown in Fig. 2. For simplicity, we focus our attention on observables defined by local geometric functionals evaluated only on the future CMC slice Σ_+. Specifically, the observables (11) are evaluated with $F_1 = 0 = G_1$, leaving

$$\mathcal{C}_{\text{CMC}} = \frac{1}{G_N L}\int_{\Sigma_+} d^d\sigma \sqrt{h}\, F_{\text{CMC}}(g_{\mu\nu}, X^\mu_+). \tag{14}$$

The value of the extrinsic curvature (13) is varied to control the distance between the CMC slice and the singularity at late times. This allows us to probe the geometry near the singularity by studying the late-time growth of these CMC observables. Considering both AdS Schwarzschild and AdS Reissner-Nordström (RN) black holes, we demonstrate how various properties of the spacetime geometry can be systematically revealed using this class of gravitational observables.

We commence our analysis by considering the AdS Schwarzschild black hole, described by the metric in Eq. (7) with

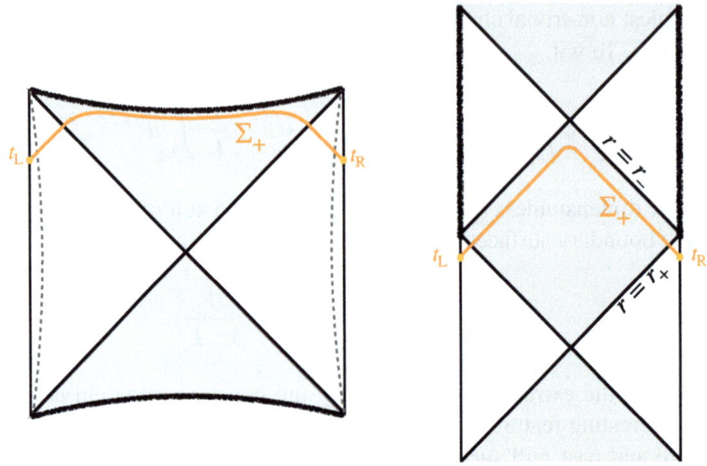

Fig. 2 The left and right panels correspond to the Penrose diagrams for the AdS Schwarzschild and AdS Reissner-Nordström black hole, respectively. Orange curves denote late-time CMC slices Σ_+ with $\alpha_+ \ll 1$

$$f(r) = k + \frac{r^2}{L^2} - \frac{\omega^{d-2}}{r^{d-2}}\,, \tag{15}$$

where the parameter ω determines the mass of the AdS black hole. At late times, the extremal surface Σ_+ approaches a constant radial slice, denoted by $r = r_f$. In the limit $\alpha_+ \ll 1$, the CMC slice approaches the future singularity with

$$\lim_{\alpha_+ \to 0} r_f \simeq \left(\frac{d^2 L^2 \omega^{d-2}}{4}\right)^{1/d} \alpha_+^{2/d}\,. \tag{16}$$

Given this surface, we evaluate the CMC observable (14) with different choices of the geometric functional F_{CMC}. Specifically, we consider:

$$F_{\text{CMC}} = 1\,, \qquad F_{\text{CMC}} = -L\,K\,, \qquad F_{\text{CMC}} = L^4 C^2\,, \tag{17}$$

where as above, K denotes the trace of the extrinsic curvature and $C^2 = C_{\mu\nu\rho\sigma}C^{\mu\nu\rho\sigma}$ is the square of the Weyl tensor. The growth rate at late times associated with these three observables can be derived as

$$\lim_{\tau\to\infty} \frac{d\mathcal{C}_{\text{CMC}}}{d\tau} \simeq \begin{cases} \frac{8\pi d}{(d-1)}\, M\, \alpha_+\,, & F_{\text{CMC}} = 1\,, \\ \frac{8\pi d}{(d-1)}\, M\,, & F_{\text{CMC}} = -L\,K\,, \\ \frac{128\pi(d-1)(d-2)}{d^2}\, \frac{M}{\alpha_+^3}\,, & F_{\text{CMC}} = L^4 C^2\,. \end{cases} \tag{18}$$

In summary, the above results reveal different behaviours for the various observables as the CMC slice approaches the singularity. Of particular note is the unbounded growth rate (with $1/\alpha_+^3$) in the case of L^4C^2, providing evidence that the extremal surface Σ_+ is close to a curvature singularity. For comparison purposes with the following recall that $\frac{d}{d-1}\,M = S\,T$ where S and T denote the entropy and temperature of the black hole, respectively.

Having investigated the behaviour of various CMC observables near the singularity of the AdS Schwarzschild black hole, we contrast these results with those for the AdS Reissner-Nordström black hole—see the right panel of Fig. 2. The metric for the latter is still expressed by Eq. (7), where the blackening factor is instead given by

$$f_{\text{RN}}(r) = k + \frac{r^2}{L^2} - \frac{\omega^{d-2}}{r^{d-2}} + \frac{q^2}{r^{2(d-2)}}\,. \tag{19}$$

Unlike the AdS Schwarzschild geometry, the AdS RN black hole features a timelike singularity and inner/outer horizons at $r = r_\pm$, where $f(r_\pm) = 0$. Consequently, the timelike singularity remains beyond the region probed by the CMC slices anchored to the asymptotic boundaries. The CMC slices effectively probe the black hole interior only up to the inner horizon located at $r = r_-$. It proves advantageous to express the resulting quantities in terms of the Bekenstein-Hawking entropy and Hawking temperature associated with the inner horizon,

$$S_- = \frac{\Omega_{k,d-1} r_-^{d-1}}{4G_{\text{N}}} \quad \text{and} \quad T_- = \frac{\left|f'_{\text{RN}}(r_-)\right|}{4\pi}\,. \tag{20}$$

For example, in the regime of large extrinsic curvature (i.e., $\alpha_+ \ll 1$), the position of the final slice $r = r_f$ is approximately located at

$$r_f \simeq r_- + 4\pi L^2 T_-\,\alpha_+^2 \quad \text{with} \quad \alpha_+ \ll 1\,. \tag{21}$$

Similar to the previous analysis, the late-time growth rate for the three observables (17) can be expressed as

$$\lim_{\tau\to\infty} \frac{d\mathcal{C}_{\text{CMC}}}{d\tau} \simeq \begin{cases} 8\pi S_- T_-\,\alpha_+ \to 0\,, & F_{\text{CMC}} = 1\,, \\ 8\pi S_- T_-\,, & F_{\text{CMC}} = -L\,K\,, \\ \# S_- T_-\,\alpha_+ \to 0\,, & F_{\text{CMC}} = L^4 C^2\,, \end{cases} \tag{22}$$

where # in the last line is a complicated function of ω, q and k. The vanishing of the growth rate for $F_{\text{CMC}} = 1$ or C^2L^4 can be attributed to the vanishing of the volume element as the CMC slice approaches a null surface (i.e., the inner horizon), while the Weyl-squared term remains finite there.

4 Summary and Outlook

In summary, our brief review has delved into the recent "complexity equals anything" proposal [1–3] for holographic complexity. These investigations have revealed an infinite class of gravitational observables defined for codimension-one or -zero subregions of the bulk spacetime. Somewhat surprisingly, (almost) each of these holographic observables stands as an equally plausible candidate for the gravitational dual of the circuit complexity of the corresponding boundary state. A key lesson is that the freedom in constructing gravitational observables is an innate feature of holographic complexity, rather than a shortcoming. As we illustrated, this freedom leads to families of gravitational observables which allow for a systematic study of the black hole interior in a way that would be impossible with, e.g., the CV or CA proposals alone. Furthermore, it seems to mimic the ambiguities found in the definition of circuit complexity in the boundary theory. Of course, one of the interesting future research directions would be to determine a more explicit relation between the ambiguities found in the bulk and the boundary constructions. This would undoubtedly lead to progress in understanding the precise interpretation of the gravitational observables in terms of the observables constructed in the boundary theory. Undoubtedly, this new proposal for holographic complexity points to numerous new opportunities for research at the intersection of quantum gravity and quantum information.

Acknowledgements RM presented this material at *Gravity, Strings and Fields: A conference in honour of Gordon Semenoff*, held July 24–28, 2023 at the Centre de Recherches Mathématiques. We would like to wish Gordon a happy birthday and to congratulate him on his enormous successes and contributions to theoretical physics. We would also like to thank Alex Belin, Eivind Jørstad, Gabor Sárosi and Antony Speranza for enjoyable collaborations on the material discussed here. Research at Perimeter Institute is supported in part by the Government of Canada through the Department of Innovation, Science and Economic Development Canada and by the Province of Ontario through the Ministry of Colleges and Universities. RCM is supported in part by a Discovery Grant from the NSERC of Canada, and by the BMO Financial Group. RCM and SMR are supported by the Simons Foundation through the "It from Qubit" collaboration. SMR is also supported by MEXT-JSPS Grant-in-Aid for Transformative Research Areas (A) "Extreme Universe", No. 21H05187 and by JSPS KAKENHI Research Activity Start-up Grant Number JP22K20370.

References

1. A. Belin, R.C. Myers, S.M. Ruan, G. Sárosi, A.J. Speranza, Does complexity equal anything? Phys. Rev. Lett. **128**(8), 081602 (2022). https://doi.org/10.1103/PhysRevLett.128.081602 [arXiv:2111.02429 [hep-th]]
2. A. Belin, R.C. Myers, S.M. Ruan, G. Sárosi, A.J. Speranza, Complexity equals anything II. JHEP **01**, 154 (2023). https://doi.org/10.1007/JHEP01(2023)154. ([arXiv:2210.09647 [hep-th]].)
3. E. Jørstad, R.C. Myers, S.M. Ruan, Complexity=anything: singularity probes. JHEP **07**, 223 (2023). https://doi.org/10.1007/JHEP07(2023)223. ([arXiv:2304.05453 [hep-th]].)

4. S. Ryu, T. Takayanagi, Holographic derivation of entanglement entropy from AdS/CFT. Phys. Rev. Lett. **96**, 181602 (2006). https://doi.org/10.1103/PhysRevLett.96.181602. ([arXiv:hep-th/0603001 [hep-th]].)
5. S. Ryu, T. Takayanagi, Aspects of holographic entanglement entropy. JHEP **08**, 045 (2006). https://doi.org/10.1088/1126-6708/2006/08/045. ([arXiv:hep-th/0605073 [hep-th]].)
6. L. Susskind, Entanglement is not enough. Fortsch. Phys. **64**, 49–71 (2016). https://doi.org/10.1002/prop.201500095[arXiv:1411.0690 [hep-th]]
7. L. Susskind, Computational complexity and Black Hole Horizons. Fortsch. Phys. **64**, 24–43 (2016). https://doi.org/10.1002/prop.201500092[arXiv:1403.5695 [hep-th]]
8. R. Jefferson, R.C. Myers, Circuit complexity in quantum field theory. JHEP **10**, 107 (2017). https://doi.org/10.1007/JHEP10(2017)107. ([arXiv:1707.08570 [hep-th]].)
9. S. Chapman, M.P. Heller, H. Marrochio, F. Pastawski, Toward a definition of complexity for quantum field theory states. Phys. Rev. Lett. **120**(12), 121602 (2018). https://doi.org/10.1103/PhysRevLett.120.121602[arXiv:1707.08582 [hep-th]]
10. P. Caputa, N. Kundu, M. Miyaji, T. Takayanagi, K. Watanabe, Phys. Rev. Lett. **119**(7), 071602 (2017). https://doi.org/10.1103/PhysRevLett.119.071602
11. D. Stanford, L. Susskind, Complexity and Shock Wave Geometries. Phys. Rev. D **90**(12), 126007 (2014). https://doi.org/10.1103/PhysRevD.90.126007 [arXiv:1406.2678 [hep-th]]
12. A.R. Brown, D.A. Roberts, L. Susskind, B. Swingle, Y. Zhao, Holographic Complexity Equals Bulk Action? Phys. Rev. Lett. **116**(19), 191301 (2016). https://doi.org/10.1103/PhysRevLett.116.191301 [arXiv:1509.07876 [hep-th]]
13. J. Couch, W. Fischler, P.H. Nguyen, Noether charge, black hole volume, and complexity. JHEP **03**, 119 (2017). https://doi.org/10.1007/JHEP03(2017)119. ([arXiv:1610.02038 [hep-th]].)

Superluminal Signalling in Double Holography

Dominik Neuenfeld and Manu Srivastava

Abstract We discuss claims in the literature which suggest that double holography is in tension with causality and argue that this tension is resolved if the theory on the brane is correctly treated as an effective field theory. In order to discuss the causal structure of the theory on the brane the notion of domain of dependence needs to be replaced and we propose possible candidate regions. This contribution to the proceedings is based on [1].

Keywords AdS/CFT · AdS/BCFT · Holography · Effective field theory

1 Introduction

Bottom-up models of AdS/BCFT [2, 3] capture essential features of holographic boundary conformal field theories in a dual, higher dimensional AdS spacetime. At the same time they are sufficiently simple such that many computations can be explicitly performed. Recently, these models and their connection to Randall-Sundrum braneworlds [4, 5] were revisited to construct models in which a gravitating AdS space can exchange matter excitations of a holographic large-c CFT with a non-gravitating bath [6–9]. The latter construction makes it possible for large AdS black holes to evaporate and thus to study the black hole information paradox in a controlled setting.

Through AdS/BCFT a gravitating spacetime in d dimensions coupled to a d-dimensional bath is dual to a BCFT$_d$, where the gravitating part of the spacetime is obtained by dualizing the d-1-dimensional boundary. Choosing the matter to be a holographic CFT$_d$, the complicated task of calculating the entropy of Hawking radiation becomes a simple geometric problem in a dual, $d+1$-dimensional spacetime

D. Neuenfeld (✉)
University of Würzburg, Würzburg, Germany
e-mail: dominik.neuenfeld@uni-wuerzburg.de

M. Srivastava
Center for Theoretical Physics, MIT, Cambridge, MA, USA
e-mail: manu_sr@mit.edu

M. Paranjape et al. (eds.), *Proceedings of Gravity, Strings and Fields*,
CRM Series in Mathematical Physics,
https://doi.org/10.1007/978-3-031-91266-5_12

by means of the HRT formula [10, 11], which relates subregion entropy in (B)CFTs to the area of extremal surface in the dual spacetime. Due to the presence of two layers of holography this construction is referred to as *double holography*.

Using doubly-holographic models it has been argued [6] that in order to compute the true, fine-grained entropy of a bath subregion R one needs to replace the usual *von Neumann entropy* S_{vN}

$$S_{\mathrm{vN}}(R) = -\operatorname{tr} \rho_R \log \rho_R, \tag{1}$$

where ρ_R is the reduced density matrix on the region R, by another notion of fine-grained entropy which is computed using the *quantum extremal island formula*,

$$S(R) = \min_I \operatorname{ext} \left(\frac{\mathrm{Area}(\partial I)}{4G_N} + S_{\mathrm{vN}}(R \cup I) \right), \tag{2}$$

which allows for additional, dynamically determined subregions I, so-called *islands*. Applying this prescription to evaporating lack holes produces a unitary page curve providing some progress in our understanding of the black hole information paradox. For a nice review, see [12].

In higher dimensions doubly-holographic derivations [7–9] are so far the only formal way of arriving at Eq. (2). However, the physics of doubly-holographic models is not well-understood and there are crucial differences to more standard theories of semi-classical gravity, for example the existence of massive gravitons. In fact, it has been argued that such differences prevent carrying over the insights obtained from the quantum extremal island formula to flat space [13]. Moreover, understanding the precise relation of bottom-up models of BCFTs to full top-down constructions in supergravity is an open problem, calling into question the validity of results only obtained in bottom-up models.

We thus face several challenges to a full understanding of the reliablity of bottom-up models: Are they internally consistent and can we constraint their parameters? Can lessons from these models be carried over to more general situations such as flat space? Can bottom-up models be derived in a controlled way from top-down constructions in supergravity?

In this note, which is based on [1], we address one particular challenge to the consistency of double holography which was pointed out by the authors of [14]—apparent superluminal signalling. We will argue that the superluminal signalling observed in [14] can be understood as an effective field theory effect in AdS. Moreover, we will argue that doubly-holographic models must be understood in a low-energy effective field theory sense, with a cutoff scale that is set by the $d+1$-dimensional AdS scale. Finally, we will discuss how this affects the causal structure of double holography and what might replace the classical notion of causal structure.

2 Double Holography from the Bottom up

The most basic bottom-up model used in double holography is a bulk AdS spacetime which is partially cut off by a time-like codimension-one surface called an *end-of-the-world (ETW) brane* or *Planck brane*. The action describing these models is the Einstein-Hilbert action in a $d+1$ dimensional bulk spacetime with negative cosmological constant Λ together a Gibbons-Hawking-York term and an action for the brane,

$$\begin{aligned} I = \frac{1}{16\pi G_N}\int d^{d+1}x\sqrt{-g}(R-2\Lambda) - \frac{1}{8\pi G_N}\int_{\mathcal{Q}\cup\mathcal{M}} d^d x\sqrt{-h}K \\ - \frac{T}{8\pi G_N}\int_{\mathcal{Q}} d^d x\sqrt{-h}. \end{aligned} \tag{3}$$

The Gibbons-Hawking-York term has support on both the asymptotic boundary $\mathcal{M}$ and the ETW brane $\mathcal{Q}$. In the present paper, the last term (the action for the ETW brane) simply consists of a constant tension term. Taking the point of view that action Eq. (3) is only an effective theory describing a much more complicated spacetime, this tension term is only the leading piece in a derivative expansion of possible brane actions.

The choice of boundary conditions is Dirichlet on $\mathcal{M}$ and Neumann on $\mathcal{Q}$ [2]. Imposing Neumann boundary conditions on the brane sets the coefficient of metric fluctuations on the brane to zero, implying that

$$K_{\mu\nu} = -\frac{T}{(d-1)}h_{\mu\nu}, \tag{4}$$

where $K_{\mu\nu}$ is the extrinsic curvature of the brane and $h_{\mu\nu}$ is the projector onto the brane. This equation has solutions for all values of T. Here, we will consider a particular class of solutions for which $0 < T < T_{\rm crit} = (d-1)/L$ (L is the bulk AdS length) and the bulk solution is the global vacuum expressed in Poincaré coordinates. The bulk metric can be expressed as a warped product of AdS$_d$ and an interval $\mu \in (-\frac{\pi}{2}, \mu_*]$, such that the ETW brane $\mathcal{Q}$ is located at $\mu = \mu_*$,

$$ds^2 = \frac{L^2}{\cos^2\mu}\left(d\mu^2 + ds^2_{\text{AdS}_d}\right) = \frac{L^2}{\cos^2\mu}\left(d\mu^2 + \frac{dy^2 - dt^2 + d\xi^i d\xi_i}{y^2}\right). \tag{5}$$

The location of $\mathcal{Q}$ is controlled by the brane tension via $T\,L/(d-1) = \sin\mu_*$.

The dual description of this model is given by a BCFT at the asymptotic boundary with the BCFT boundary located at the intersection $\mathcal{Q}\cap\mathcal{M}$, in our case $y=0$. This model reproduces many characteristic properties of BCFTs, such as the form of the boundary entropy, the g-theorem, and—if we allow for couplings to bulk scalars—the behaviour of non-trivial one-point functions.

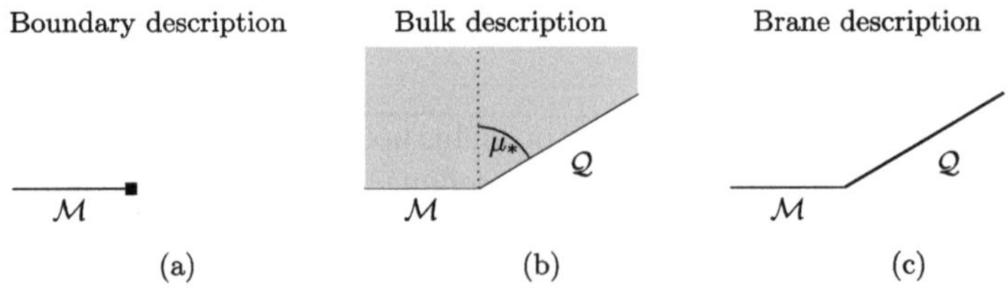

Fig. 1 The three description in double holography

The crucial aspect of those models, which lies at the heart of the derivation of Eq. (2), is that in addition to the two dual *bulk* and *boundary* descriptions, there is a third, emergent, way of thinking about the model as $0 < T - T_{\text{crit}} \ll L^{-1}$. We will call this third perspective the *brane perspective*, c.f., Fig. 1. It is obtained by integrating out the bulk between the brane and the asymptotic boundary. The brane perspective consists of a holographic large-c CFT on the asymptotic boundary which is glued to a large-c CFT on a gravitating spacetime given by the brane world-volume. It is this duality (or rather triality) frame in which the theory takes the form as the union of a gravitating region and a non-gravitating bath.

3 An Apparent Inconsistency of the Brane Perspective

It is well known that information cannot escape black holes in local quantum field theory. However, Eq. (2) computes a true, quantum-gravitational entropy of black holes (or their radiation) which demonstrates that black hole evaporation is in fact unitary and information leaves the black hole—an effect simply not captured by the semi-classical approximation to quantum gravity.

An obvious requirement for accepting this lesson more generally is self-consistency of the brane description of double holography. However, in [14] it was observed that that the brane perspective seems to allow for superluminal signalling, calling into question whether these models are local and/or causal.

Consider signalling between two points, p and q, where q is on the asymptotic boundary and p is on the brane, Fig. 2. Using the metric Eq. (5) it is easy to compute the time Δt it takes a light-like signal to either travel through the bulk, Δt_{bulk}, or around the asymptotic boundary and the brane Δt_{brane}. A bit of algebra reveals that

$$\Delta t_{\text{brane}}^2 - \Delta t_{\text{bulk}}^2 = 2 y_p y_q (1 - \sin \mu_*) \leq 0, \tag{6}$$

where y_p and y_q are the coordinate locations of p and q in the metric Eq. (5). This demonstrates that even in this simplest example, the signal between q and p is faster through the bulk than around the boundary.

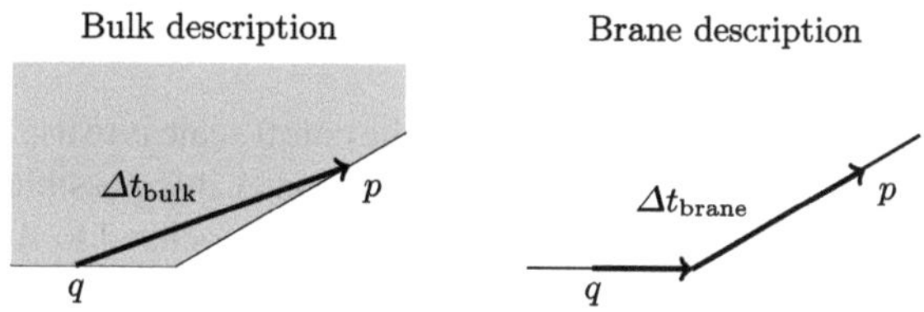

Fig. 2 The two ways of sending a signal between p and q

This observation is, at least naively, in conflict with microcausality in the brane description, since two operators which are spacelike separated in the brane description can still be causally connected through the bulk.

In AdS/CFT this possible inconsistency is absent due to the Gao-Wald theorem [15] which states that the signalling time between two points on the asymptotic boundary of AdS is lower bounded by the signalling time around the boundary. This theorem does not hold in the case of double holography since the endpoints of signalling are not necessarily at an *asymptotic* boundary.

One possible resolution of this issue was discussed in [16], where the authors considered a top-down model in supergravity. They found that asking whether the causal structure for points in the brane perspective is respected by the bulk geometry is an ill-posed question. The reason is that unlike in bottom-up models, in the top-down model considered no canonical mapping between the brane and the bulk exists.

We will now argue that bottom-up models need to be treated as effective field theories. If done correctly, superluminal signalling can never be observed and the problem is resolved without the need for referring to a top-down construction.

4 Bottom-Up Braneworld Models are EFTs

To understand how the inconsistency discussed above gets resolved in bottom-up models, consider the gravitational theory in the brane perspective [8],

$$\begin{aligned} I_{\text{brane}} = \frac{1}{16\pi G_N^{\text{brane}}} \int d^d x \sqrt{-g} \Big(\mathcal{R} + 2\Lambda_b \\ + \frac{L^2}{(d-4)(d-2)} \left(\mathcal{R}^{ij}\mathcal{R}_{ij} - \frac{d}{4(d-1)}\mathcal{R}^2 \right) + \dots \Big) + W_{CFT}[g]. \end{aligned} \tag{7}$$

Due to the presence of higher derivative terms in the gravitational action, we should expect a local theory only above a length scale set by the dimensionful coupling of the first higher derivative term. Here, the cutoff scale is set by L, the bulk AdS length.

We can arrive at the same conclusion by another argument. In AdS/CFT, the location along the holographic direction in the bulk is related to an energy scale in the boundary description. By imagining extending the bulk spacetime beyond the ETW brane at μ_* all the way to $\mu = \pi$ we see that the brane effectively cuts off part of the spacetime close to the boundary, i.e., it regulates the UV of the CFT living at an *imagined boundary*.

One way of making precise the cutoff scale is to use bulk reconstruction, see Fig. 3. Given access to a boundary subregion A it is possible to reconstruct bulk operators within the entanglement wedge (EW) associated to A [17]. The EW of A is defined as the bulk domain of dependence of a spacelike region bounded by A and the HRT surface associated to A.

Subregions in the imagined part of the asymptotic boundary whose EW are behind the brane do not encode operators in the physical bulk spacetime. Thus, there is a minimal size of subregions in the CFT such that operators in front of the brane can be encoded. We can take the size of those regions as a definition of the UV cutoff for the imagined CFT. This UV cutoff is related to a UV cutoff on the brane by a Weyl rescaling which maps the boundary metric to the brane metric.

Applying this procedure to a constant cutoff, we find that the CFT cutoff is set by z_c the location of the brane and the Weyl rescaled cutoff on the brane is L, c.f. Fig. 3a. In the case at hand, i.e., for a brane at a constant value of μ as shown in Fig. 3b, this procedure gives

$$L_{\text{cutoff}} = \frac{L}{\sin \mu_*}. \tag{8}$$

Now that we have established a cutoff scale on the brane, we can revisit the computation of [14] and check if the causality violation can be seen at scales larger than Eq. (8). Naively, it seems that the causality violation in Eq. (6) can appear at all scales, since it linearly depends on the coordinate distance between p and q. However, we are saved by the relation between proper and coordinate distance in AdS.

Consider Fig. 4a. The point p is obtained by signalling for a time Δt from q through the bulk. We arrive at p' by sending a signal around the boundary and brane for the same amount of time. The coordinate location of $y_{p'}$ is give by

$$y_{p'} = \sqrt{(y_p + y_q)^2 - 2y_p y_q (1 - \sin \mu_*)} - y_q. \tag{9}$$

In order for the advance of signals through the bulk to be visible to an observer in the EFT, the physical distance between the two points must be larger than Eq. (8). Maximizing over all points q the largest coordinate distance between p and p' is

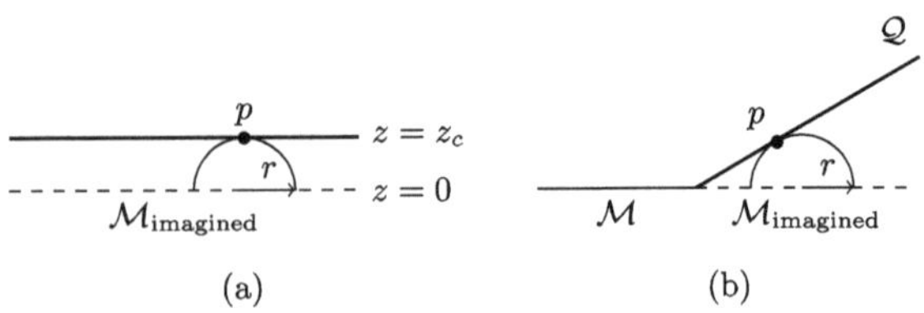

Fig. 3 **a** The case of a brane at $z = z_c$. **b** The case of a brane at $\mu = \mu_*$

$$\left(y_p - y_{p'}\right)_{\max} = y_p(1 - \sin \mu_*), \tag{10}$$

from which we can easily compute the proper distance $s/2$ between both points,

$$\frac{s}{2} = -\frac{L}{\cos \mu_*} \log(\sin \mu_*) \leq \frac{1}{2} L_{\text{cutoff}}, \tag{11}$$

which is below the cutoff scale. Thus, the faster-than-light signalling cannot be observed in the EFT on the brane which has a cutoff given by Eq. (8).

5 Causality at the Cutoff

Usually, the causal structure of a background spacetime is discussed in terms of *domains of dependence*. The domain of dependence $\mathcal{D}(A)$ of a set A consists of all points p such that any past or future directed, inextendable causal (i.e., time- or lightlike) curve through p must intersect A. In local theories the state of the physical system in $\mathcal{D}(A)$ is determined by specifying initial values, e.g., a density matrix ρ_A, on A.

In the case of semi-classical gravity the natural cutoff scale is usually assumed to be the Planck scale where one expects the notion of geometry and therefore also the notion of domains of dependence to break down. In contrast, in double-holography the background geometry is well-defined in the large N limit which suppresses fluctuations of the geometry in the bulk such that the cutoff scale L can remain large compared to fluctuations of the geometry. Because of this we should still be able to define a causal structure in a geometric way for the brane.

However, in double holography data on A is not sufficient to determine the physical state in all of $\mathcal{D}(A)$ as can easily be seen from Fig. 4b. Given the naive domain of dependence $\mathcal{D}(B)$ of a subregion B of AdS space, an excitation with a finite proper width s that starts at a point p between the asymptotic boundary and B and propagates at the speed of light towards B will enter $\mathcal{D}(B)$ although it initially started out at

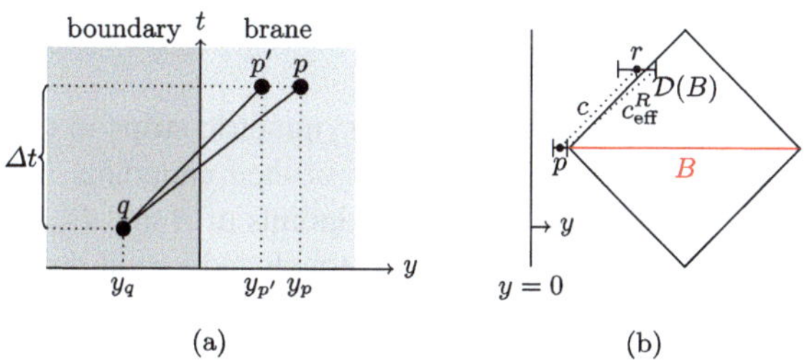

Fig. 4 **a** This figure shows the propagation of signals in double holography through the bulk $q \to p$ and around the boundary $q \to p'$. **b** If excitations of fixed proper width propagate on AdS space, the leading edge of the excitation can move faster than the speed of light

spacelike separation. The reason for this behaviour lies in the geometry of AdS space. As the excitation propagates, the coordinate size grows while the proper size of the excitation remains constant. As a result, the leading edge of the signal propagates faster than the speed of light, with $c_{\text{eff}}^R = \frac{c}{\sin \mu_*}$. A similar argument shows that the signal propagation speed for signals to the left is unchanged, $c_{\text{eff}}^L = c$.

Any region that replaces the domain of dependence must respect this behaviour of AdS and thus need to be contained in $\mathcal{D}_{\text{eff}}$ which is defined with respect to c_{eff}^R and c_{eff}^L. We will now discuss three different, geometric notions of domain of dependence which respect this bound. More details about the computations and more examples can be found in [1].

5.1 Unitary Domain of Dependence

Our first alternative to the domain of dependence is defined as the largest region for which time-evolution can be implemented by a unitary operation. To this end consider a density matrix $\rho(0) = \rho_B$ on a subregion B at constant time t_0. A state $\rho(s_0) = \rho_{B'}$ on a partial Cauchy slice B' with has $\mathcal{D}(B) = \mathcal{D}(B')$ and contains the same information can be obtained if there exist unitary operators $U(s)$ with $0 \leq s \leq s_0$ such that

$$\rho_{B'} = U(s_0)\rho_B U^\dagger(s_0). \tag{12}$$

Definition Given a partial Cauchy slice B at a constant Poincaré time t_0, the *unitary domain of dependence* $\mathcal{C}(B)$ is the union of all points p such that there exists a Cauchy slice B' with $\mathcal{D}(B) = \mathcal{D}(B')$ passing through p and the state on B' is obtained from ρ_B by the action of unitaries.

Note that condition Eq. (12) implies that $S(B) = S(B')$ (here, we consider situations where no islands arise and thus Eqs. (1) and (2) agree). We can then consider different Cauchy slicings B and B', split the region B' into two subregions B'_1 and B'_2 and use subadditivity of entropy to find

$$S(B) \leq S(B'_1) + S(B'_2). \tag{13}$$

We choose the subregions to be translation-symmetric strips in the ξ^i direction so that they are fully specified by the y-location of their endpoints. The left endpoints of B and B'_1 are the same, as are the right endpoints of B and B'_2. We then consider the remaining free point at which B'_1 and B'_2 touch and vary it throughout the (y, t) plane. For every point we can test if Eq. (13) holds by computing the contributions separately using the HRT prescription introduced in the introduction. For a discussion of the validity of the HRT formula in braneworlds consider [1]. This way we can obtain an upper bound on the size of the unitary domain of dependence. Examples in $d = 2$ and $d = 3$ are shown in Fig. 5.

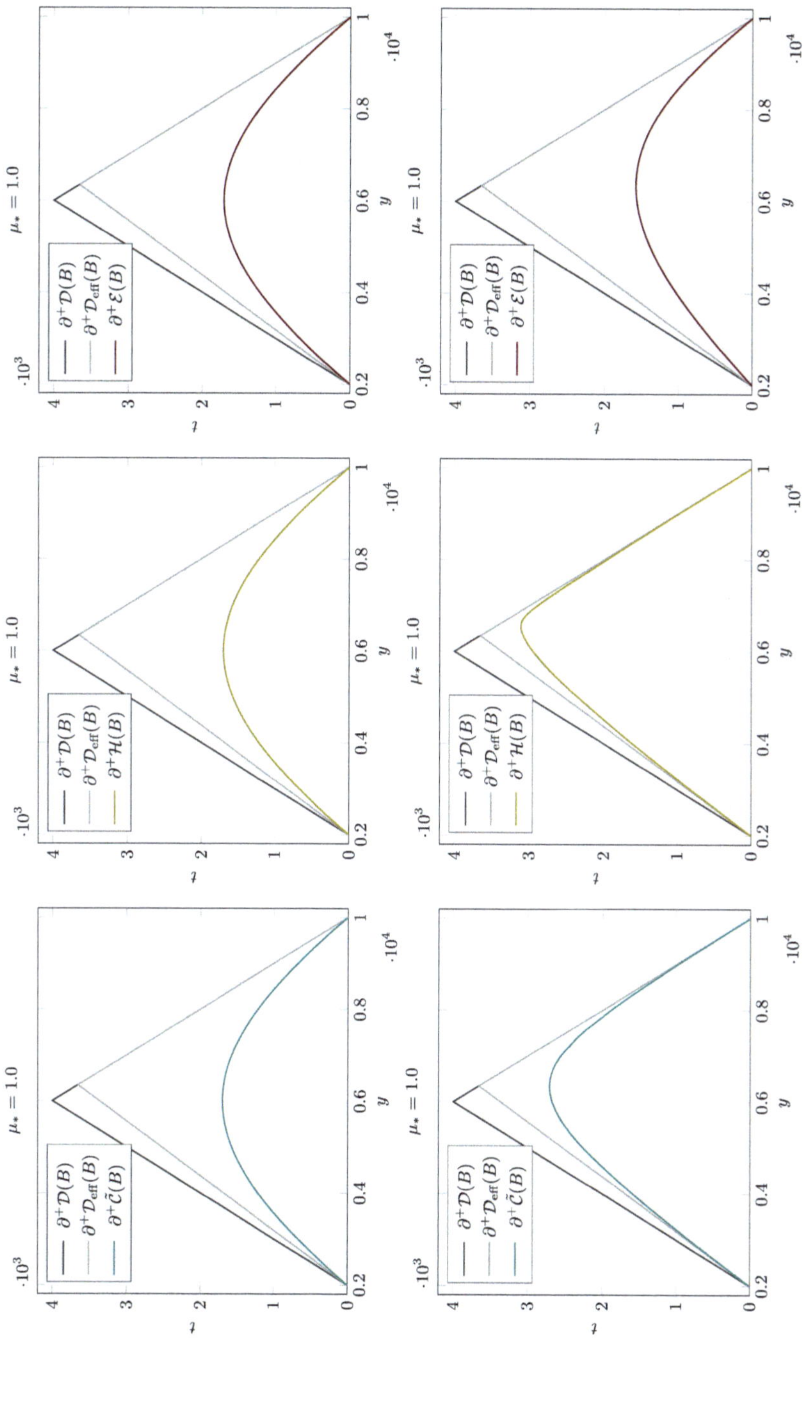

Fig. 5 Upper bounds on the domain of dependence (dark gray), the effective domain of dependecen (light gray), the unitary domain of dependence (blue, left), the holographic domain of dependence (yellow, middle), and the EFT domain of dependence (red, right) for $d = 2$ (top) and $d = 3$ (bottom)

5.2 Holographic Domain of Dependence

Similarly to unitarity, we can use holography as a guide to define a replacement for the domain of dependence on the brane. In AdS/CFT, the entanglement wedge associated to a boundary subregion A is the bulk region that is encoded in the reduced density matrix on A. From the bulk point of view, it is the domain of dependence of the partial Cauchy slice bounded by A and the HRT surface associated to A. In AdS/CFT with an asymptotic boundary, the intersection of the entanglement wedge of A with the asymptotic boundary agrees with $\mathcal{D}(A)$. Assuming that this is still true for holography at a cutoff, we have another candidate domain of dependence, which we will call the *holographic domain of dependence*.

Definition Given a partial Cauchy slice B at constant Poincaré time, the *holographic domain of dependence* $\mathcal{H}(B)$ is the intersection of the entanglement wedge associated to B and the brane, i.e., $\mathcal{H}(B) = \mathcal{W}_{\mathcal{E}}[B] \cap \mathcal{Q}$ with $\mathcal{Q}$ and $\mathcal{W}_{\mathcal{E}}[B]$ being the brane and the entanglement wedge associated to B respectively.

Here, we associate an entanglement wedge $\mathcal{W}_{\mathcal{E}}[B]$ to B in the obvious way, now using the HRT prescription for a subregion on the brane. The holographic domain of dependence is a natural replacement for the domain of dependence from the holographic point of view leveraging the duality between the brane and the bulk perspectives in double holography. Note that $\mathcal{H}(B)$ and $\mathcal{C}(B)$ are defined independently and there is no natural relationship between them. The region that this prescription yields is shown in Fig. 5.

5.3 EFT Domain of Dependence

We might also define a *EFT Domain of Dependence* as the spacetime region in which we believe effective field theory to remain valid. In the presence of gravity there are different competing proposals of how such a region should be characterized. Here, we will take the nice-slice criterion proposed in [18], which states that all local curvatures of a Cauchy slice must be smaller than the cutoff scale. While there does not seem to be a formal definition in the literature, we will take it to mean that

$$|\mathfrak{K}(y, \xi^i)| \leq \frac{1}{L_{\text{cutoff}}}, \tag{14}$$

where $\mathfrak{K}(y, \xi^i)$ is the extrinsic curvature scalar of the Cauchy slice evaluated at (y, ξ^i). The intrinsic and extrinsic curvatures are related by the Gauss-Codazzi equations. Since we are in AdS vacuum, the Gauss-Codazzi equations together with Eq. (14) imply that the intrinsic curvature also cannot be bigger than $\mathcal{O}(1/L_{\text{cutoff}})$.

Definition The *EFT domain of dependence* $\mathcal{E}(B)$ of a partial Cauchy slice B located at constant Poincaré time is the union of all points p such that there exists a Cauchy slice B' with $D(B) = D(B')$ passing through p and whose extrinsic curvature obeys Eq. (14).

This definition captures the region in which we usually assume validity of an EFT description and is shown in Fig. 5. In examples shown in Fig. 5 and additional examples [1], the EFT domain of dependence is at least as small as the other two notions of domain of dependence. This implies that at least in the cases studied, a breakdown of local physics will not be visible outside the spacetime region in which we deem EFT to be valid based on the nice-slice criterion.

6 Conclusion

Here we have argued that the seemingly acausal effects discussed in [14] have a UV origin and should be understood as artifacts of the limit of the effective field theory of bottom-up models of double holography. If one wants to discuss causality these models one needs to replace the naive domain of dependence with some other region. We have discussed different prescriptions that have the potential to define a notion of causality in brane worlds. While, somewhat miraculously, they agree in two dimensions, they are different in higher dimensions with the example of $d = 3$ given here. We consider the examples given here as evidence that bottom-up models of double holography are self-consistent if understood as an effective field theory.

References

1. D. Neuenfeld, M. Srivastava, JHEP **10**, 164 (2023)
2. T. Takayanagi, Phys. Rev. Lett. **107**, 101602 (2011)
3. M. Fujita, T. Takayanagi, E. Tonni, JHEP **11**, 043 (2011)
4. L. Randall, R. Sundrum, Phys. Rev. Lett. **83**, 4690–4693 (1999)
5. A. Karch, L. Randall, JHEP **05**, 008 (2001)
6. A. Almheiri, R. Mahajan, J. Maldacena, Y. Zhao, JHEP **03**, 149 (2020)
7. H. Geng, A. Karch, JHEP **09**, 121 (2020)
8. H.Z. Chen, R.C. Myers, D. Neuenfeld, I.A. Reyes, J. Sandor, JHEP **10**, 166 (2020)
9. H.Z. Chen, R.C. Myers, D. Neuenfeld, I.A. Reyes, J. Sandor, JHEP **12**, 025 (2020)
10. S. Ryu, T. Takayanagi, Phys. Rev. Lett. **96**, 181602 (2006)
11. V.E. Hubeny, M. Rangamani, T. Takayanagi, JHEP **07**, 062 (2007)
12. A. Almheiri, T. Hartman, J. Maldacena, E. Shaghoulian, A. Tajdini, Rev. Mod. Phys. **93**(3), 035002 (2021)
13. H. Geng, A. Karch, C. Perez-Pardavila, S. Raju, L. Randall, M. Riojas, S. Shashi, JHEP **01**, 182 (2022)
14. H. Omiya, Z. Wei, JHEP **07**, 128 (2022)
15. S. Gao, R.M. Wald, Class. Quant. Grav. **17**, 4999–5008 (2000)
16. A. Karch, H. Sun, C.F. Uhlemann, JHEP **10**, 012 (2022)
17. B. Czech, J.L. Karczmarek, F. Nogueira, M. Van Raamsdonk, Class. Quant. Grav. **29**, 155009 (2012)
18. D.A. Lowe, J. Polchinski, L. Susskind, L. Thorlacius, J. Uglum, Phys. Rev. D **52**, 6997–7010 (1995)

Operator Products in the SU(∞) Principal Chiral Model

Peter Orland

Abstract The SU(N) principal chiral model is asymptotically free and integrable in $1+1$ dimensions. In the large-N limit, there is no scattering, but correlation functions are *not* those of a free field theory. We briefly review the derivation of form factors for local operators. Two-point functions for such operators are known exactly. The two-point function of scaling-field operators has the short-distance behavior expected from the renormalization group. We briefly discuss non-vacuum operator products. The ultimate goal is to derive the Lagrangian field theory from this axiomatic quantum-field-theory formalism.

Keywords Large-N · Planar diagrams · Integrability · Form factors

1 The Principal Chiral Model

Although it is a pleasure to present this work to celebrate Gordon Semenoff's seventieth birthday, I have an ulterior motive. Problems remain in the subject, calling out for more expertise and physical intuition: Gordon, help!

The SU(N) principal chiral model (PCM) [1] in 1+1 dimensions is considered a signpost for quantum chromodynamics. The PCM is an asymptotically-free matrix quantum field theory. It is known to have a mass gap, but this is not analytically proved. Integrability [2] yields the exact S matrix for all N [3].

One method to study integrable quantum models is the form factor program, dating from the 1970's [4], formalized by Smirnov [5] (see Ref. [6] for a clear introduction). Unfortunately, finding many form factors of the PCM is hard, due

---

Contribution to the proceedings volume for *Gravity, Strings and Fields: A conference in honor of Gordon Semenoff.*

P. Orland (✉)
Baruch College, CUNY, New York, USA
e-mail: orland@nbi.dk; peter.orland@baruch.cuny.edu

The Graduate School, University Center, CUNY, New York, NY, USA

M. Paranjape et al. (eds.), *Proceedings of Gravity, Strings and Fields*,
CRM Series in Mathematical Physics,
https://doi.org/10.1007/978-3-031-91266-5_13

to bound-state poles. The large-N method is also limited. At large-N, the O(N) and $\mathbb{CP}(N{-}1)$ models have bubble diagrams, with a linear structure; but Feynman diagrams of the large-N PCM are planar. *Combining* the large-N and form-factor approaches, however, yields results [7–9].

Form factors of local operators and exact expressions for correlation functions have been found. The short-distance two-point function of the scaling field [10, 11], matches the renormalization group result [12]. The short-distance forms of other quantities are known, although the derivations are not yet published.

These methods lie in the realm of axiomatic field theory. We hope to use the operator-product expansion to address the most pregnant question raised by this work: whether there is a proof of equivalence with Lagrangian field theory. If so, the model would be truly solved at large-N, firmly establishing the existence of the mass gap. This is tough, Gordon, and I need your guidance.

1.1 *The Basic Features of the PCM*

The Lagrangian of the SU(N) PCM is, with ultraviolet cut-off Λ implicit,

$$\mathcal{L} = \frac{N}{2g_0^2}\mathrm{Tr}\,\partial_\mu U^\dagger \partial^\mu U. \tag{1}$$

Here g_0 is the bare coupling in the 't Hooft limit $N \to \infty$. The global symmetry is $U(x) \to V_L U(x) V_R$, where $V_L,\, V_R \in \mathrm{SU}(N)$ are the left- and right-transformations, respectively. Local operators include the scaling field, $\Phi(x)$, its Hermitian conjugate $\Phi(x)^\dagger$, the left and right Noether currents $j^{\mathrm{L}}_\mu(x)$, $j^{\mathrm{R}}_\mu(x)$, respectively, and the stress-energy-momentum tensor (or stress tensor) $T_{\mu\nu}(x)$.

The scaling field $\Phi(x)$ is a complex $N \times N$ matrix field related to the bare field by

$$\langle \ldots \Phi(x) \ldots . \rangle = Z^{-1/2} \langle \ldots U(x) \ldots . \rangle, \tag{2}$$

where Z is a Λ-dependent real renormalization factor. The field $\Phi(x)^\dagger$ is defined similarly. The scaling field is neither unitary nor of unit determinant, in general, but (2) is well-defined as $\Lambda \to \infty$.

The currents are

$$j^{\mathrm{L}}_\mu(x)_b = \mathrm{iTr}\; t_b\, \partial_\mu U(x)\, U(x)^\dagger \;, \quad j^{\mathrm{R}}_\mu(x)_b = \mathrm{iTr}\; t_b\, U(x)^\dagger \partial_\mu U(x) \;, \tag{3}$$

where the generators $t_a,\ a = 1, \ldots, N^2 - 1$, of SU($N$) satisfy $\mathrm{Tr} t_a t_b = \delta_{ab}$.

The stress tensor $T_{\mu\nu}(x)$ is the symmetric operator $T_{\nu\mu}(x) = T_{\mu\nu}(x)$, which is divergenceless $\partial^\mu T_{\mu\nu}(x) = 0$, and has vanishing vacuum expectation value.

The beta function and the anomalous dimension of the scaling field are, respectively,

$$\beta(g_0) = \frac{\partial g_0^2}{\partial \ln \Lambda} = -\beta_1 g_0^4 + \cdots, \quad \beta_1 = \frac{1}{8\pi},$$

$$\gamma(g_0) = \frac{\partial}{\partial \ln \Lambda} \ln \frac{1}{N} \langle \mathrm{Tr} \Phi(x) \Phi(0)^\dagger \rangle = \gamma_1 g_0^2 + \cdots, \quad \gamma_1 = \frac{N^2 - 1}{4\pi N^2}.$$

The asymptotically-free short-distance behavior of the scaling field is

$$\frac{1}{N} \langle \mathrm{Tr}\, \Phi(x) \Phi(0)^\dagger \rangle \simeq C \ln^{\gamma_1/\beta_1}(|mx|^{-1}) \longrightarrow C \ln^2(m|x|). \tag{4}$$

where m is the mass gap and the limit is taken as $N \to \infty$ [12]. We recovered this result from entirely nonperturbative methods [10]. With Katzav [11], we found the universal constant C in (4), which is impossible to calculate with perturbation theory.

Form factors of currents and the stress tensor have also been found (including some form factors for currents for finite N) [9, 13, 14]. There were errors in Ref. [13], some of which were corrected in [14].

1.2 The S Matrix and the 't Hooft Limit

Completely integrable quantum field theories [2] have infinitely many conservation laws. These imply that for any scattering process of particles, the sets of incoming and outgoing momenta are the same, i.e., $\{p_1, \ldots, p_n\}_{\mathrm{out}} = \{p_1, \ldots, p_n\}_{\mathrm{in}}$. This implies that the result of scattering is only exchange of momenta and there is no creation or destruction of excitations. Another consequence is that the S matrix factorizes into two-body S matrices, for in-states (defined as $|p_1, \ldots, p_n\rangle$, with $p_1 > p_2 > \cdots > p_n$):

$$S^n(p_1, \ldots, p_n) = \prod_{j=1}^{n-1} \left[\prod_{l=j+1}^{n} S^2(p_j, p_l) \right]. \tag{5}$$

If (5) holds for $n = 3$, it generally holds for any larger value of n. Consistency of the $n = 3$ case implies the Yang-Baxter relation

$$S^2(p_j, p_l) S^2(p_j, p_m) S^2(p_l, p_m) = S^2(p_l, p_m) S^2(p_j, p_m) S^2(p_j, p_l). \tag{6}$$

We now discuss the example of the S matrix of the chiral Gross-Neveu model with symmetry SU(N). The rapidity θ of a mass-m excitation parametrizes energy and momentum by, respectively, $m \cosh \theta_j = E_j$, $m \sinh \theta_j = p_j$. Rapidity differences are denoted by $\theta_{jk} = \theta_j - \theta_k$.

An S matrix with SU(N) invariance has asymptotic particle states, labeled by $r = 1, \ldots, N-1$ with corresponding masses obeying the sine law (this is a general kinematic feature of integrable quantum field theories [15]) $\mu_r = \mu_1 \sin \frac{\pi r}{N} / \sin \frac{\pi}{N}$. These are bound states of r elementary particles (whose label is $r = 1$). The $r = N-1$ case is the antiparticle of the $r = 1$ particle. The chiral-Gross-Neveu S-matrix element of two such elementary particles [16], each with a single color quantum number a, c is

$$ {}_{\text{out}}\langle p_1', a'; p_2' c' | p_1, a; p_2, c\rangle_{\text{in}} = \langle a', c' | S_{\text{CGN}}(\theta) | a, c\rangle \delta(p_1' - p_1)\delta(p_2' - p_2), \quad (7) $$

where

$$ S_{\text{CGN}}(\theta) = \frac{\Gamma(\mathrm{i}\theta/2\pi + 1)\Gamma(-\mathrm{i}\theta/2\pi - 1/N)}{\Gamma(\mathrm{i}\theta/2\pi + 1 - 1/N)\Gamma(-\mathrm{i}\theta/2\pi)} \left(1\!\!1 - \frac{2\pi \mathrm{i}}{N\theta} P \right), \quad (8) $$

where $1\!\!1$ keeps colors the same in the initial and final states, P exchanges the two initial colors in the final state, and where $\theta = \theta_{12}$. There is only forward scattering in (7). The bound-state poles of this S matrix have residues of order $1/N$ [17].

The mass spectrum of the PCM also obeys a sine law $m_r = m_1 \sin \frac{\pi r}{N} / \sin \frac{\pi}{N}$. The rules for the spectrum are similar to the chiral-Gross-Neveu case. The main difference is that the excitations are color dipoles. Again, there is only forward scattering. The S matrix of two elementary ($r = 1$) particles contains two factors of (8) [3]:

$$ S_{\text{PP}}(\theta) = \frac{\sin(\theta/2 - \pi \mathrm{i}/N)}{\sin(\theta/2 + \pi \mathrm{i}/N)} S_{\text{CGN}}(\theta) \otimes S_{\text{CGN}}(\theta) \quad (9) $$

Note that the binding energy of bound states with $r = 2, \ldots, N-2$ vanish in the 't Hooft limit. As is the case for the chiral Gross-Neveu model, the residues of the bound-state poles vanish. Thus only the elementary particle ($r = 1$) and antiparticle ($r = N-1$) survive as $N \to \infty$. Consequently, for the rest of our analysis, we need only (9) and its crossed version $S_{PA}(\theta)$, representing the scattering of an elementary antiparticle with an elementary particle. Crossing is implemented by exchanging the s- and t-channels, and replacing θ by $\hat{\theta} = \pi \mathrm{i} - \theta$. The $1/N$-expansion of the particle-particle S-matrix element (9) is

$$ S_{\text{PP}}(\theta) = \left[1 + O\left(\frac{1}{N^2}\right) \right] \left[1\!\!1 \otimes 1\!\!1 - \frac{2\pi \mathrm{i}}{N\theta} P \otimes 1\!\!1 - \frac{2\pi \mathrm{i}}{N\theta} 1\!\!1 \otimes P - \frac{4\pi^2}{N^2\theta^2} P \otimes P \right]. \quad (10) $$

The scattering matrix of one particle and one antiparticle $S_{PA}(\theta)$ is obtained by crossing (10) from the s-channel to the t-channel:

$$S_{PA}(\theta)^{d_2c_2;c_1d_1}_{a_1b_1;b_2a_2} = \left[1 + O(1/N^2)\right]\left[\delta^{d_2}_{b_2}\delta^{c_2}_{a_2}\delta^{c_1}_{a_1}\delta^{d_1}_{b_1}\right.$$
$$\left. - \frac{2\pi\mathrm{i}}{N\hat{\theta}}\left(\delta_{a_1a_2}\delta^{c_1c_2}\delta^{d_2}_{b_2}\delta^{d_1}_{b_1} + \delta^{c_2}_{a_2}\delta^{c_1}_{a_1}\delta_{b_1b_2}\delta^{d_1d_2}\right) - \frac{4\pi^2}{N^2\hat{\theta}^2}\delta_{a_1a_2}\delta^{c_1c_2}\delta_{b_1b_2}\delta^{d_1d_2}\right]. \quad (11)$$

2 Large-N Form Factors of the PCM

2.1 Rules for Exact Form Factors

The form factor of a local operator $\mathcal{O}(x)$ is an expression of the form given by

$$f^{\mathcal{O}}(\theta_1, a_1; \ldots; \theta_n, a_n) = \langle 0|\mathcal{O}(0)|\theta_1, a_1; \ldots; \theta_n, a_n\rangle_{\text{in}}, \quad (12)$$

where $\theta_1, \ldots, \theta_n$ are the external rapidities of excitations and $a_1, \ldots a_n$ are the corresponding quantum numbers. Other matrix elements, besides (12), may be found by crossing particle states from the ket to the bra.

To determine form factors, it is convenient to use Smirnov's axioms [5, 6]:

1. Scattering Axiom

$$f^{\mathcal{O}}(\theta_1, a_1; \ldots; \theta_{i-1}, a_{i-1}; \theta_{i+1}, a_{i+1}; \theta_i, a_i; \theta_{i+2}, a_{i+2}; \ldots; \theta_n, a_n)$$
$$= S^{k_ik_{i+1}}_{a_ia_{i+1}}(\theta_i - \theta_{i+1}) f^{\mathcal{O}}(\theta_1, a_1; \ldots; \theta_{i-1}, a_{i-1}; \theta_i, a_i; \theta_{i+1}, a_{i+1}; \theta_{i+2}, a_{i+2} \ldots; \theta_n, a_n).$$

This axiom is a generalization of Watson's theorem.

2. Periodicity Axiom

$$f^{\mathcal{O}}(\theta_1, a_1; \ldots; \theta_n, a_n) = \gamma^{\mathcal{O}}_{a_n} f^{\mathcal{O}}(\theta_n - 2\pi\mathrm{i}, a_n; \theta_1, a_1; \ldots; \theta_{n-1}, a_{n-1}),$$

where the phase factor $\gamma^{\mathcal{O}}_a$ appears in the equal-time commutation relation

$$\mathcal{O}_a(x)\mathcal{O}(y) = \gamma^{\mathcal{O}}_a \mathcal{O}(y)\mathcal{O}_a(x), \quad x^0 = y^0, \; x^1 > y^1,$$

for arbitrary $\mathcal{O}(y)$. In our case, the excitations are bosons, so we shall take $\gamma^{\mathcal{O}}_n = 1$.

3. Annihilation-Pole Axiom

$$\mathrm{i}\,\mathrm{Res} f^{\mathcal{O}}(\theta_1, a_1; \ldots; \theta_n, a_n)\big|_{\theta_n = \theta_{n-1} + \pi\mathrm{i}} = f^{\mathcal{O}}(\theta_1, a_1; \ldots, \theta_{n-2}; a_{n-2})\, C_{a_{n-1}a_n}$$
$$- S^{b_{n-1}b_1}_{c_1a_1}(\theta_1 - \theta_{n-1}) S^{c_1b_2}_{c_2a_2}(\theta_2 - \theta_{n-1}) \cdots S^{c_{n-4}b_{n-3}}_{c_{n-3}a_{n-3}}(\theta_{n-3} - \theta_{n-1})$$
$$\times S^{b_{n-3}b_{n-2}}_{a_{n-1}a_{n-2}}(\theta_{n-2} - \theta_{n-1}) f^{\mathcal{O}}(\theta_1, a_1; \ldots; \theta_{n-2}, a_{n-2})\, C_{b_{n-1}a_n},$$

where C is the charge-conjugation matrix. The normalization of the left-hand side follows from the standard state normalization, e.g. $<\theta|\theta'> = 4\pi\delta(\theta - \theta')$. This axiom and the previous one are special cases of "generalized crossing", discussed in detail by Babujian et.al. [6]. See also Appendix B in Ref. [5].

4. Lorentz-Invariance Axiom If an operator $\mathcal{O}(x)$ carries Lorentz spin s (not the center-of-mass energy squared s), the form factors must transform under a boost $\theta_j \to \theta_j + \alpha$ for all $j = 1, \ldots, n$ as

$$f^{\mathcal{O}}(\theta_1 + \alpha, a_1; \ldots; \theta_n + \alpha, a_n) = e^{s\alpha} f^{\mathcal{O}}(\theta_1, a_1; \ldots; \theta_n, a_n) ,$$

5. Bound-State Pole Axiom Poles appear in the form factors, at differences of rapidities corresponding to the formation of bound states. The rules for the residues of the poles depend on the nature of the model under consideration. In the large-N limit these poles vanish, hence we will not require further details.

6. Minimality Axiom In the spirit of the bootstrap, we assume form factors have as much analyticity as possible. It can sometimes be checked that this is consistent with some perturbative method, which means either standard covariant perturbation theory or the $1/N$-expansion. In cases where we can find form factors, minimality stands up very well. If a first guess for the form factor $f^{\mathcal{O}}(\theta_1, \ldots, \theta_n)$ satisfies the first four axioms, then so does

$$f^{\mathcal{O}}_{\text{minimal}}(\theta_1, \ldots, \theta_n) = f^{\mathcal{O}}(\theta_1, \ldots, \theta_n) \frac{P_n(\{\cosh(\theta_j - \theta_k)\})}{Q_n(\{\cosh(\theta_j - \theta_k)\})} ,$$

where P_n and Q_n are symmetric polynomials. In this way, we can eliminate all the poles in form factors, except those required by Axiom 3. or corresponding to bound states. For Axiom 3 to be satisfied:

$$P_n|_{\theta_n = \theta_{n-1} + \pi \mathrm{i}} = P_{n-2} , \quad Q_n|_{\theta_n = \theta_{n-1} + \pi \mathrm{i}} = Q_{n-2} .$$

The overall normalization of form factors can often be found through physical or mathematical considerations.

Although the S matrix becomes unity as $N \to \infty$, the correlation functions are not those of a free field, as is clear from (4). Technically, the nontriviality of our form factors is due to the fact that factors of $1/N$ in the particle-antiparticle S-matrix elements (11) are cancelled by traces appearing in the application of Axiom 2 above.

2.2 *The Scaling Field of the SU(∞) PCM*

We use the shorthand $P_j = P, \theta_j, a_j, b_j$, for particle states and $A_j = A, \theta_j, b_j, a_j$ for antiparticle states (note that we have reversed the order of the left color and the right color in this last expression). The form factors were found by applying the scattering axiom to the first external excitation (with rapidity θ_1) repeatedly, pushing this excitation past all the others, finally comparing with the periodicity axiom [7, 8].

The form factor of the scaling field, for an $M-1$-antiparticle and M-particle in-state is [7, 8]

$$\begin{aligned}&\langle 0|\Phi(0)_{b_0 a_0}\,|A_1;\ldots;A_{M-1};P_M;\ldots;P_{2M-1}\rangle\\&=N^{-M+1/2}\sum_{\sigma,\tau\in S_M}F_{\sigma\tau}(\theta_1,\theta_2,\ldots,\theta_{2M-1})\prod_{j=0}^{M-1}\delta_{a_j\,a_{\sigma(j)+M}}\delta_{b_j\,b_{\tau(j)+M}}\,,\end{aligned}\tag{13}$$

where S_M is the permutation group of M objects, and $F_{\sigma\tau}=F^0_{\sigma\tau}+O\left(\frac{1}{N}\right)$, with

$$F^0_{\sigma\tau}(\theta_1,\theta_2,\ldots,\theta_{2M-1})=\frac{(-4\pi)^{M-1}\prod_{l=1}^{M}\left[1-\delta_{\sigma(l)\tau(l)}\right]}{\prod_{j=1}^{M-1}[\theta_j-\theta_{\sigma(j)+M}+\pi\mathrm{i}][\theta_j-\theta_{\tau(j)+M}+\pi\mathrm{i}]}.\tag{14}$$

The expression (14) has no double poles. The simple poles occur where differences of rapidities are equal to $\pm\pi\mathrm{i}$. The annihilation-pole axiom was used to find the overall normalization.

Substituting the $M=2$ form factor into the LSZ reduction formula yields the $1/N$ expansion of the two-particle S matrix [7].

2.3 Current Operators

Dropping the superscript L from the left current (3), its form factor for one particle and one antiparticle [9] is

$$\begin{aligned}\langle 0|j_\mu(0)_{a_3 a_0}|P_1,A_2\rangle_{\mathrm{in}}&=\frac{2\pi\mathrm{i}(p_1-p_2)_\mu}{\theta_{12}+\pi\mathrm{i}}\delta_{a_3 a_1}\delta_{a_0,a_2}\delta_{b_1 b_2}\\&=\frac{2\pi\mathrm{i}\epsilon_{\mu\nu}\eta^{\nu\lambda}(p_1+p_2)_\lambda}{\theta_{12}+\pi\mathrm{i}}\tanh\frac{\theta_{12}}{2}\delta_{a_3 a_1}\delta_{a_0,a_2}\delta_{b_1 b_2}\,.\end{aligned}\tag{15}$$

The first expression on the right-hand side clearly satisfies the Lorentz-invariance axiom. The latter form is more convenient for checking the annihilation-pole axiom of form factors with more excitations. The normalization is fixed by requiring the charge operator $Q_0=\int j_0 dx$ to satisfy the SU(N) commutation relations [9].

The generalization to the form factor with M antiparticles and M particles, $M=1,2\ldots$, in the in-state was found by Cortés Cubero [14]:

$$\begin{aligned}&\langle 0|j_\mu(0)_{a_{2M+1}a_o}|P_1;\ldots;P_M;A_{M+1};\ldots;A_{2M}\rangle_{\mathrm{in}}=\frac{(4\pi)^M\mathrm{i}\tilde{P}_\mu}{2N^{M-1}}\\&\times\sum_{\substack{\sigma\in S_{M+1}\\ \tau\in S_M}}\left[\prod_{l=0}^{M}\delta_{a_l\,a_{M+\sigma(l)+1}}\right]\left[\prod_{l=0}^{M}\delta_{b_l\,b_{M+\tau(l)}}\right]J_{\sigma\tau}(\theta_1,\ldots,\theta_{2M}),\end{aligned}\tag{16}$$

where $\tilde{P}_\mu = \epsilon_{\mu\nu}\eta^{\nu\alpha}(p_1 + \cdots + p_{2M})$ and $J_{\sigma\tau} = J^0_{\sigma\tau} + O\left(\frac{1}{N}\right)$, with

$$J^0_{\sigma\tau}(\theta_1, \ldots, \theta_{2M}) = \tanh \frac{\theta_{M+1+\sigma(0)} - \theta_{\sigma^{-1}(M)}}{2} \times \frac{\prod_{l=1}^{M}\left[1 - \delta_{1+\sigma(l)\,\tau(l)}\right]}{\left\{\prod_{j=1}^{M} 1_{\sigma(j)\neq M}[\theta_j - \theta_{\sigma(j)+M+1} + \pi \mathrm{i}]\right\}\prod_{j=1}^{M}[\theta_j - \theta_{\tau(j)+M} + \pi \mathrm{i}]}. \tag{17}$$

2.4 The Stress Tensor

The large-N stress tensor form factors can be found using the same methods. The result is

$$\langle 0|T_{\mu\nu}(0)|P_1; \ldots; P_M; A_{M+1}; \ldots; A_{2M}\rangle_{\text{in}} = -\frac{2^{M-1}\pi^{M+1}\tilde{P}_\mu \tilde{P}_\nu}{N^{M-1}} \times \sum_{\sigma,\tau\in S_M}\left[\prod_{l=0}^{M}\delta_{a_l\, a_{M+\sigma(l)}}\right]\left[\prod_{l=0}^{M}\delta_{b_l\, b_{M+\sigma(l)}}\right] T_{\sigma\tau}(\theta_1, \ldots, \theta_{2M}), \tag{18}$$

where $T_{\sigma\tau} = T^0_{\sigma\tau} + O\left(\frac{1}{N}\right)$, with

$$T^0_{\sigma\tau} = \tanh^2 \frac{\theta_{2M} - \theta_1}{2} \prod_{l=1}^{M} \frac{1 - \delta_{\sigma(l)\,\tau(l)}}{[\theta_j - \theta_{\sigma(j)+M} + \pi \mathrm{i}][\theta_j - \theta_{\tau(j)+M} + \pi \mathrm{i}]}. \tag{19}$$

3 Correlation Functions

3.1 Closed-Form Expressions

The Wightman function of the scaling field can be written as a series by expanding in terms of a complete set of states

$$\mathcal{W}_N(x) = \frac{1}{N}\langle 0|\mathrm{Tr}\Phi(0)\Phi(x)^\dagger|0\rangle = \frac{1}{N}\sum_{a_0,b_0}\sum_X \left|\langle 0|\Phi(0)_{b_0 a_0}|X\rangle_{\text{in}}\right|^2 e^{\mathrm{i}p_X\cdot x},$$

where X denotes an arbitrary choice of particles, momenta and colors and where p_X is the momentum eigenvalue of the state $|X\rangle$. This series is, for $N \to \infty$, explicitly (here we have reordered the numbering of the rapidities)

$$\mathcal{W}(x) = \int \frac{d\theta}{4\pi} e^{\mathrm{i}x\cdot p} + \frac{1}{4\pi}\sum_{l=1}^{\infty}\int d\theta_1 \cdots d\theta_{2l+1} e^{\mathrm{i}x\cdot(p_1+\cdots p_{2l+1})}\prod_{j=1}^{2l}\frac{1}{(\theta_j - \theta_{j+1})^2 + \pi^2}, \tag{20}$$

where $l = M - 1$. For further details on how (20) was obtained, see Ref. [8].

We can find a similar result for the current-current correlation function [14],

$$\lim_{N\to\infty}\frac{1}{N}\langle\ 0|j_\mu^a(0)j_\nu^b(x)|0\rangle = -\frac{\delta^{ab}\eta_{\mu\nu}}{4}\sum_{M=1}^{\infty}\int d\theta_1\cdots d\theta_{2M}e^{\mathrm{i}x\cdot(p_1+\cdots p_{2M})}$$
$$\times(p_1+\cdots+p_{2M})^2\tanh^2\frac{\theta_{2M}-\theta_1}{2}\prod_{l=1}^{2M-1}\frac{1}{(\theta_l-\theta_{l+1})^2+\pi^2}. \quad (21)$$

Large-N factorization implies that all Wightman functions of scaling fields are products of vacuum expectation values of traces. Thus if all the traces in a product are only over one factor of Φ and one of $\Phi^\dagger$, the vacuum expectation value is a product of two-point functions (20). Since $\mathcal{W}(x)\to 0$ as $\eta^{\mu\nu}x^\mu x^\nu\to\infty$, the cluster property must hold. We have not considered, however, the cluster property for Wightman functions with traces over larger products, e.g., $\mathrm{Tr}\,\Phi(0)\Phi(x)^\dagger\Phi(y)\Phi(z)^\dagger$.

3.2 *Short-Distance Asymptotics*

An older short-distance method exploits a Wick rotation to Euclidean space [18]. Here is a Minkowski-space version: defining $\zeta=\tanh^{-1}x^1/x^0$, we have

$$x\cdot p_j = R\cosh(\theta_j-\zeta),\quad R=\sqrt{x\cdot x}.$$

Lorentz boosting by ζ causes the changes $\theta_j\to\theta_j+\zeta$, $x^0\to R$, $x^1\to 0$, and

$$e^{\mathrm{i}x\cdot(p_1+\cdots p_{2l+1})}\to\exp\left[\mathrm{i}mR(\cosh\theta_1+\cdots+\cosh\theta_{2l+1})\right].$$

For small R, the *distribution* $\exp mR\cosh\theta$ is asymptotically 1, for $|\theta|<L\equiv|\ln(mR)|$, and 0 for $|\theta|>L$ (where this function oscillates rapidly). Writing $u_j=\theta_j/L$,

$$\mathcal{W}(x)=\frac{L}{2\pi}+\frac{L}{4\pi}\sum_{l=1}^{\infty}\int_{-1}^{1}du_1\cdots\int_{-1}^{1}du_{2l+1}\prod_{j=1}^{2l}\frac{1}{L[(u_j-u_{j+1})^2+(\pi/L)^2]}. \quad (22)$$

Recently we learned that, more than seven decades ago, Kac and Pollard [19] found the same result for the expression (22) as that in Refs. [10, 11]. To sum (22) we need properties of the square root of the Laplacian on functions $f(u)$, with $f(\pm1)=0$ [20]:

$$\Delta^{1/2}f(u)=\sqrt{-\frac{d^2}{du^2}}f(u)=\frac{1}{\pi}\int_{-1}^{1}du'\,\mathrm{PV}\,\frac{f(u')-f(u)}{(u'-u)^2}, \quad (23)$$

where PV denotes the principal value. The Green's function is [19–21]

$$K(u,v) = \Delta^{-1/2}\delta(u-v) = \frac{1}{2\pi}\ln\frac{1-uv+(1-u^2)^{1/2}(1-v^2)^{1/2}}{1-uv-(1-u^2)^{1/2}(1-v^2)^{1/2}}. \quad (24)$$

Now [10, 11, 19]

$$\frac{1}{L[(u-u')^2+(\pi/L)^2]} = \langle u'|e^{-\frac{\pi}{L}\Delta^{1/2}+O(1/L)}|u\rangle. \quad (25)$$

Defining $\Delta^{1/2}\phi_n(u) = \lambda_n\phi_n(u)$, $u \in (-1,1)$, we can sum the series (22):

$$\mathcal{W}(x) = \frac{L}{4\pi}\sum_n \left|\int_{-1}^{1} du\ \varphi_n(u,L)\right|^2 \frac{1}{1-e^{-2\pi\lambda_n/L+O(1/L^2)}} \simeq CL^2, \quad (26)$$

for small L, where

$$C = \frac{1}{8\pi^2}\sum_n \left|\int_{-1}^{1} du\ \varphi_n(u)\right|^2 \lambda_n^{-1} = \frac{1}{16\pi}. \quad (27)$$

agreeing with (4). Our nonperturbative calculation has revealed the asymptotic freedom of the PCM. To date, this is the only method yielding the coefficient C [11].

In work to appear, we found the short-distance forms for correlation functions of currents and the stress tensor. For example, the former is

$$\frac{1}{N}\langle 0|j_\mu^A(0)j_\nu^B(x)|0\rangle \simeq \frac{\eta_{\mu\nu}\delta^{AB}}{16x^\mu x_\mu}, \quad (28)$$

which is consistent with currents having no anomalous dimensions.

4 General Matrix Elements of Operator Products

The work in this section describes an open problem. We wish to make the connection between our results and Lagrangian field theory. For example, (16), (17) imply that currents are conserved: $\partial^\mu j_\mu = 0$. The remaining classical equation of motion is $[\partial_0 + j_0, \partial_1 + j_1] = 0$. We would expect to recover a corrected version of this equation of motion. Another problem is to find the effective coupling $g(R)$, where again $R = \sqrt{x\cdot x}$, we would need to check each side of

$$T_{\mu\nu}(0) = \frac{g(R)^2}{N}\left[j_\mu\!\left(-\frac{x}{2}\right)j_\nu\!\left(\frac{x}{2}\right) - \frac{1}{2}\eta_{\mu\nu}j^\alpha\!\left(-\frac{x}{2}\right)j_\alpha\!\left(\frac{x}{2}\right)\right] \quad (29)$$

The left-hand side has known matrix elements, e.g.,

$$\langle P|T_{\mu\nu}(0)|P'\rangle = -\frac{2\pi^2}{(2\pi\mathrm{i}+\theta_{PP'})^2}(p+p')_\mu(p+p')_\nu\delta_{aa'}\delta_{bb'}. \tag{30}$$

To solve either of these problems requires finding non-vacuum matrix elements of bilinears of current operators, that is

$${}_{\mathrm{in}}\langle E_1,\ldots,E_m|j_\mu(0)j_\nu(x)|E_1',\ldots,E_n'\rangle_{\mathrm{in}}. \tag{31}$$

These can be written as integrals, just as the vacuum expectation values can. The integrals are harder to evaluate, however. For large L, with $\tanh\zeta = x^1/x^0$, as before,

$$\langle P|[j_\mu(x)j_\nu(y)]_{a_1c_1}|P'\rangle = \delta_{a_1a}\delta_{c_1a'}\delta_{bb'}\frac{\partial^2}{\partial x^\mu\partial y^\nu}\frac{1}{2}\int_{-1}^{1}du\int_{-1}^{1}dv K(u,v)$$
$$\times\coth\frac{Lu-\theta-\zeta}{2}\coth\frac{Lu-\theta'-\zeta}{2}\frac{1}{v-(\theta+\zeta-2\pi\mathrm{i})/L}\frac{1}{v-(\theta'+\zeta+2\pi\mathrm{i})/L} \tag{32}$$

where $K(u,v)$ is the Green's function (24). We have similar results for multi-excitation states (31), but have not yet been able to simplify such expressions.

Acknowledgements This work was supported in part by a grant from the PSC-CUNY.

References

1. F. Gürsey, On the symmetries of strong and weak interactions. Il Nuovo Cimento. **16**, 230–240 (1960). https://doi.org/10.1007/BF02860276
2. A.B. Zamolodchikov, A.B. Zamolodchikov, Factorized S-matrices in two dimensions as the exact solutions of certain relativistic quantum field theories. Ann. Phys. **120**, 253–291 (1979). https://doi.org/10.1016/0003-4916(79)90391-9
3. A.M. Polyakov, P.B. Wiegmann, Theory of nonabelian goldstone bosons in two dimensions. Phys. Lett. **131**B, 121–126 (1983). https://doi.org/10.1016/0370-2693(83)91104-8. P.B. Wiegmann, On the theory of nonabelian goldstone bosons in two dimensions: exact solution of the SU(N) × SU(N) nonlinear σ model. Phys. Lett. **141**B, 217–222 (1984). https://doi.org/10.1016/0370-2693(84)91256-5. E. Abdalla, M.C.B. Abdalla, A. Lima-Santos, On the exact S matrix of the principal chiral model. Phys. Lett. **140**B, 71–75 (1984). https://doi.org/10.1016/0370-2693(84)91050-5. P.B. Wiegmann, Exact factorized S matrix of the chiral field in two dimensions. Phys. Lett. **142**B, 173–176 (1984). https://doi.org/10.1016/0370-2693(84)91256-5
4. S.N. Vergeles, V.M. Gryanik, Two-dimensional quantum field theories which admit exact solutions. Sov. J. Nucl. Phys. **23**, 704–709 (1976). P.H. Wiesz, Exact quantum sine-Gordon form factors. Phys. Lett. B**67**, 179–182 (1977). https://doi.org/10.1016/0370-2693(77)90097-1. M. Karowski, P. Wiesz, Exact form factors in (1+1)-dimensional field theoretic models with soliton behaviour. Nucl. Phys. B**139**, 455–476 (1978). https://doi.org/10.1016/0550-3213(78)90362-0

5. F.A. Smirnov, Form factors in completely integrable models of quantum field theory. Adv. Ser. Math. Phys. **14** (1992). World Scientific
6. H. Babujian, A. Fring, M. Karowski, A. Zapletal, Exact form factors in integrable quantum field theories: the sine-Gordon model. Nucl. Phys. B **538** [FS], 535–586 (1999). https://doi.org/10.1016/S0550-3213(98)00737-8
7. P. Orland, Summing planar diagrams by an integrable bootstrap. Phys. Rev. D **84** 105005 (2011). https://doi.org/10.1103/PhysRevD.84.105005
8. P. Orland, Summing planar diagrams by an integrable bootstrap.II. Phys. Rev. D **86**, 045023 (2012). https://doi.org/10.1103/PhysRevD.86.045023
9. A. Cortés Cubero, Multiparticle form factors of the principal Chiral Model At Large N. Phys. Rev. D **86**, 025025 (2012). https://doi.org/10.1103/PhysRevD.86.025025
10. P. Orland, Seeing asymptotic freedom in an exact correlator of a large-N matrix field theory. Phys. Rev. D **90** 125038 (2014). https://doi.org/10.1103/PhysRevD.90.125038
11. E. Katzav, P. Orland, Universal coefficient of the exact correlator of a large-N matrix field theory. Phys. Rev. D **94**, 065041 (2016). https://doi.org/10.1103/PhysRevD.94.065041
12. P. Rossi, M. Campostrini, E. Vicari, The large-N expansion of unitary-matrix models. Phys. Rep. **302**, 143–209 (1998). https://doi.org/10.1016/S0370-1573(98)00003-9
13. A. Cortés Cubero, P. Orland, Correlation functions of the SU(∞) principal chiral model. Phys. Rev. D **88**, 025044 (2013). https://doi.org/10.1103/PhysRevD.88.025044
14. A. Cortés Cubero, (2+1)-dimensional Yang-Mills theory and form factor perturbation theory. Phys. Rev. D **90**, 065002 (2014). https://doi.org/10.1103/PhysRevD.90.065002
15. B. Schroer, T.T. Truong, P. Weisz, Towards an explicit construction of the sine-Gordon field theory. Phys. Lett. **63B**, 422–424 (1976). https://doi.org/10.1016/0370-2693(76)90386-5
16. B. Berg, M. Karowski, V. Kurak, P. Weisz, Factorized U(n) symmetric S matrices in two-dimensions. Nucl. Phys. B **143**, 125–132 (1978). https://doi.org/10.1016/0550-3213(78)90489-3
17. V. Kurak, J.A. Swieca, Antiparticles as bound states of particles in the factorized S-matrix framework. Phys. Lett. **82B**, 289–291 (1978). https://doi.org/10.1016/0370-2693(79)90758-5
18. J.L. Cardy, G. Mussardo, Form-factors of descendent operators in perturbed conformal field theories. Nucl. Phys. B **340** 387–402 (1990). https://doi.org/10.1016/0550-3213(90)90452-J. V.P. Yurov, A.B. Zamolodchikov, Correlation functions of integrable 2-D models of relativistic field theory. Ising model. Int. J. Mod. Phys. A **6** 3419–3440 (1991). https://doi.org/10.1142/S0217751X91001660
19. M. Kac, H. Pollard, The distribution of the maximum of partial sums of independent random variables. Canad. J. Math. **2**, 375–384 (1950). https://doi.org/10.4153/CJM-1950-034-9
20. M. Riesz, Intégrales de Riemann-Liouville et potentiels. Acta Szeged Sect. Math. **9** 1–42 (1937–1938). https://doi.org/10.1007/978-3-642-37535-4_38. Rectification, ibid., Acta Szeged sect. Math. **9**, 116–118 (1937–1938). acta.bibl.u-szeged.hu/13497
21. R.M. Blumenthal, R.K. Getoor, D.B. Ray, On the distribution of first hits for symmetric stable processes. Trans. Am. Math. Soc. **99**, 540–554 (1961). https://doi.org/10.2307/1993561

Puzzles with Negative Mass

M. B. Paranjape

Abstract The Schwarzschild metric is a solution of the vacuum Einstein equations for any value of the mass, which can be positive or negative. The negative mass solution has a naked singularity, there is no event horizon. It was shown [1–3] how to fill in a "negative mass star" so that the singularity is smoothed out, and the energy-momentum satisfies the dominant energy condition and can even gives rise to a dynamically stable configuration. However, some puzzles persist. Firstly, the exchange of a spin two graviton normally provides an attractive force between sources, while a simple calculation of the geodesic motion in the presence of a negative mass star shows repulsion. Secondly, how does a gas of negative mass particles behave thermally? If the energy is negative, the partition function makes no sense, it certainly does not converge. Should the temperature also be taken negative? The solution of these puzzles is still at large.

1 Introduction

Negative mass in an asymptotically flat spacetime requires energy-momentum sources which do not satisfy the dominant energy condition. This is because of the positive energy theorem [4–6], which states that if the dominant energy condition is satisfied, implies the ADM must be positive. For a negative mass in an asymptotically flat spacetime, the dominant energy condition must be violated. We can imagine introducing an energy-momentum source in a sphere centred on where the singularity would occur, with exterior metric given by the negative mass Schwarzschild solution; a negative mass star, analogous to what is done for the positive mass metric. The interior metric would of course no longer be the Schwarzschild metric and would in principle be something non-singular that depended on the type of matter introduced. Such a metric would have a negative ADM mass, disallowing the possibility that the introduced energy-momentum satisfies the dominant energy condition.

M. B. Paranjape (✉)
Groupe de physique des particules, Département de physique, Centre de recherche mathématiques and Institut Courtois, Université de Montréal, Montréal, Québec, Canada
e-mail: paranj@lps.umontreal.ca

M. Paranjape et al. (eds.), *Proceedings of Gravity, Strings and Fields*,
CRM Series in Mathematical Physics,
https://doi.org/10.1007/978-3-031-91266-5_14

The dominant energy condition is the most physical of the so-called energy conditions, it asserts that the energy density is positive and that the flow of the energy-momentum current as seen by any inertial observer is inside the local, future-directed directed light-cone. Hence such a star, although eluding a singularity, would still be an unsatisfactory example of a physical, negative mass star.

However, we can evade the positive energy theorem by considering spacetimes that are not asymptotically flat. Indeed, de Sitter spacetime is an ideal candidate, it does not have an asymptotic region, it is exponentially expanding everywhere. There is no positive energy theorem for matter in de Sitter spacetime. There exists an exact solution of the Einstein equations with a cosmological constant, which corresponds to a mass, sitting in de Sitter spacetime, the Schwarzschild-de Sitter spacetime. The cosmological constant can be thought of as originating from the matter fields in the system, through a vacuum expectation value of a scalar field, for example. This seems to be physically relevant, both for the inflationary scenario but also for the present universe where observationally there seems to be a significant amount of dark energy, both of which correspond to a cosmological constant. The mass parameter in Schwarzschild-de Sitter spacetime can have any sign, positive or negative. This metric then gives a model spacetime for describing negative mass. The Schwarzschild-de Sitter metric is singular, however the singularity can be removed by excising a spherical region around the singularity and filling it in with matter, and this can be done respecting the dominant energy condition. Then, negative mass corresponds to a local under-density in the local energy-momentum density given by the cosmological constant, while a positive mass corresponds to an over-density. It is clear that the negative mass here is relative to the background cosmological constant.

The Schwarzschild-de Sitter metric is given by (in so-called Schwarzschild coordinates)

$$d\tau^2 = \left(1 - \tfrac{2GM}{r} - \tfrac{\Lambda}{3}r^2\right)dt^2 - \frac{1}{\left(1 - \tfrac{2GM}{r} - \tfrac{\Lambda}{3}r^2\right)}dr^2 - r^2 d\Omega^2 \tag{1}$$

with $\Lambda > 0$ which satisfies the Einstein equation in the presence of a cosmological constant Λ, the origin thereof taken to be a physical condensate. Such a condensate continues to satisfy the dominant energy condition. The metric is deformed to

$$d\tau^2 = \left(1 - \tfrac{\Lambda(r)}{3}r^2\right)dt^2 - \frac{1}{\left(1 - \tfrac{\Lambda(r)}{3}r^2\right)}dr^2 - r^2 d\Omega^2 \tag{2}$$

where

$$\Lambda(r) = \begin{cases} \lambda(r) & \text{for} \quad r < r_0 \\ \frac{6GM}{r^3} + \Lambda & \text{for} \quad r > r_0 \end{cases} \tag{3}$$

where $\lambda(r_0) = \frac{6GM}{r_0^3} + \Lambda$ and $\lim_{r \to 0} \lambda(r) \to 0$ and r_0 is interpreted as the radius of the "star". A fluctuation of the density in a spherical region, so that Λ and M remain constants in the exterior, describes a particle of mass M, sitting in de Sitter spacetime. For negative M, the fluctuation corresponds to an under density, and if it does not burrow deeper than Λ, the dominant energy condition is satisfied [1–3]. The exterior geometry with negative M corresponds to a negative mass particle.

2 Why Do Negative Masses Repel?

The geodesics for a test mass in de Sitter spacetime are easily calculated. As the spacetime is homogeneous, we need only find the geodesic that passes through the origin, and then a simple coordinate transformation will give the geodesic that has a small transversal displacement, *i.e.*, a small impact parameter. Adding a mass at the origin, *i.e.*, going to Schwarzschild-de Sitter spacetime, will give a perturbation relative to the pure de Sitter spacetime geodesic. For positive mass, obviously we expect an attraction, while for negative mass, we expect a repulsion. We will see this explicitly.

2.1 *Repulsion from Negative Mass*

For a metric of the form

$$d\tau^2 = B(r)dt^2 - A(r)dr^2 - r^2 d\Omega^2 \tag{4}$$

corresponding to the Schwarzschild-de Sitter metric (2), the angular deflection is easily expressed as an integral

$$\varphi(r) = \int dr \frac{A^{1/2}}{r^2\left(\frac{1}{J^2 B} - \frac{E}{J^2} - \frac{1}{r^2}\right)^{1/2}}. \tag{5}$$

Here J is a parameter representing the angular momentum and E is the energy of the scattered particle. Replacing for $B = A^{-1} = 1 - \frac{2GM}{r} - \frac{\Lambda}{3}r^2 = B_0 + \delta B$ where $\delta B = -\frac{2GM}{r}$ we find [7], assuming the M is small, it is clear that we will get an expression is linear the M:

$$\begin{aligned}
\varphi(r) &= \int dr \frac{1}{r^2}\left(\frac{1}{J^2} - \left(\frac{E}{J^2} + \frac{1}{r^2}\right)B\right)^{-1/2} \\
&= \int dr \frac{1}{r^2}\left(\frac{1}{J^2} - \left(\frac{E}{J^2} + \frac{1}{r^2}\right)(B_0 + \delta B)\right)^{-1/2} \\
&\approx \int dr \frac{1}{r^2}\left(\frac{1}{J^2} - \left(\frac{E}{J^2} + \frac{1}{r^2}\right)B_0\right)^{-1/2}\left(1 + \frac{1}{2}\frac{\left(\frac{E}{J^2} - \frac{1}{r^2}\right)\delta B}{\frac{1}{J^2} - \left(\frac{E}{J^2} + \frac{1}{r^2}\right)B_0}\right)
\end{aligned}$$

$$= \pi + \int dr \frac{1}{r^2}\left(\frac{1}{J^2} - \left(\frac{E}{J^2} + \frac{1}{r^2}\right) B_0\right)^{-3/2} \frac{1}{2}\left(\frac{E}{J^2} - \frac{1}{r^2}\right)\left(-\frac{2GM}{r}\right) \tag{6}$$

Pure de Sitter space being a homogenous spacetime means that all geodesics are essentially identical. They are simply translations or rotations of each other, just as in flat space. Thus an obvious geodesic orbit corresponds to a straight line passing through the origin. We can obtain the corresponding geodesic that passes by the origin with an arbitrary impact parameter by a simple coordinate translation. The change in the azimuthal angle φ for an undeflected geodesic, for pure de Sitter space is obviously π. The correction comes from the mass, and for small mass M, the correction is obviously linear in the mass. Hence a deflection that attracts for small positive M implies a deflection that repels for small negative mass M. Therefore, negative mass repels.

2.2 *Puzzle with One Graviton Exchange*

Repulsion by negative mass is in contradiction to what is expected from the exchange of a spin 2 graviton. Two particles experience a force or potential through the exchange of mediating particle. This force or potential is directly obtained from the propagator of the field of the mediating particle. In a relativistically invariant quantum field theory, when the mediating particle is of even spin, its propagator contains an even number of metric tensors $\eta_{\mu\nu}$, [8]. The particles which feel a force or potential between them couple to the mediating particle with a "charge". The force or potential is bilinear in the charges of the particle. For charges of the same sign, (like charges), a bilinear expression in the charges always has positive sign. The force or potential then is attractive or repulsive based solely on the properties of the propagator of the mediating particle. For the exchange of a spin zero, there are no factors of the metric, and the effective interaction is attractive between like charges. For spin 1, the propagator contains 1 factor of $\eta_{\mu\nu}$ and the details show that the minus signs in the metric conspire to give a repulsion between like charges. Hence two particles of the same charge repel in this case, as is familiar in electromagnetism. However, opposite charges do attract in electromagnetism. Continuing to the exchange of a spin two particle, as in the case of the graviton, the propagator contains two factors of $\eta_{\mu\nu}$, cancelling the sign and giving rise to attraction between like charges. In this case the "charge" is in fact the mass. Hence two particles of positive mass attract. Correspondingly, two particles of negative mass should also attract, as the net negative sign cancels out. However, our analysis in the previous section shows that this is in fact not the case, negative masses repel. This is simply a puzzle that needs to be understood, why do negative mass particles repel when exchange of a spin 2 graviton would indicate that they should attract.

A simple analysis of the Newton equations, [10], gives for a particle of mass m in the gravitational field of a particle of mass M,

$$\dot{\vec{p}} = -\nabla V \tag{7}$$

$$\vec{p} = m\vec{v} \tag{8}$$

$$V = -\frac{GMm}{r} \tag{9}$$

where $\vec{p}$ is the momentum of the particle of mass m, $\vec{v}$ is its velocity and r is the separation between the two masses. Although we allow both M and m to be of any sign, we find that m simply cancels out in the Newton equation (7). The limit where $|m| \ll |M|$, corresponds to the case where m is a test mass essentially following geodesic motion.

Thus an obvious resolution of the puzzle is the fact that the potential and the momentum of the particle, in which the potential induces a change, are both proportional to the mass of the particle, and that mass then cancels out. This is in contradistinction, say, to the electromagnietic case, where the potential is proportional to the charge of the particle but not its mass. Attraction or repulsion are determined only by the mass creating the gravitational potential and not by the mass experiencing the gravitational potential. This understanding corresponds to the application of the equivalence principle, all particles behave the same way in a gravitational field.

Taking the case of two negative mass particles, the sign of the potential, (9), is negative consistent with one graviton exchange, (and which normally gives rise to attraction). In the Newton law (7), the mass m cancels from both sides, changing the sign of the putative effective force on the particle. The two negative mass particles move away from one another corresponding to repulsion. However, the momentum of the negative mass particle is opposite to its motion, hence keeping consistency with the sign of the potential, when it is defined as including the product of both masses. For the case of a positive mass particle $M > 0$ excercising a force on a negative mass particle $m < 0$, again, m cancels from the Newton law (7), giving rise to effectively attraction. The particles now move together, however, the momentum of the negative mass particle is away from direction of motion, again keeping consistency with the propagator from one gravitaon exchange. All other combinations of masses can be easily analyzed. The upshot is that a negative mass particle repels all particles of either sign while a positive mass particle attracts particles of either sign. However, crucially, the momentum of the negative mass particle is in the opposite direction to its velocity.

Hence the puzzle shifts from whether negative mass particles attract or repel, in accordance with one graviton exchange, to the question of how do negative mass particles have momentum in the opposite direction to their motion. This is still a puzzle. Any models for negative mass are seen as cavities in an ambient fluid [1–3]. Examination of this in detail requires an analysis of the model that one uses for the fluid. The mass flow due to the motion of a bubble (cavity) in an ambient fluid is obviously in the opposite direction to the motion [9], as the actual momentum is only in the fluid motion. The total momentum depends on the particular fluid that is considered. Understanding this in complete detail is still a puzzle that needs to be properly understood.

3 Puzzle with the Statistical Distribution of a Gas of Negative Mass Particles

We can imagine that there is a gas of negative mass particles in the early universe. Actually, there should be a gas of both, positive mass particles and negative mass particles, all relative to the background energy density. The background density is chosen such that there is no net positive or negative mass relative to the background. Such a gas might correspond to over-densities for positive mass particles and under-densities for negative mass particles, relative to the background total energy density. The exterior metric to such fluctuations would be Schwarzschild-de Sitter for the simplest case.

The simplest distribution that one could imagine that the particles are subjected to, would be the Boltzmann distribution, especially if they are non-relativistic. This works fine for the positive mass particles. But how does it work for the negative mass particles? The Boltzmnn distribution says that particles are distributed according to

$$F(\vec{p}) = \mathcal{N} e^{-E/k_B T} \tag{10}$$

where $E = |\vec{p}|^2/2M$ is the particle's energy, nominally being the kinetic energy. For negative mass particles, it is consistent to ascribe a negative kinetic energy with their motion [9]. But this simply makes no sense in the Boltzmann distribution, as it becomes an increasing exponential, and does not give rise to a reasonable statistical distribution. The distribution will be unstable to populating more and more particles of larger and larger negative mass. This instability does not seem to be physically reasonable.

One could imagine that negative temparture should be used to describe an ensemble of negative mass particles. However, to make a consistent system of positive mass particles with positive temperature interaction with a system of negative mass particles with negative temperature seems problematic. So how do negative mass fluctuations, relative to the background, distribute themselves statistically?

It seems likely that fluctuations around the background, will be equally positive and negative. It is also clear that the positive energy fluctuations should be considered as being regions of positive mass with positive kinetic energy. But what about the negative energy fluctuations? They should be considered on the same level as the positive energy fluctuations, and the question is, how should we consider the energy of the fluctuations which correspond to under densities? It seems reasonable that use the distribution function

$$F(\vec{p}) = \mathcal{N} e^{-|E|/k_B T} \tag{11}$$

would be a possible alternative. However there is no proof that this is actually how they will be distributed. It would be interesting to find the dynamics within the context of quantum field theory that would explain the statistical distribution of positive and negative mass.

4 Conclusion

Intrinsic negative mass particles are almost surely an unphysical, inherently unstable concept. If there were negative mass particles, the universe would be unstable to creating unlimited numbers of negative mass particles and in the process, lower its energy. However, relative negative mass is perfectly physical, [1–3] in certain space-times that are not asymptotically flat, and shares some of the counterintuitive aspects of negative mass in gravitational theories [10]. Understanding the gravitational effects and behaviour of negative mass within the context of relativistic quantum field theory seems like one of the most puzzling and interesting aspects of general relativity.

Acknowledgements I thank Gordon Semenoff for all the help and support he has given to me over the past 45 years. I wish him well and many more years of intellectual achievement. I also thank the NSERC of Canada for financial support.

References

1. J. Belletête, M.B. Paranjape, On negative mass. Int. J. Mod. Phys. D **22**, 1341017 (2013). https://doi.org/10.1142/S0218271813410174
2. S. Mbarek, M.B. Paranjape, Negative mass bubbles in de Sitter spacetime. Phys. Rev. D **90**(10), 101502 (2014). https://doi.org/10.1103/PhysRevD.90.101502
3. M.C. Johnson, M.B. Paranjape, A. Savard, N. Tapia-Arellano, Stable, thin wall, negative mass bubbles in de Sitter space-time. Gen. Rel. Grav. **52**(8), 80 (2020). https://doi.org/10.1007/s10714-020-02732-9
4. R. Schon, S.-T. Yau, On the Proof of the positive mass conjecture in general relativity. Commun. Math. Phys. **65**, 45–76 (1979). https://doi.org/10.1007/BF01940959
5. R. Schon, S.-T. Yau, Proof of the positive mass theorem. 2. Commun. Math. Phys. **79**, 231–260 (1981). https://doi.org/10.1007/BF01942062
6. E. Witten, A simple proof of the positive energy theorem. Commun. Math. Phys. **80**, 381 (1981). https://doi.org/10.1007/BF01208277
7. S. Weinberg, *Gravitation and Cosmology: Principles and Applications of the General Theory of Relativity* (Wiley, New York, 1972). ISBN 978-0-471-92567-5, 978-0-471-92567-5
8. M.E. Peskin, D.V. Schroeder, *An Introduction to Quantum Field Theory* (Addison-Wesley, Reading, USA, 1995). 978-0-201-50397-5
9. M.B. Paranjape, Negative mass in the universe is relative, repulsive, and real, *Peter Suranyi 87th Birthday Festschrift* (World Scientific Press, 2022), pp. 243–251
10. J.M. Luttinger, On negative mass in the theory of gravitation. *Gravity Research Foundation Essay Competition* (1951). https://tinyurl.com/yc8exf3e

Gravity of String Theory

Jeong-Hyuck Park

Abstract What is the gravity that string theory predicts? Is it different from General Relativity? After presenting an affirmative answer to this question, I will discuss the condition that the gravity of string theory is compatible with observations: solar system test and supernova data.

Keywords String theory · Gravity · Non-riemannian geometry · Test

1 Prologue

This is going to be my first writing on physics where I refer to myself as *I*, which I hope to be appropriate. I first met Gordon around 2000 right after finishing my PhD. Despite its elegant mathematical structure, I was unhappy with the physics I had been doing [1], since I failed to see any (immediate) connection to the real world. Surely, it was due to the lack of my ability to do so, and I was the one to blame as I did not decide to go for hep-ph nor cond-mat physics. The unhappiness was then serious to me, as I had chosen to do physics not for fun but for a sort of duty to human kind. Given this personal situation, Gordon appeared to me as an ideal role model of a prominent physicist who does high-standard formal theory that is relevant to real nature. It was almost a shock to me to learn about Gordon's pioneering work on *graphene* [2].

Gordon has been a kind mentor to me whenever we met through his frequent visits to Korea and also at various conferences including the series of PIMS-APCTP summer schools, which deeply influenced my works on gauged matrix models and (with controversy) on the ideal Bose gas [3]. But this is not the subject I am going to discuss in this writing.

J.-H. Park (✉)
Department of Physics, Sogang University, Mapo-gu, Seoul, Korea
e-mail: park@sogang.ac.kr

M. Paranjape et al. (eds.), *Proceedings of Gravity, Strings and Fields*,
CRM Series in Mathematical Physics,
https://doi.org/10.1007/978-3-031-91266-5_15

2 Three Questions

Since 2010, I have been occupied with three questions. I state them below with possible answers. Explanation will follow in next sections.

Q1. *What is the gravity that string theory predicts?*
String theory predicts its own gravity with its own *Einstein Equation* [4]

$$G_{AB} = T_{AB} \,, \tag{1}$$

where A, B are $\mathbf{O}(D, D)$ vector indices with spacetime dimension D.

Q2. *Is it different from General Relativity?*
Yes, because it is inevitably a modified gravity and because it allows novel spacetime geometries where a Riemannian metric is not defined even locally.

Q3. *Is the gravity of string theory compatible with experiments?*
It can pass a solar system test and further can be consistent with late-time cosmological data, if the following inequality holds for ordinary matter [5, 6]

$$\left| \frac{\int_{\text{baryon}} d^3x \, \sqrt{-g}\, e^{-2\phi}\big(K_\mu{}^\mu - T_{(0)}\big)}{\int_{\text{baryon}} d^3x \, \sqrt{-g}\, e^{-2\phi} K_t{}^t} \right| \lesssim 10^{-5} \,, \tag{2}$$

which involves the components of the energy-momentum tensor in (1), to be explained below. I do not know whether or not ordinary baryonic matter satisfies this inequality. I merely hope one day the broad readership of the present proceedings in honor of Gordon can answer.

3 Gravity of String Theory

3.1 Gravity of Closed String Massless Sector

It was often said that string theory implies General Relativity (GR), as the quantisation of a closed string upon a flat spacetime background produces a metric $g_{\mu\nu}$ which is the only fundamental and geometric variable in GR. It is also widely thought that string theory can be relevant to real world only at high energy scale. However, this should be true for massive states of strings. Further, the closed string massless sector contains not only the metric, but also a skew-symmetric B-field and a scalar dilaton ϕ. The effective description of the sector is traditionally given by

$$\int \mathrm{d}^D x \, \sqrt{-g} e^{-2\phi} \big(R + 4\partial_\mu \phi \partial^\mu \phi - \tfrac{1}{12} H_{\lambda\mu\nu} H^{\lambda\mu\nu} \big) \,, \tag{3}$$

where $H_{\lambda\mu\nu} = \partial_\lambda B_{\mu\nu} + \partial_\mu B_{\nu\lambda} + \partial_\nu B_{\lambda\mu}$ is the field strength of the B-field, dubbed H-flux. Riemannian geometry was naturally adopted to write the action, which is the underlying differential geometry of GR that is the gravity of point particle. Novel formulation of the action was developed under the nomenclature, double field theory (DFT) [7–10], where the spacetime dimension is formally doubled[1] and the $\mathbf{O}(D, D)$-symmetry secretly hidden in (3) becomes manifest. The $\mathbf{O}(D, D)$ rotates the triplet $\{g_{\mu\nu}, B_{\mu\nu}, \phi\}$ into one another and thus geometrises the Buscher rule [13, 14]. The original DFT Lagrangian proposed in [10] consists of six terms given by the derivatives of generalised metric $\mathcal{H}_{AB}$ and dilaton d,

$$\mathcal{H}_{AB} = \begin{pmatrix} g^{-1} & -g^{-1}B \\ Bg^{-1} & g - Bg^{-1}B \end{pmatrix}, \qquad e^{-2d} = \sqrt{-g}e^{-2\phi} . \tag{4}$$

It appeared as if one writes the Ricci scalar curvature explicitly in terms of a metric. With Imtak Jeon and Kanghoon Lee, I looked for the underlying differential geometry which must be based on $\mathbf{O}(D, D)$ symmetry principle and beyond Riemannian: Initially without the dilaton d [15] and later incorporating both $\mathcal{H}_{AB}$ and d as the fundamental variables, we managed to construct the relevant covariant derivatives with a novel Christoffel connection and consequently identified the three terms in the action (3) as a single (generalised) scalar curvature [16]. That is to say, the action (3) corresponds to the 'pure gravity' of string theory where the whole closed string massless sector is to be gravitational and geometric. The covariant derivatives acting on tensors [16] and spinors [17–19] then naturally realise the notion of $\mathbf{O}(D, D)$-symmetric minimal coupling to other sectors in string theory [20], or (phenomenologically) to the Standard Model [21]. In this way, the $\mathbf{O}(D, D)$ symmetry principle dictates unambiguously how each component field of the closed string massless sector, $\{g_{\mu\nu}, B_{\mu\nu}, \phi\}$, should couple to various types of matter, see Sect. 2 of [4] for various examples. In particular, a point particle couples to the metric $g_{\mu\nu}$ only, and thus the Equivalence Principle holds in the string frame rather than in Einstein's [22].

Based on these works, in 2018 with Stephen Angus and Kyungho Cho, I arrived at the 'Einstein Double Field Equation', $G_{AB} = T_{AB}$ (1) [4]. Parallel to GR, our Einstein curvature G_{AB} is off-shell conserved $\nabla_A G^{AB} = 0$ [23] in terms of the covariant derivative previously constructed in [16], while the energy-momentum tensor is on-shell conserved. For a self-contained exposition of the formalism, I refer again the Sect. 2 of [4].

[1] In fact, the theory assumes a *doubled-yet-gauged* coordinate system where a gauge orbit corresponds to a single physical point [11]. The proper distance measures the distance between gauge orbits [12].

3.2 Comparison with General Relativity

With the parametrisation of the generalised metric and dilaton (4), the Einstein Double Field Equation (1) decomposes into three equations [4],

$$\begin{aligned} R_{\mu\nu} + 2\nabla_\mu(\partial_\nu\phi) - \tfrac{1}{4}H_{\mu\rho\sigma}H_\nu{}^{\rho\sigma} &= K_{(\mu\nu)}\,, \\ \tfrac{1}{2}e^{2\phi}\nabla^\rho\big(e^{-2\phi}H_{\rho\mu\nu}\big) &= K_{[\mu\nu]}\,, \\ R + 4\Box\phi - 4\partial_\mu\phi\partial^\mu\phi - \tfrac{1}{12}H_{\lambda\mu\nu}H^{\lambda\mu\nu} &= T_{(0)}\,, \end{aligned} \tag{5}$$

which are now spelled in terms of the familiar Riemannian language. While $K_{(\mu\nu)}$, $K_{[\mu\nu]}$, and $T_{(0)}$ appearing on the right hand sides of the equalities are various components of the $\mathbf{O}(D, D)$-symmetric energy-momentum tensor T_{AB} [4], given the unequivocal coupling of the closed string massless sector to matter, each of them may be conveniently obtained in an alternative manner[2] by the variation of the matter Lagrangian with respect to the component fields, $g_{\mu\nu}$, $B_{\mu\nu}$, and $d \equiv \phi - \frac{1}{2}\ln\sqrt{-g}$. Note that the variation of the metric is to be taken with fixed d rather than ϕ, and thus the Einstein curvature $G_{\mu\nu}$ has not immediately appeared in (5). That was also the reason why we chose a letter $K_{\mu\nu}$ in (5) different from the conventional one *i.e.* $T_{\mu\nu}$.

Although (5) might appear complicated in comparison with the original Einstein equation, Wheeler's insight on gravity is still working, "Matter tells spacetime how to curve. Spacetime tells matter how to move": (at least perturbatively) $K_{[\mu\nu]}$, $K_\mu{}^\mu - T_{(0)}$, and $K_{(\mu\nu)}$ determine in sequence the H-flux, dilaton ϕ, and the metric. Note for the middle step,

$$\Box\left(e^{-2\phi}\right) = \left(K_\mu{}^\mu - T_{(0)} + \frac{1}{6}H_{\lambda\mu\nu}H^{\lambda\mu\nu}\right)e^{-2\phi}\,. \tag{6}$$

Minimally coupled to the metric, a particle then follows a geodesic in string frame.

- *Newtonian Limit.*
  The Newtonian limit can be taken for the geodesic motion. From the temporal component of the metric, one can read off the Newtonian potential $-MG/r$ with the Newtonian mass [4, 5, 24],

$$M = \frac{1}{4\pi G}\int dx^3\sqrt{-g}e^{-2\phi}\left(-K_t{}^t - \frac{1}{2}H_{tij}H^{tij}\right). \tag{7}$$

  This result itself is in contrast to GR where the Newtonian mass is given, with the normalisation of the Einstein equation $G_{\mu\nu} = T_{\mu\nu}$, by

$$M = \frac{1}{8\pi G}\int dx^3\sqrt{-g}\left(-T_t{}^t + \sum_{i=1}^{3} T_i{}^i\right). \tag{8}$$

[2] Yet, for fermionic spinors, they are equivalent only on-shell, see (2.116) of [4].

Beside the electric H-flux contribution in (7), the two formulas (7) and (8) are generically inequivalent. For example, pressureless dusts set $K_{\mu\nu} = T_{\mu\nu}$ and $T_i{}^i = 0$. Then there appears a factor two difference: $4\pi G$ *vs.* $8\pi G$. Further, for a QCD baryon, electric gluon field contributes to the mass in (7) but not the magnetic one. In contrast, in (8) both do so.

3.3 Non-riemannian Geometry

Another characteristic of DFT is that it can enlarge the notion of spacetime or string vacua beyond the Riemannian paradigm, by admitting non-Riemannian geometries where an invertible Riemannian metric $g_{\mu\nu}$ cannot be defined even locally. This is possible when we view the generalised metric $\mathcal{H}_{AB}$ and the DFT-dilaton d as the fundamental variables of DFT: *A priori* they should be better given not by the parametrisations (4) but through their own defining properties. Namely the DFT-dilaton exponentiates to a unit-weight scalar density, e^{-2d}, and the generalised metric is defined to be a symmetric $\mathbf{O}(D, D)$ element. It turns out that (4) is not the most general parametrisation of the generalised metric: The upper-left $D \times D$ block where the inverse of the metric g^{-1} is positioned can be degenerate, such that no invertible Riemannian metric is associated.

- The first non-Riemannian example was obtained rather accidentally in [25], through an $\mathbf{O}(D, D)$ rotation over the two isometric directions of a well-known macroscopic fundamental string configuration [26].
- With Kevin Morand [27], I have classified all possible DFT geometries by two non-negative integers $(n, \bar{n})$ with $n + \bar{n} \leq D$. Only the $(0, 0)$ is Riemannian and others are non-Riemannian. However, BRST quantum consistency on string worldsheet imposes $n = \bar{n}$ in the familiar critical dimension, $D = 10$ or $D = 26$ [28]. Supersymmetric DFTs [18, 20] and doubled Green–Schwarz or RNS string actions [29, 30] readily admit such non-Riemannian geometries.
- The perfectly symmetric vacua of DFT are the maximally non-Riemannian cases of $(D, 0)$ and $(0, D)$ where the generalised metric coincides with the $\mathbf{O}(D, D)$ invariant constant metric up to an overall sign factor. They preserve the full $\mathbf{O}(D, D)$ symmetry to feature infinitely many isometries (Killing vectors) [31]. Consequently, they do not admit any moduli nor string mode [32]. Riemannian spacetime emerges after the spontaneous symmetry breaking of $\mathbf{O}(D, D)$, where the massless Nambu–Goldstone bosons are $g_{\mu\nu}$ and $B_{\mu\nu}$ [33].
- Singularities appearing in a wide class of supergravity solutions [34–36] are actually non-Riemannian geometries [37]. Through both coordinate and B-field gauge transformations, the fundamental variables of DFT, $\mathcal{H}_{AB}$ and d, can be mapped to finitely regular $(1, 1)$ non-Riemannian geometry. Consequently, DFT curvatures

remain finite and the sugra curvature singularities are now identified as DFT coordinate singularities. Further, geodesics of [34–36] turn out to be all complete. That is to say, singularities in GR can be mere artifact of Riemannian geometry. They are regular non-Riemannian.

- Non-Riemannian geometries of $(1,0)$, $(1,1)$, or $(D-1,0)$ types have been also identified in [27, 30, 33, 38] as the non-relativistic Newton–Cartan or ultra-relativistic Carrollian gravities/strings of [39–45]. An undoubled differential tool kit (connection and curvature) was proposed in [46] for the unifying undoubled description of the $(n,\bar{n})$ non-Riemannian geometries.
- Non-Riemannian geometry realises the fracton physics of [47–51] in a geometric manner [52]. On a generic $(n,\bar{n})$ background, particle becomes immobile over $n+\bar{n}$ spacetime directions, while string becomes chiral over n directions and anti-chiral over $\bar{n}$ directions. It follows that string is immobile too. The infinite-dimensional isometries produce infinitely many Noether currents, which might shed light on the information loss paradox in Riemannian singular geometries. Taking into account the immobility, it may be worth while to examine non-Riemannian geometry as an internal space.

4 Test of String Gravity

Given the rigid and unambiguous coupling of the string gravity to matter, it should be possible to test the theory conclusively. It is often said that the presence of a massless scalar like the dilaton would break the Equivalence Principle, acting as a "fifth force". However, as mentioned earlier, the principle can be saved in the string frame, at least for the geodesics of a point particle. Yet, the dilaton does couple to the gauge bosons in the Standard Model, $-\frac{1}{4\alpha'}\sqrt{-g}e^{-2\phi}F_{\mu\nu}F^{\mu\nu}$ [21] (*c.f.* [53]), and thus potentially affects the coupling constant, or the fine structure constant. The variation of the dilaton needs to be dynamically suppressed, which I sketch below.

4.1 Astrophysical Test: Solar System

A modern method to test any (modified) gravity is the parametrised post-Newtonian (PPN) analysis [54], which contains two key quantities, $\beta_{\rm PPN}$, $\gamma_{\rm PPN}$. The current stringent observational bounds of them come from the solar gravity (Shapiro time-delay by Cassini spacecraft) [54, 55] and perihelion shifts of Mercury or Mars [54, 56], which require both $\beta_{\rm PPN}$ and $\gamma_{\rm PPN}$ to be close to unity within the error of 10^{-5}.

With Kangsin Choi, I have performed a PPN analysis of the $D=4$ string gravity [5]. We identified $\beta_{\rm PPN}-1$ as the electric H-flux and $\gamma_{\rm PPN}-1$ as a certain equation of state parameter (a fraction of energy-momentum tensor integrals),

$$H=\pm 2\sqrt{\beta_{\rm PPN}-1}MG\,dt\wedge d(\cos\vartheta)\wedge d\varphi\,,\qquad \gamma_{\rm PPN}-1=\frac{\displaystyle\int_{\rm Sun}d^3x\;e^{-2d}\big(T_{(0)}-K_\mu{}^\mu\big)}{\displaystyle\int_{\rm Sun}d^3x\;e^{-2d}K_t{}^t}\,. \tag{9}$$

The integral is taken over the interior of the star or Sun. We showed, assuming the weak energy condition $-K_t{}^t\geq 0$, that the electric H-flux should be trivial and hence rather precisely $\beta_{\rm PPN}=1$.

We turn to $\gamma_{\rm PPN}-1$ which ought to be negligible as 10^{-5} (2). The fractional expression (9) itself does not tell us whether it is so or not. The answer should depend on the properties of the matter that constitutes the Sun.

- The Sun consists of atoms. If we model the atoms as featureless point particles, we have $T_{(0)}=0$ (minimal coupling) and $K_\mu{}^\mu/K_t{}^t\simeq 1-v^2$ where v is the averaged speed which should be small for non-relativistic particles. Thus, we would have $\gamma_{\rm PPN}\simeq v^2\simeq 0$, and the model would not pass the test.
- However, atoms are not certainly the ultimate building block of the Sun. The integrals in (9) are to be localised at baryons (enormously many of them) which have rich subhadronic structure. We expect a hierarchical reduction,

$$\frac{\displaystyle\int_{\rm Sun}d^3x\;e^{-2d}\big(K_\mu{}^\mu-T_{(0)}\big)}{\displaystyle\int_{\rm Sun}d^3x\;e^{-2d}K_t{}^t}\simeq\frac{\displaystyle\int_{\rm baryon}d^3x\;e^{-2d}\big(K_\mu{}^\mu-T_{(0)}\big)}{\displaystyle\int_{\rm baryon}d^3x\;e^{-2d}K_t{}^t}\,. \tag{10}$$

  Intriguingly, a recent experiment has revealed pressure inside protons as high as $10^{35}\,{\rm Pa}\sim 0.005\,{\rm GeV}^4$ [57] which is comparable to their mass density, $10^{18}\,{\rm kg/m}^3\sim 0.004\,{\rm GeV}^4\sim(\Lambda_{\rm QCD})^4$ [58].
- If we assume an $\mathbf{O}(D,D)$-symmetric, chiral effective field theory for hadrons,

$$S_{\rm eff.}=-\int d^4x\,\sqrt{-g}e^{-2\phi}g^{\mu\nu}\partial_\mu\Phi^I\partial_\nu\Phi^J\mathcal{G}_{IJ}(\Phi)\,, \tag{11}$$

  we get precisely $T_{(0)}=K_\mu{}^\mu=g^{\mu\nu}\partial_\mu\Phi^I\partial_\nu\Phi^J\mathcal{G}_{IJ}$ and thus obtain $\gamma_{\rm PPN}=1$.
- QCD relates $\gamma_{\rm PPN}-1$ to the gluon and quark condensates,

$$\gamma_{\rm PPN}-1\simeq\frac{\displaystyle\int_{\rm baryon}d^3x\;\big[e^{-2d}{\rm Tr}(B^2-E^2)-m\bar\psi\psi\big]}{\displaystyle\int_{\rm baryon}d^3x\;\big[e^{-2d}{\rm Tr}(E^2)+i\bar\psi\gamma^t D_t\psi\big]}\,, \tag{12}$$

  which may vanish according to 'pseudo-conformal' scenarios [59–61]. The solar system experiment appears to be telling us about the gravitational form factor inside hadrons, or the right hand side of the equality $\simeq$ in (12).

4.2 *Late-Time Cosmological Test: Supernova and Quasar*

Beside the above test at the astrophysical scale, in collaboration with Lee et al. [6], I have confronted the 'Friedmann equations' of the string gravity [62] against the late-time cosmological data of type Ia supernovae [63, 64] and absorption spectra of quasars [65–68]. The former probes the evolution of the Hubble parameter and the latter imposes strict limits on the temporal variation of the fine-structure constant. We showed that an *open* Universe is in admirable agreement with the observations without any dark energy nor dark matter, free of any coincidence problem. The Universe should be open to stabilise the dilaton. However, ordinary baryonic matter, even though it turns out to be minute as $\Omega_{\text{matter}} \approx 0.01$, still needs to satisfy the suppression (2). Thus, the observations at two distinct scales commonly point to it. We call for further verifications.

5 Epilogue

In a nutshell, string theory predicts its own gravity constituted by the closed string massless sector, admits novel non-Riemannian geometry, modifies GR in a concrete manner, and calls for test.

Non-relativistic particles are traditionally treated as pressureless dusts. Yet, it should be the thermal pressure that is negligible. The dusts may hide ultra-relativistic high pressure inside them where the energy is confined. The string gravity then probes both of them through the equation of state parameter (9).

Lastly, as Gordon emphasizes to me, the theory also needs to be quantized, desirably in an $\mathbf{O}(D, D)$-symmetric manner.

Acknowledgements I would like to thank the organizers of the conference in Honour of Gordon Semenoff on Gravity, Strings and Fields, held in Montreal over July 24–28, 2023, where I got helpful feedback on my talk and hence this proceeding. This work is supported by Basic Science Research Program through the National Research Foundation of Korea (NRF) Grants, NRF-2023R1A2C200536011, NRF-2020R1A6A1A0304787721 (Center for Quantum Spacetime: CQUeST), and NRF-2023K2A9A1A0109874011 (Korea–Canada Cooperative Program).

References

1. J.H. Park, N = 1 superconformal symmetry in four-dimensions. Int. J. Mod. Phys. A **13**, 1743–1772 (1998). [arXiv:hep-th/9703191 [hep-th]]; J.H. Park, Superconformal symmetry in six-dimensions and its reduction to four-dimensions. Nucl. Phys. B **539**, 599–642 (1999). [arXiv:hep-th/9807186 [hep-th]]
2. G.W. Semenoff, Condensed matter simulation of a three-dimensional anomaly. Phys. Rev. Lett. **53**, 2449 (1984)

3. J.H. Park, S.W. Kim, Thermodynamic instability and first-order phase transition in an ideal Bose gas. Phys. Rev. A **81**, 063636 (2010). [arXiv:0809.4652 [cond-mat.stat-mech]]. Click https://drive.google.com/file/d/1I4MKH7nUg6UHHg4mN5VVYxVkyE83N7mP/view?usp=sharing LINK for one-page summary with further references)
4. S. Angus, K. Cho, J.H. Park, Einstein double field equations. Eur. Phys. J.C **78**(6), 500 (2018). [arXiv:1804.00964 [hep-th]]
5. K.S. Choi, J.H. Park, Post-Newtonian feasibility of the closed string massless sector. Phys. Rev. Lett. **129**(6), 061603 (2022). [arXiv:2202.07413 [hep-th]]
6. H. Lee, J.H. Park, L. Velasco-Sevilla, L. Yin, *Late-time Cosmology without Dark Sector but with Closed String Massless Sector*. [arXiv:2308.07149 [hep-th]]
7. W. Siegel, Two vierbein formalism for string inspired axionic gravity. Phys. Rev. D **47**, 5453 (1993). ([hep-th/9302036])
8. W. Siegel, Superspace duality in low-energy superstrings. Phys. Rev. D **48**, 2826 (1993). ([hep-th/9305073])
9. C. Hull, B. Zwiebach, Double field theory. JHEP **0909**, 099 (2009). ([arXiv:0904.4664 [hep-th]])
10. O. Hohm, C. Hull, B. Zwiebach, Generalized metric formulation of double field theory. JHEP **08**, 008 (2010). ([arXiv:1006.4823 [hep-th]])
11. J.H. Park, Comments on double field theory and diffeomorphisms. JHEP **06**, 098 (2013). ([arXiv:1304.5946 [hep-th]])
12. J.H. Park, Stringy gravity: solving the dark problems at 'short' distance. EPJ Web Conf. **168**, 01010 (2018). ([arXiv:1707.08961 [hep-th]])
13. T.H. Buscher, A symmetry of the string background field equations. Phys. Lett. B **194**, 59 (1987)
14. T.H. Buscher, Path integral derivation of quantum duality in nonlinear sigma models. Phys. Lett. B **201**, 466 (1988)
15. I. Jeon, K. Lee, J.H. Park, Differential geometry with a projection: application to double field theory. JHEP **1104**, 014 (2011). ([arXiv:1011.1324 [hep-th]])
16. I. Jeon, K. Lee, J.H. Park, Stringy differential geometry, beyond Riemann. Phys. Rev. D **84**, 044022 (2011). ([arXiv:1105.6294 [hep-th]])
17. I. Jeon, K. Lee, J.H. Park, Incorporation of fermions into double field theory. JHEP **11**, 025 (2011). ([arXiv:1109.2035 [hep-th]])
18. I. Jeon, K. Lee, J.H. Park, Phys. Rev. D **85**, 081501 (2012) [erratum: Phys. Rev. D **86**, 089903 (2012)] [arXiv:1112.0069 [hep-th]]
19. I. Jeon, K. Lee, J.H. Park, Ramond-Ramond Cohomology and O(D, D) T-duality. JHEP **09**, 079 (2012). ([arXiv:1206.3478 [hep-th]])
20. I. Jeon, K. Lee, J.H. Park, Y. Suh, Stringy unification of type IIA and IIB supergravities under N = 2 D = 10 supersymmetric double field theory. Phys. Lett. B **723**, 245–250 (2013). ([arXiv:1210.5078 [hep-th]])
21. K.S. Choi, J.H. Park, Standard model as a double field theory. Phys. Rev. Lett. **115**(17), 171603 (2015). [arXiv:1506.05277 [hep-th]]
22. S.M. Ko, J.H. Park, M. Suh, The rotation curve of a point particle in stringy gravity. JCAP **06**, 002 (2017). ([arXiv:1606.09307 [hep-th]])
23. J.H. Park, S.J. Rey, W. Rim, Y. Sakatani, $\mathbf{O}(D, D)$ covariant Noether currents and global charges in double field theory. JHEP **1511**, 131 (2015). ([arXiv:1507.07545 [hep-th]])
24. K. Cho, K. Morand, J.H. Park, Stringy Newton Gravity with H-flux. Phys. Rev. D **101**(6), 064020 (2020). [arXiv:1912.13220 [hep-th]]
25. K. Lee, J.H. Park, Covariant action for a string in "doubled yet gauged" spacetime. Nucl. Phys. B **880**, 134–154 (2014). ([arXiv:1307.8377 [hep-th]])
26. A. Dabholkar, G.W. Gibbons, J.A. Harvey, F. Ruiz Ruiz, Superstrings and solitons. Nucl. Phys. B **340**, 33–55 (1990)
27. K. Morand, J.H. Park, Classification of non-Riemannian doubled-yet-gauged spacetime. Eur. Phys. J. C **77**(10), 685 (2017). [arXiv:1707.03713 [hep-th]]

28. J.H. Park, S. Sugimoto, String theory and non-riemannian geometry. Phys. Rev. Lett. **125**(21), 211601 (2020). [arXiv:2008.03084 [hep-th]]
29. J.H. Park, Green-Schwarz superstring on doubled-yet-gauged spacetime. JHEP **11**, 005 (2016). ([arXiv:1609.04265 [hep-th]])
30. C.D.A. Blair, A worldsheet supersymmetric Newton-Cartan string. JHEP **10**, 266 (2019). ([arXiv:1908.00074 [hep-th]])
31. C.D.A. Blair, G. Oling, J.H. Park, Non-riemannian isometries from double field theory. JHEP **04**, 072 (2021). ([arXiv:2012.07766 [hep-th]])
32. K. Cho, K. Morand, J.H. Park, Kaluza–Klein reduction on a maximally Non-riemannian space is moduli-free. Phys. Lett. B **793**, 65–69 (2019). ([arXiv:1808.10605 [hep-th]])
33. D.S. Berman, C.D.A. Blair, R. Otsuki, Non-riemannian geometry of M-theory. JHEP **07**, 175 (2019). ([arXiv:1902.01867 [hep-th]])
34. E. Witten, On string theory and black holes. Phys. Rev. D **44**, 314–324 (1991)
35. C.P. Burgess, R.C. Myers, F. Quevedo, On spherically symmetric string solutions in four-dimensions. Nucl. Phys. B **442**, 75–96 (1995). ([arXiv:hep-th/9410142])
36. G.T. Horowitz, A. Strominger, Black strings and P-branes. Nucl. Phys. B **360**, 197–209 (1991)
37. K. Morand, J.H. Park, M. Park, Identifying Riemannian singularities with regular non-riemannian geometry. Phys. Rev. Lett. **128**(4), 041602 (2022). [arXiv:2106.01758 [hep-th]]
38. S.M. Ko, C. Melby-Thompson, R. Meyer, J.H. Park, Dynamics of Perturbations in double field theory and non-relativistic string theory. JHEP **12**, 144 (2015). ([arXiv:1508.01121 [hep-th]])
39. J. Gomis, H. Ooguri, Nonrelativistic closed string theory. J. Math. Phys. **42**, 3127 (2001). ([hep-th/0009181])
40. U.H. Danielsson, A. Guijosa, M. Kruczenski, IIA/B, wound and wrapped. JHEP **10**, 020 (2000). ([arXiv:hep-th/0009182 [hep-th]])
41. R. Andringa, E. Bergshoeff, J. Gomis, M. de Roo, "Stringy" Newton-Cartan Gravity. Class. Quant. Grav. **29**, 235020 (2012). ([arXiv:1206.5176 [hep-th]])
42. E. Bergshoeff, J. Gomis, G. Longhi, Dynamics of Carroll particles. Class. Quant. Grav. **31**(20), 205009 (2014). [arXiv:1405.2264 [hep-th]]
43. C. Duval, G.W. Gibbons, P.A. Horvathy, P.M. Zhang, Carroll versus Newton and Galilei: two dual non-Einsteinian concepts of time. Class. Quant. Grav. **31**, 085016 (2014). ([arXiv:1402.0657 [gr-qc]])
44. X. Bekaert, K. Morand, Connections and dynamical trajectories in generalised Newton-Cartan gravity II. An ambient perspective. J. Math. Phys. **59**(7), 072503 (2018). [arXiv:1505.03739 [hep-th]]
45. T. Harmark, J. Hartong, N.A. Obers, Nonrelativistic strings and limits of the AdS/CFT correspondence. Phys. Rev. D **96**(8), 086019 (2017). [arXiv:1705.03535 [hep-th]]
46. K. Cho, J.H. Park, Remarks on the Non-riemannian sector in Double Field Theory. Eur. Phys. J. C **80**(2), 101 (2020). [arXiv:1909.10711 [hep-th]]
47. C. Chamon, Quantum glassiness. Phys. Rev. Lett. **94**(4), 040402 (2005).[arXiv:cond-mat/0404182 [cond-mat.str-el]]
48. J. Haah, Local stabilizer codes in three dimensions without string logical operators. Phys. Rev. A **83**(4), 042330 (2011). [arXiv:1101.1962 [quant-ph]]
49. S. Vijay, J. Haah, L. Fu, A new kind of topological quantum order: a dimensional hierarchy of quasiparticles built from stationary excitations. Phys. Rev. B **92**(23), 235136 (2015). [arXiv:1505.02576 [cond-mat.str-el]]
50. S. Vijay, J. Haah, L. Fu, Fracton topological order, generalized lattice gauge theory and duality. Phys. Rev. B **94**(23), 235157 (2016). [arXiv:1603.04442 [cond-mat.str-el]]
51. N. Seiberg, Field theories with a vector global symmetry. SciPost Phys. **8**(4), 050 (2020). [arXiv:1909.10544 [cond-mat.str-el]]
52. S. Angus, M. Kim, J.H. Park, Fractons, non-Riemannian geometry, and double field theory. Phys. Rev. Res. **4**(3), 033186 (2022). [arXiv:2111.07947]
53. I. Jeon, K. Lee, J.H. Park, Double field formulation of Yang-Mills theory. Phys. Lett. B **701**, 260–264 (2011). ([arXiv:1102.0419 [hep-th]])

54. C.M. Will, The confrontation between general relativity and experiment. Living Rev. Rel. **17**, 4 (2014). ([arXiv:1403.7377 [gr-qc]])
55. B. Bertotti, L. Iess, P., Tortora, A test of general relativity using radio links with the Cassini spacecraft. Nature **425**, 374–376 (2003)
56. A. Konopliv, S. Asmar, W. Folkner, O. Karatekin, D. Nunes, S. Smrekar, C. Yoder, M. Zuber, Mars high resolution gravity fields from MRO, Mars seasonal gravity, and other dynamical parameters. Icarus **211**(1), 401–428 (2011)
57. V.D. Burkert, L. Elouadrhiri, F.X. Girod, The pressure distribution inside the proton. Nature **557**(7705), 396–399 (2018)
58. X.D. Ji, A QCD analysis of the mass structure of the nucleon. Phys. Rev. Lett. **74**, 1071–1074 (1995). [arXiv:hep-ph/9410274 [hep-ph]]
59. H.K. Lee, Y.L. Ma, W.G. Paeng, M. Rho, Cusp in the symmetry energy, speed of sound in neutron stars and emergent pseudo-conformal symmetry. Mod. Phys. Lett. A **37**(03), 2230003 (2022).[arXiv:2107.01879 [nucl-th]]
60. Y. Fujimoto, K. Fukushima, L.D. McLerran, M. Praszalowicz, Trace anomaly as signature of conformality in neutron stars. Phys. Rev. Lett. **129**(25), 252702 (2022). [arXiv:2207.06753 [nucl-th]]
61. M. Rho, Pseudo-Conformal sound speed in the core of compact stars. Symmetry **14**(10), 2154 (2022). [arXiv:2209.02327 [nucl-th]]
62. S. Angus, K. Cho, G. Franzmann, S. Mukohyama, J.H. Park, $\mathbf{O}(D, D)$ completion of the Friedmann equations. Eur. Phys. J. C **80**(9), 830 (2020). [arXiv:1905.03620 [hep-th]]
63. D. Scolnic, D. Brout, A. Carr, A.G. Riess, T.M. Davis, A. Dwomoh, D.O. Jones, N. Ali, P. Charvu, R. Chen, et al., The Pantheon+ Analysis: the full data set and light-curve release. Astrophys. J. **938**(2), 113 (2022). [arXiv:2112.03863 [astro-ph.CO]]
64. A.G. Riess, W. Yuan, L.M. Macri, D. Scolnic, D. Brout, S. Casertano, D.O. Jones, Y. Murakami, L. Breuval, T.G. Brink, et al., A comprehensive measurement of the local value of the hubble constant with 1 $\mathrm{kms}^{-1}\mathrm{Mpc}^{-1}$ uncertainty from the hubble space telescope and the SH0ES team. Astrophys. J. Lett. **934**(1), L7 (2022). [arXiv:2112.04510 [astro-ph.CO]]
65. J.A. King, J.K. Webb, M.T. Murphy, V.V. Flambaum, R.F. Carswell, M.B. Bainbridge, M.R. Wilczynska, F.E. Koch, Spatial variation in the fine-structure constant—new results from VLT/UVES. Mon. Not. Roy. Astron. Soc. **422**, 3370–3413 (2012). ([arXiv:1202.4758 [astro-ph.CO]])
66. M.R. Wilczynska, J.K. Webb, J.A. King, M.T. Murphy, M.B. Bainbridge, V.V. Flambaum, A new analysis of fine-structure constant measurements and modelling errors from quasar absorption lines. Mon. Not. Roy. Astron. Soc. **454**(3), 3082–3093 (2015). ([arXiv:1510.02536 [astro-ph.CO]])
67. C.J.A.P. Martins, A.M.M. Pinho, Stability of fundamental couplings: a global analysis. Phys. Rev. D **95**(2), 023008 (2017). [arXiv:1701.08724]
68. M.R. Wilczynska, J.K. Webb, M. Bainbridge, S.E.I. Bosman, J.D. Barrow, R.F. Carswell, M.P. Dkabrowski, V. Dumont, A.C. Leite, C.C. Lee, et al., Four direct measurements of the fine-structure constant 13 billion years ago. Sci. Adv. **6**(17), eaay9672 (2020). [arXiv:2003.07627 [astro-ph.CO]]

Holographic Mean Field Theory and Kondo Condensation

Sang-Jin Sin

Abstract In this talk, we give a report on recent development of the holographic mean field theory for strongly interacting fermion systems and its application to the recent discovery of the Kondo condensation phenomena. First, we investigate various types of the symmetry-breakings and their effect on the spectral function. After classifying the Green functions according to the spectral shapes and singularity types from the analytic Green's function, we give the density of state of the fermion when it couples to the complex scalar without the coupling to the gauge field. The resulting DOS agrees with the measured small gap in highly P doped Si system. We interpreted this as the Kondo condensation.

Keywords Holography · Mean field theory · Kondo condensation

1 Introduction

Study on strongly interacting system has been the frontier of physics for more than half century. It has been the continuous source of motivation for a new theoretical paradigm. Along this line, the gravity dual picture [1–6] has been suggested as a new paradigm to understand strange metals [7, 8] and quantum critical points [9–11] of strongly coupled systems. A few experimental data were also compared with the theory showing agreements, especially in quark gluon plasma [12], in clean graphene [13–17] and surface of topological insulators [18, 19].

On the other hand, the QCP itself is a phenomena at zero temperature taking a measure zero sector of the phase diagram. Therefore for the theory to be compared with experiment, it is essential to consider the symmetry broken phases as well as the unbroken phase near the QCP. Such symmetry breaking is an essential step for a new collective phenomena, because a phenomenon associated with manybody collective phenonena is possible only when we have an order that leads to a gap which protect the new ground state from the excitations causing disorder. Otherwise, what we

S.-J. Sin (✉)
Department of Physics, Hanyang University, Seoul, South Korea
e-mail: sangjin.sin@gmail.com

M. Paranjape et al. (eds.), *Proceedings of Gravity, Strings and Fields*,
CRM Series in Mathematical Physics,
https://doi.org/10.1007/978-3-031-91266-5_16

will have is just fermi liquid. While gap calculation is usually a consequence of a mean field theory which is seeking for value of the condensation corresponding to the certain fermion bilinear, its validity is limited to the weakly interacting cases. Therefore a dream for theorists is to develop a mean field theory which is valid for the strongly interacting system maintaining the spirit of Landau-Ginzburg and Wilson.

In our recent works [20], the fermion spectral function (SF) [21–24] of each broken symmetry phase was calculated and classified. Subsequently, we realized that [25] the gapless mode in the scalar coupling is the AdS analogue of the surface mode of topological insulator whose topological stability might be related to amazing stability of the Fermi Liquid. The holographic mean field theory has universal structure so that it has the power to accommodate various different phenomena in the same fashion. Just as the superconductivity is described by the condensation of itinerant electrons $\Delta_S \sim cc$, one may also expect a new quantum ground state of manybody coming from the condensation of the itinerant-localized electrons with spin. Such many Kondo phenomena would be also described by considering the scalar condensation $\Delta_K \sim f^\dagger c$ if the mean field theory works for strongly interacting case like Kondo system. Recently, we realized such idea in the holographic set-up [26] by finding that the Kondo condensation Δ_K produces a gap, which was also observed experimentally [26].

Here, we report our theory [27] to the simplest case where the order parameter is the complex scalar which represent the condensation of the $f^\dagger c$, where c denotes the itinerant fermion and f denotes the impurity fermion which does not move. As a consequence the condensation is a scalar field but it is complex field but it does not carry any charge. Therefore it can not couple to the $U(1)$ charge. The resulting fermion spectral function gives a density of state which is consistent with the oberved DOS measured in highly P doped Si system, which has a small gap. We interpreted this as the Kondo condensation which is a magnetic analogue of BCS system.

2 Symmetry Breaking and Holographic Mean Field Theory

Our holographic mean field theory has four components: 1st, bulk fermions ψ which are dual to the boundary fermions with strong interactions. 2nd, order parameter fields Φ^I describing the condensations of fermion bilinear which is the analogue of the Hubbard-Statonovich field of the usual mean field theory. 3rd, the gravity describing the interactions between elctrons, and finally the gauge field which should be included to describe the density or chemical potential of the fermions. In this work, we do not include the vector for the analytic work but we can do it when we work numerically. For simplicity we assume that they can be described by local fields in the bulk not in the boundary, because we should allow the strongly interacting system in the boundary be described by nonlocal field theory. The total action is given by [24]

$$S_{total} = S_\psi + S_{bdy} + S_{g,\Phi} + S_{int}, \tag{1}$$

$$S_\psi = \int d^5x \sum_{j=1}^{2} \sqrt{-g}\, \bar{\psi}^{(j)}\Big(\tfrac{1}{2}(\overrightarrow{\not{D}} - \overleftarrow{\not{D}}) - m^{(j)}\Big)\psi^{(j)}, \tag{2}$$

$$S_{g,\Phi} = \int d^5x\sqrt{-g}\Big(R - 2\Lambda + |D_M\Phi_I|^2 - m_\Phi^2|\Phi|^2\Big), \tag{3}$$

$$S_{bdy} \quad = \tfrac{i}{2}\int_{bdy} d^4x\sqrt{-h}\Big(\bar{\psi}^{(1)}\psi^{(1)} \pm \bar{\psi}^{(2)}\psi^{(2)}\Big), \tag{4}$$

$$S_{int} \quad = \int d^5x\sqrt{-g}\Big(\bar{\psi}^{(1)}\Phi\cdot\Gamma\psi^{(2)} + h.c\Big). \tag{5}$$

where $\not{D} = \Gamma^M(\partial_M + \frac{1}{4}\omega_{M\alpha\beta}\Gamma^{\alpha\beta})$, $\omega_{M\alpha\beta}$ is the spin connection, $\Phi\cdot\Gamma = \Gamma^{\underline{\mu_1\mu_2\cdots\mu_I}}\Phi_{\underline{\mu_1\mu_2\cdots\mu_I}}$. Φ^I is the order parameter field which couples with bilinear spinor in the bulk, leading to the symmetry breaking under the presence of the source or its condensation. Additionally, we will turn on just one component of field Φ to calculate the spectral function. The gamma matrix convention and the geometry are chosen and given as follows,

$$\Gamma^{\underline{t}} = \sigma_1\otimes i\sigma_2, \quad \Gamma^{\underline{x}} = \sigma_1\otimes\sigma_1, \quad \Gamma^{\underline{y}} = \sigma_1\otimes\sigma_3, \quad \Gamma^{\underline{z}} = \sigma_2\otimes\sigma_0, \quad \Gamma^{\underline{u}} = \sigma_3\otimes\sigma_0 \tag{6}$$

$$ds^2 = \frac{1}{u^2}(dt^2 + \sum_{i=1}^{3} d\mathbf{x_i}^2 + du^2), \quad f(u) = 1, \quad h = gg^{uu}, \quad u_h = \infty, \tag{7}$$

where the underlined indices represent tangent space ones. Under this convention, the boundary locates at $u = 0$.

Notice that in AdS_5, including two-flavors of fermions is mendatory because holography projects out half of the fermion degrees of freedom while we need a full 4 component spinor in the 4 dimensional boundary. On the other hand, in AdS_4, considering one flavor is still allowed since the boundary is of 2+1 dimension where spinors are of two components.

We will analytically determine and analyze the fermions' Green's function in the presence of an order parameter field. So, we consider the absence of gauge field for simplicity. Furthermore, we will begin our analysis in the probe limit and later we will eventually calculate the spectral function in the full back-reacted background. We will compare it with the probe limit analytic results to check the reliability of the latter.

We now perform the variational analysis to show the boundary fermions in different quantization choices. The standard-standard (SS) and standard-alternative (SA) quantization can be distinguished by the sign of the boundary action (4). We first simplify the action by introducing $\zeta^{(j)}$,

$$\psi^{(j)} = (-gg^{uu})^{-1/4}\zeta^{(j)}e^{-i\omega t + ik_x x + ik_y y + ik_z z}. \tag{8}$$

Then, the variation of bulk fermions action (2), after the equation of motion is imposed can be written as a boundary term given below.

$$\delta S_{bulk} = \frac{i}{2}\sum_{i=1}^{2}\int d^4x\Big[\bar\zeta_-^{(i)}\delta\zeta_+^{(i)} - \bar\zeta_+^{(i)}\delta\zeta_-^{(i)} - \delta\bar\zeta_-^{(i)}\zeta_+^{(i)} + \delta\bar\zeta_+^{(i)}\zeta_-^{(i)}\Big]. \quad (9)$$

If we add the variation of boundary action with sign $\pm$ (4) depending on SS and SA quantization respectively, the variation of total action is given by

$$\delta S_{tot}^{(SS)} = \frac{i}{2}\int_{bdy} d^4x\Big(\bar\zeta_-^{(1)}\delta\zeta_+^{(1)} + \delta\bar\zeta_+^{(1)}\zeta_-^{(1)} + \bar\zeta_-^{(2)}\delta\zeta_+^{(2)} + \delta\bar\zeta_+^{(2)}\zeta_-^{(2)}\Big), \quad (10)$$

$$\delta S_{tot}^{(SA)} = \frac{i}{2}\int_{bdy} d^4x\Big(\bar\zeta_-^{(1)}\delta\zeta_+^{(1)} + \delta\bar\zeta_+^{(1)}\zeta_-^{(1)} - \bar\zeta_+^{(2)}\delta\zeta_-^{(2)} - \delta\zeta_-^{(2)}\zeta_+^{(2)}\Big). \quad (11)$$

From this expression, we see what are chosen as independent degree of freedom to makes the variation of total action zero. We call such independent fermions as the source fermions. We define a 4 component spinor ξ's by

$$\xi_S^{(SS)} := (\zeta_+^{(1)}, \zeta_+^{(2)}), \quad \text{and} \quad \xi_C^{(SS)} := (\zeta_-^{(1)}, \zeta_-^{(2)}), \quad (12)$$

as the boundary spinors for SS-quantization. The indices S, C in ξ_S and ξ_C are adopted since they correspond to the source and condensation terms. The ξ_C's supposed to be determined by ξ_S's. Similarly, for the SA-quantization we define

$$\xi_S^{(SA)} := (\zeta_+^{(1)}, \zeta_-^{(2)}), \quad \text{and} \quad \xi_C^{(SA)} := (\zeta_-^{(1)}, \zeta_+^{(2)}). \quad (13)$$

One should remember that all $\zeta_\pm$ are two component spinors while ξ's are 4 component ones. The extension $\xi(u)$ of these fermions to the bulk of the AdS can also be considered by the original $\zeta(u)$'s whose boundary values were used above as ζ, so that

$$\xi(u) = u^{m_s}\xi_S + u^{m_c}\xi_C + \cdots . \quad (14)$$

We can now rewrite the on-shell effective action as

$$S_{tot}^{(SS)} = \frac{1}{2}\int_{bdy} d^4x\Big(\xi_S^{(SS)\dagger}(-\sigma_0\otimes\sigma_2)\xi_C^{(SS)} + h.c\Big), \quad (15)$$

$$S_{tot}^{(SA)} = \frac{1}{2}\int_{bdy} d^4x\Big(\xi_S^{(SA)\dagger}(-\sigma_3\otimes\sigma_2)\xi_C^{(SA)} + h.c\Big). \quad (16)$$

Green's function

We can determine the boundary Green's functions from the the effective actions (15)-(16) and the definition of source and condensation fermions. Since we have 4 components of $\xi(u)$, there must be 4 independent solutions $\Psi_i(u)$ which can span the space of spinor solutions. Our $\xi(u)$ with prescribed boundary value should be a linear combination of these, so that

$$\xi(u) = \sum_i c^i \Psi_i(u). \tag{17}$$

By taking the a-th component of this equation, we have $\xi^a(u) = \Psi_i^a(u)c^i$, which can be written as a matrix equation $\xi(u) = \Psi(u)\mathbf{c}$. Here $\Psi_i^a(u)$ is the a-th component of i-th solution and we considered $\xi(u)$ and $\mathbf{c}$ as column matrices. By expanding the matrix $\Psi(u)$ near the boundary,

$$\Psi(u) = u^{m_s}\mathbb{S} + u^{m_c}\mathbb{C} + \cdots \tag{18}$$

We now can write the source and condensation fermions depending on the quantization choice as follows,

$$\xi_S^{(\mathbb{Q})} = \mathbb{S}^{(\mathbb{Q})}\mathbf{c}, \quad \xi_C^{(\mathbb{Q})} = \mathbb{C}^{(\mathbb{Q})}\mathbf{c}. \tag{19}$$

where $\mathbb{Q}$ stands for quantization choice. From eqs. (15)-(16) and using (19),

$$S_{total}^{(\mathbb{Q})}\Big|_{bdy} = \frac{1}{2}\int_{bdy} d^4x\left(\xi_S^\dagger \Gamma_{bdy}\xi_C + h.c\right)^{(\mathbb{Q})}, \tag{20}$$

$$= \frac{1}{2}\int_{bdy} d^4x(\xi_S^\dagger \Gamma_{bdy}\mathbb{C}\mathbb{S}^{-1}\xi_S + h.c)^{(\mathbb{Q})}, \tag{21}$$

$$= \frac{1}{2}\int_{bdy} d^4x(\xi_S^\dagger \mathbb{G}\xi_S + h.c)^{(\mathbb{Q})}. \tag{22}$$

the boundary Green's function can be defined as follow

$$\mathbb{G}^{(SS)} = -(\sigma_0 \otimes \sigma_2)\mathbb{C}\mathbb{S}^{-1}, \tag{23}$$

$$\mathbb{G}^{(SA)} = -(\sigma_3 \otimes \sigma_2)\mathbb{C}\mathbb{S}^{-1}. \tag{24}$$

Notice that the definition of Green's function remains valid even for the zero bulk fermion mass. Consequently, as far as we can extract the leading-order terms of the bulk fermions near the boundary, the Green's function calculation remains solvable. This is helpful because, in the zero fermion mass, we will be able to obtain Green's function analytically for all Lorentz symmetry breaking interaction with proper choice of scaling dimension of the order parameter field. The term suitable here refers to choosing scaling dimensions for the source term, in which any u-dependence in the interacting term will be eliminated after fully expressing the vierbein and spin connection. This ensures the results remain u-independent at the interaction terms and allows solvable Dirac equations.

3 Kondo Condensation and Observation of a New Quantum Manybody Ground State

When a magnetic moment is embedded in a metal, it captures itinerant electrons to form the Kondo cloud, which can spread out over a few micro-meters [28]. For a metal with dense magnetic impurities such that Kondo clouds overlap with each other, correlated ground states are formed. When the impurities form a regular lattice, the result is a heavy fermion or anti-ferromagnetic order depending on the dominant interaction. Even in the case of random impurities, overlapping Kondo clouds are expected to form a coherent ground state. Here, we examine this issue by performing electrical transport and high- precision tunnelling density-of-states (DOS) spectroscopy measurements in a highly P- doped crystalline silicon metal where disorder-induced localized magnetic moments exist. Recently, a small gap in the conductivity of the Si metal below 2 K was detected. See Fig. 1. For several years it was a mystery and finally it was interpreted as the Kondo condensation, the formation of a correlated ground state of overlapping Kondo clouds and a detailed theoretical analysis using a holographic method reproduces the unusual DOS spectra [26]. See Fig. 2. Here we briefly report this development as an application of the holographic mean field theory sketched above.

The basic idea is as follows. We take the order parameter to be the complex scalar Φ representing the condensation of the

$$f_{\downarrow}^{\dagger} c_{\uparrow},$$

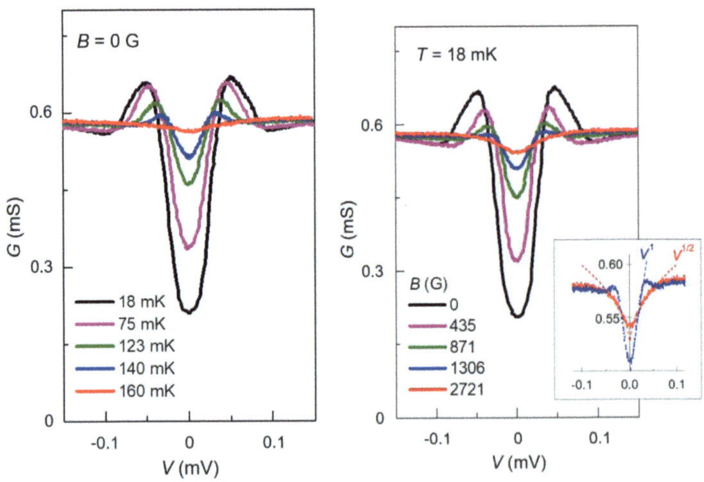

(a) fixed B (b) fixed T

Fig. 1 Observation of a tiny gap in heavily P doped Si : **a** for fixed B **b** for fixed T

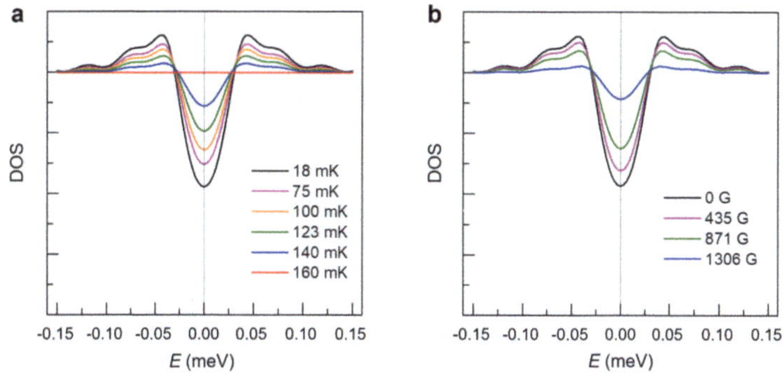

Fig. 2 Theoretical fit by holography calculation: **a** for fixed B **b** for fixed T

where c denotes the itinerant fermion and f denotes the impurity fermion which does not move. The condensation is a complex scalar field but it does not carry any charge. Therefore it can not couple to the $U(1)$ charge. The resulting fermion spectral function gives a density of state which is consistent with the oberved DOS measured in highly P doped Si system, which has a small gap. We interpreted this as the Kondo condensation which is a magnetic analogue of BCS system.

3.1 *Physical Picture*

Observation of the abnormal electronic DOS spectra in the Si:P metal and their B-driven tuning are very surprising because both the host Si and the dopant P are not magnetic in the crystral structure, and degenerate Si:P is regarded as a metal (E_F is located within the conduction band. We consider several possible causes for the observed exotic DOS spectra. First, we rule out a Coulomb gap scenario because it could not explain the magnetic behaviour. Second, we rule out spin glass-type magnetic gaps[29], because no noticeable differences are seen between field-cooled and zero-field-cooled DOS spectra. Neither of these mechanisms would explain the existence of DOS peaks outside the gap as well as the observed non-Fermi-liquid behaviours in bulk Si:P resistance. We also exclude the emergence of superconductivity. Unlike boron-doped degenerate SiO_2, a clear signature of superconductivity (a collapse of bulk resistance) is not detected down to 10 mK in the Si:P metal. We also eliminate the possibility of the Kondo lattice gap because the local moments in our sample is randomly distributed. Indeed our DOS is highly symmetric while the DOS of the Kondo lattice system has a characteristic asymmetry. In fact Si:P is not a heavy fermion system.

The resistivity measurement clearly shows that there is a Kondo effect in low temperatures below 1 K. However, single Kondo effect can not explain the presence of the peaks near the gap because it is independent local moment's random capturing of itinerant electrons, while the peaks near the gap in DOS are consequence of the correlation between the degrees of freedom with different frequencies which is impossible without quantum coherence. Because our pseudo gap is phenomena below the Kondo temperature, therefore the Kondo effect and electron correlations must happen simultaneously. To explain the gap with side peaks, we must introduce an interaction between the local moments and itinerant electrons so that the whole electron system can form a new ground state with a small gap. The only reasonable possibility we can imagine is the Ruderman-Kittel-Kasuya-Yosida (RKKY) interaction between the under-screened local moments. Notice that the under-screend local moment can not induce the Kondo effect because the requested interaction strength to capture the second electron, should be much stronger than that of capturing the first electron.

Thus, the emerging picture is as follows: When T or B increases, the pairing between the local moments is destroyed by thermal fluctuation or by Zeeman flip, so that the gap will slowly disappear and the system undergoes from a gapped scalar ordered state, which we interpret as a gapped random singlet state, to the paramagnetic state. Notice that the physics described here is just for the localized moments which are just a tiny fraction of the whole doped phosphorous atoms. So the observed DOS can be regarded as a sum of usual metalic spectrum of the Si:P and the gapped random singlet phase of local moments. The gapped aspect of the system is due to the finiteness of the randomness. Although we could describe the experimental data quite quantitavely in terms of a Ginzberg-Landau-Wilson theory, one should notice that such theory is a macroscopic theory which is valid near the quantum critical point. Therefore despite the fact that the physics of the scalar order in our Si:P is parallel to the generic Doniach phase diagram[30], we do not fully understand the full microscopic mechanism by which the magnetic order appears in such a disordered dilute-Kondo system.

3.2 Density of States

The problem here is how to calculate the DOS of this system to compare with the observed data. The problem involves all the sources of modern many-body theory's difficulties: impurity, randomness as well as strong correlations. The hope is that one can find order parameter(s) characterizing the system near the critical point in terms of which a Ginzberg-Landau (GL) theory can be formulated such that it works even for strongly interacting system with impurities.

Notice that both the magnetization and the gap of superconductor, the two most typical order parameters, can be understood as the expectation value of fermion bilinears [31] $\langle\chi^\dagger\sigma\chi\rangle$ and $\langle\chi\chi\rangle$ of the fermion χ. When there is condensation of spin pairings, the correlator $\langle\chi_\uparrow^\dagger(x)\chi_\downarrow(y)\rangle$ with two separated positions x, y is non-zero, and in the long wave length limit where we do not need to distinguish two positions, it is a charge neutral spin 0 field and we take this as our order parameter of the system. In order words, we suggest that condensation of dimers can be described by a *complex but neutral scalar* order, which we call $\Phi_0(x)$. To encode the quantum fluctuation effect in this theory, we introduce the energy scale parameter r and a prescription how the order parameter depends on the scale. The suggestion that this can be done by the canonical field equation in the anti-deSitter space is the key of the holographic theory. From this point of view we may call the theory as Ginzberg-Landau-Wilson theory, because what quantum field theory does is to relate physics at different scales [32, 33].

We calculate the spectral function in the presence of the scalar order using the holographic theory with the following prescription. Let $\psi(r, x)$ be the holographic fermion which is dual to the actual fermion $\chi(x)$. Then use the Yukawa type interaction

$$\Phi(r, x)\bar{\psi}\psi(r, x)$$

with canonical Kinetic terms of $\psi(r, x)$ and $\Phi(r, x)$ to calculate the spectral function. More details of the calculation will be given in the supplementary material. Here we simply describe the result below. For temperature evolution, we choose the parameters of the theory such that it can fit the medium temperature 100 mK data best. For the data fitting other than the 100 K, we modeled the temperature dependence of the condensation parameter by $M(T) = M_0\sqrt{1 - T/T^*}$, which is most typical temperature dependence in many cases, and choose M_0 such that the theory fit the data at a particular temperature best, which we take T=100 mK. Then use the same parameters to see how the theory can fit the data for other temperatures. As a result, the data for $T > 100\ K$ is fitted better than that of lower temperature ones. Figure 2 demonstrates the fitting data of the temperature evolution of DOS by our theory. The theory parameters are fixed such that T=100 mK data are best fit. The normalization constant $N_0 = 2.9 \times 10^{-4}$ were chosen, where (Experimental data) = N_0 (Theory result). Notice that with this choice, higher temperature data can be fit quite well while the lower temperature ones are not so well fitted. Nevertheless we can say that the qualitative features agreement in the gap and side peaks are remarkable.

Acknowledgements This work is supported by Mid-career Researcher Program through the National Research Foundation of Korea grant No. NRF- 2021R1A2B5B02002603 and the Basic research Laboratory support program RS-2023- 00218998. We thank the APCTP for the hospitality during the focus program, where part of this work was discussed.

References

1. J.M. Maldacena, The Large N limit of superconformal field theories and supergravity. Adv. Theor. Math. Phys.**2**, 231 (1998). https://doi.org/10.4310/ATMP.1998.v2.n2.a1, arxiv:[hep-th/9711200]
2. E. Witten, Anti-de Sitter space and holography. Adv. Theor. Math. Phys. **2**, 253 (1998). arxiv: [hep-th/9802150]
3. S.S. Gubser, I.R. Klebanov, A.M. Polyakov, Gauge theory correlators from noncritical string theory. Phys. Lett. B **428**, 105 (1998). arxiv: [hep-th/9802109]

4. E. Alvarez, C. Gomez, Geometric holography, the renormalization group and the c theorem. Nucl. Phys. B **541**, 441 (1999). arxiv: [hep-th/9807226]
5. V. Balasubramanian, P. Kraus, Space-time and the holographic renormalization group. Phys. Rev. Lett. **83**, 3605 (1999). arxiv: [hep-th/9903190]
6. J. de Boer, E.P. Verlinde, H.L. Verlinde, On the holographic renormalization group. JHEP **08**, 003 (2000). arxiv: [hep-th/9912012]
7. S.A. Hartnoll, A. Lucas, S. Sachdev, *Holographic Quantum Matter* (MIT Press, 2018)
8. J. Zaanen, Y. Liu, Y.-W. Sun, K. Schalm, *Holographic Duality in Condensed Matter Physics* (Cambridge University Press, 2015)
9. S. Sachdev, *Quantum Phase Transitions*, 2nd ed. (Cambridge University Press, 2011)
10. K.G. Wilson, Renormalization group and critical phenomena. 1. Renormalization group and the Kadanoff scaling picture. *Phys. Rev. B* **4**, 3174 (1971)
11. K.G. Wilson, The renormalization group: critical Phenomena and the Kondo problem. Rev. Mod. Phys. **47**, 773 (1975)
12. P. Kovtun, D.T. Son, A.O. Starinets, Viscosity in strongly interacting quantum field theories from black hole physics. Phys. Rev. Lett. **94**, 111601 (2005). arxiv: [hep-th/0405231]
13. J. Crossno, J.K. Shi, K. Wang, X. Liu, A. Harzheim, A. Lucas et al., Observation of the dirac fluid and the breakdown of the wiedemann-franz law in graphene. Science **351**, 1058 (2016)
14. A. Lucas, J. Crossno, K.C. Fong, P. Kim, S. Sachdev, Transport in inhomogeneous quantum critical fluids and in the Dirac fluid in graphene. Phys. Rev. B **93**, 075426 (2016). arxiv: [1510.01738]
15. Y. Seo, G. Song, P. Kim, S. Sachdev, S.-J. Sin, Holography of the Dirac Fluid in Graphene with two currents. Phys. Rev. Lett. **118**, 036601 (2017). arxiv: [1609.03582]
16. Y. Seo, G. Song, S.-J. Sin, Strong correlation effects on surfaces of topological insulators via holography. Phys. Rev. B **96**, 041104 (2017). arxiv: [1703.07361]
17. Y. Seo, G. Song, C. Park, S.-J. Sin, Small fermi surfaces and strong correlation effects in dirac materials with holography. JHEP **10**, 204 (2017). arxiv: [1708.02257]
18. M. Liu, J. Zhang, C.-Z. Chang, Z. Zhang, X. Feng, K. Li et al., Crossover between weak antilocalization and weak localization in a magnetically doped topological insulator. Phys. Rev. Lett. **108**, 036805 (2012)
19. D. Zhang, A. Richardella, D.W. Rench, S.-Y. Xu, A. Kandala, T.C. Flanagan et al., Interplay between ferromagnetism, surface states, and quantum corrections in a magnetically doped topological insulator. Phys. Rev. B **86**, 205127 (2012)
20. E. Oh, Y. Seo, T. Yuk, S.-J. Sin, Ginzberg-Landau-Wilson theory for Flat band, Fermi-arc and surface states of strongly correlated systems. JHEP **01**, 053 (2021). arxiv: [2007.12188]
21. S.-S. Lee, A Non-Fermi liquid from a charged black hole: a critical Fermi Ball. Phys. Rev. D **79**, 086006 (2009). arxiv: [0809.3402]
22. M. Cubrovic, J. Zaanen, K. Schalm, String theory, quantum phase transitions and the emergent Fermi-Liquid. Science **325**, 439 (2009). arxiv: [0904.1993]
23. H. Liu, J. McGreevy, D. Vegh, Non-fermi liquids from holography. Phys. Rev. D **83**, 065029 (2011)
24. N. Iqbal, H. Liu, Real-time response in AdS/CFT with application to spinors. Fortsch. Phys. **57**, 367 (2009). arxiv: [0903.2596]
25. E. Oh, T. Yuk, S.-J. Sin, The emergence of strange metal and topological liquid near quantum critical point in a solvable model. JHEP **11**, 207 (2021). arxiv: [2103.08166]
26. H. Im et al., Observation of Kondo condensation in a degenerately doped silicon metal. Nat. Phys. **19**, 676 (2023). arxiv: [2301.09047]
27. S. Sukrakarn, T. Yuk, S.-J. Sin, Ginzburg-Landau-Wilson theory for strongly correlated systems II: analytic and backreacted results, 2311.01897
28. I.V. Borzenets, J. Shim, J. Chen, A. Ludwig, A.D. Wieck, S. Tarucha et al., Observation of the Kondo screening cloud. Nature **579**, 210 (2020)
29. R. Oppermann, B. Rosenow, Magnetic gaps related to spin glass order in fermionic systems. Phys. Rev. Lett. **80**, 4767 (1998)
30. S. Doniach, The kondo lattice and weak antiferromagnetism. Physica B+ C **91**, 231 (1977)

31. E. Fradkin, *Field Theories of Condensed Matter Physics* (Cambridge University Press, 2013)
32. K.G. Wilson, Renormalization group and critical phenomena. i. renormalization group and the kadanoff scaling picture. Phys. Rev. B**4**, 3174 (1971)
33. K.G. Wilson, The renormalization group: critical phenomena and the Kondo problem. Rev. Mod. Phys. **47**, 773 (1975)

Scale Invariance at the Edge

Gordon W. Semenoff

Abstract Some aspects of the theory of fermions living on three dimensional space-time with a flat co-dimension one boundary are discussed, particularly a case where the boundary condition preserves scale and translation invariance but violates the residual Lorentz and conformal symmetries. This case is interesting because the Dirac equation has normalizable stationary edge states which must be taken into account in the quantization. We show that a consequence of the edge states for quantization of the Dirac field is that there are no states of the Dirac field theory which are simultaneously scale, C and P invariant even when these are good symmetries of the theory. The scale invariant states of the Dirac field contain either a nonzero scale covariant momentum density or a U(1) charge density concentrated near the edge.

1 Introduction

The quantization of the Dirac theory on a space with a boundary requires boundary conditions. There is the obvious question, especially with fermions, as to whether the boundary conditions respect the symmetry of the theory. For example, the free massless Dirac theory in the bulk of 3 dimensional Minkowski space-time is a (trivial example of a) conformal field theory with the full $SO(3, 2)$ conformal symmetry group. Consider the same problem on a half-space – 3 dimensional Minkowski space with an infinite, flat, co-dimension one boundary which is itself a 2 dimensional Minkowski spacetime. One would expect that the bulk $SO(3, 2)$ invariance would be reduced the $SO(2, 2)$ symmetry of the boundary. This symmetry consists of translation in time and space along the boundary, a Lorentz transform along the direction of the boundary and scaling and a conformal transformation of the boundary.

However, the existence of this residual symmetry can of course depend on the boundary condition. There turns out to be a boundary condition which is relevant to the study of some two dimensional Dirac materials which terminate on boundaries

G. W. Semenoff (✉)
Department of Physics and Astronomy, University of British Columbia, Vancouver, British Columbia, Canada
e-mail: gordonws@phas.ubc.ca

M. Paranjape et al. (eds.), *Proceedings of Gravity, Strings and Fields*,
CRM Series in Mathematical Physics,
https://doi.org/10.1007/978-3-031-91266-5_17

where the boundary condition breaks what would be some of the residual space-time symmetry. In the example of graphene, perhaps the prototypical example of a Dirac material [1, 2], the edge that we are referring to has a zig-zag configuration [3, 4]. The appropriate boundary condition has less spacetime symmetry than is possible for a massless Dirac field [5]. It preserves only the scale and translation symmetry of the edge. It violates Lorentz and conformal symmetry. On the other hand, even the massless non-interacting Dirac Hamiltonian has interesting behaviour there. It exhibits a continuous set of bound states at the edge.

Bound states in a scale invariant field theory are unusual. They must have vanishing binding energy. Indeed they appear in the middle of the spectrum, between he positive and negative energy eigenstates of the Dirac Hamiltonian. In the following we will examine how the scale symmetry as well as the discrete spacetime symmetries are represented when these bound states are present. We will identify scale invariant states of the Dirac field theory. Those states have interesting properties in that requiring scale invariance will break some of the other symmetries of the theory, either charge conjugation C or parity P, which are perfectly good symmetries of the Hamiltonian and the boundary conditions but seem to be incompatible with scale invariance. The result, for scale invariant states will be, either a C invariant,P violating state which has a one-point function with non-vanishing momentum density $< \mathbf{T}^{02}(x) >$ or a P invariant, C violating state with a one-point function with non-vanishing U(1) charge density.

2 Dirac Theory on a 3D Half-Space

Consider the massless Dirac equation

$$\partial\!\!\!/ \psi^{(A)}(x) = 0\,, \quad A = 1, 2 \tag{1}$$

for two complex two-component spinors in 2+1-dimensions, with "valley" labels $A = 1, 2$, defined on the half-space $x^\mu = (x^0, x^1, x^2)$ with $x^1 > 0$ and with the boundary conditions

$$\lim_{x^1\to 0} \left(1 + i\gamma^0\right) \psi^{(1)}(x) = 0\,, \quad \lim_{x^1\to 0} \left(1 - i\gamma^0\right) \psi^{(2)}(x) = 0 \tag{2}$$

We are considering the case where there are two valleys, $A = 1, 2$, which have opposite boundary conditions in order to preserve some discrete spacetime symmetries, on which we will elaborate shortly. These turn out to describe the continuum limit of the tight-binding model of graphene on its hexagonal lattice with a zig-zag edge, where we have assumed that the electron is spinless.

The boundary condition is needed so that the normal component of the current vanishes at the boundary

$$\lim_{x^1 \to 0} \bar{\psi}^{(A)}(x)\gamma^1\psi^{(A)}(x) = 0\,, \quad A = 1, 2 \tag{3}$$

This is needed so that the Hamiltonian is Hermitian. An alternative, Lorentz invariant boundary condition would be $\lim_{x^1\to 0}\left(1 \pm \gamma^1\right)\psi^{(A)}(x) = 0$, similar to the MIT bag model boundary condition. It actually preserves the Lorentz and conformal symmetry of the boundary. However, it does not allow edge states. In the following we will focus on the Dirac fermion problem with edge states, that is, with the boundary condition (2).

3 Boundary Condition Breaks Lorentz and Conformal Symmetry

In the bulk of three dimensional Minkowski space, the space-time symmetries of the fermions are the full set of set of conformal transformations. In the presence of a boundary, these are reduced to the spacetime symmetries of the boundary, which are generated by spacetime Noether charges which have the algebra

$$-i\left[P_\mu,\, \psi^{(A)}(x)\right] = \partial_\mu\psi^{(A)}(x)\,, \quad \mu = 0, 2 \tag{4}$$

$$-i\left[M,\, \psi^{(A)}(x)\right] = \left(x_2\partial_0 - x_0\partial_2 + \frac{1}{2}\gamma^1\right)\psi^{(A)}(x) \tag{5}$$

$$-i\left[\Delta,\, \psi^{(A)}(x)\right] = \left(x^\mu\partial_\mu + 1\right)\psi^{(A)}(x) \tag{6}$$

$$-i\left[K_\lambda,\, \psi^{(A)}(x)\right] = \left\{\left(x_\lambda x^\mu - \frac{x^2}{2}\delta^\mu_\lambda\right)\partial_\mu + \frac{1}{2}\epsilon_{\lambda\rho\sigma}x^\rho\gamma^\sigma + x_\lambda\right\}\psi^{(A)}(x),\ \lambda = 0, 2 \tag{7}$$

However, we can see that the spin parts of the Lorentz transformation in Eq. (5) and of the conformal transformation in Eq. (7) are incompatible with the boundary conditions in Eq. (2). This further reduces the symmetries to translations and scale transformations generated by P^0, P^2 and Δ.

4 CPT

Consider the explicit representation of the 2×2 Dirac matrices matrices

$$\gamma^0 = -i\sigma^3\,,\ \gamma^1 = \sigma^1\,,\ \gamma^2 = \sigma^2 \tag{8}$$

where σ^a are the Pauli matrices. With this choice of Dirac matrices, the discrete spacetime symmetries C, P and T are the transformations

$$\mathrm{C}: (\psi^{(1)}(x), \psi^{(2)}(x)) \rightarrow (\gamma^1\psi^{*(2)}(x), \gamma^1\psi^{*(1)}(x)) \tag{9}$$

$$\mathrm{P}: (\psi^{(1)}(x), \psi^{(2)}(x)) \rightarrow (i\gamma^2\psi^{(2)}(x'), -i\gamma^2\psi^{(1)}(x')) \quad x' = (x_0, x_1, -x_2) \tag{10}$$

$$\mathrm{T}: (\psi^{(1)}(x), \psi^{(2)}(x)) \rightarrow (\mathcal{T}\gamma^2\psi^{(2)}(\tilde{x}), \mathcal{T}\gamma^2\psi^{(1)}(\tilde{x})) \quad \tilde{x} = (-x_0, x_1, x_2) \tag{11}$$

We note that there is no valley symmetry in this model. The interchange of the two valleys is not a symmetry because the valleys have different boundary conditions. Also, note that the valleys are interchanged under any one of the C, P or T transformations. This means that, if we had a single valley, rather than two valleys, none of C, P or would be symmetries. and CPT would be a symmetry. This is consistent with the fact that the theory is not Lorentz invariant, since a Lorentz transformation would have to preserve the valley index, whereas CPT made from the transformations above necessarily interchanges the valleys.

5 Explicit Solution

With the choice of Dirac gamma-matrices in Eq. (8), the explicit forms of the Dirac Eq. (1) and boundary conditions (2) are

$$\begin{bmatrix} -i\partial_0 & \partial_1 - i\partial_2 \\ \partial_1 + i\partial_2 & i\partial_0 \end{bmatrix} \begin{bmatrix} u^{(A)}(x) \\ v^{(A)}(x) \end{bmatrix} = 0 \,, \; A = 1, 2 \tag{12}$$

$$u^{(1)}(x^1 \to 0) = 0 \,, \; v^{(2)}(x^1 \to 0) = 0 \tag{13}$$

The Eq. (12) and the boundary conditions (3) have the explicit solutions

$$\psi^{(1)}_{\omega k\ell}(x) = \frac{e^{-i\omega x^0 + ikx^2}}{\sqrt{2}\pi} \begin{bmatrix} \sin \ell x^1 \\ \frac{k}{\omega} \sin \ell x^1 + \frac{\ell}{\omega} \cos \ell x^1 \end{bmatrix} \tag{14}$$

$$\psi^{(2)}_{\omega k\ell}(x) = \frac{e^{-i\omega x^0 + ikx^2}}{\sqrt{2}\pi} \begin{bmatrix} \frac{k}{\omega} \sin \ell x^1 - \frac{\ell}{\omega} \cos \ell x^1 \\ \sin \ell x^1 \end{bmatrix} \tag{15}$$

$$\ell > 0 \,, \; \omega = \pm\sqrt{k^2 + \ell^2} \tag{16}$$

$$\psi^{(1)}_k(x) = \sqrt{\frac{k}{\pi}} \begin{bmatrix} 0 \\ e^{-k(x^1 - ix^2)} \end{bmatrix}, \; k > 0 \tag{17}$$

$$\psi^{(2)}_k(x) = \sqrt{\frac{-k}{\pi}} \begin{bmatrix} e^{k(x^1 + ix^2)} \\ 0 \end{bmatrix}, \; k < 0 \tag{18}$$

The solutions $\psi^{(A)}_{\omega k\ell}(x)$ are Dirac spinors which propagate in the bulk of the half-space and we shall refer to them as "bulk states". They have two signs of the frequency, ω and the wave-number k. Their plane-wave parts $\sim e^{-i\omega x^0 + ikx^2}$ carry a representation of translations in time and along the x^2 direction, parallel to the edge.

The solutions $\psi_k^{(A)}(x)$ which are localized near the edge of the half-plane at $x^1 = 0$ are referred to as "edge states". They also carry a representation of translations parallel to the edge $\sim e^{ikx^2}$, and time translation where they have zero frequency. They have the peculiar feature that their momentum quantum numbers, k, for edge states in a given valley have a the same sign for the entire set of edge wave-functions.

The bulk and edge states obey a completeness relation

$$\sum_\omega \int dkd\ell\psi^{(A)}_{\omega k\ell}(x^0,x^1,x^2)\psi^{(A)\dagger}_{\omega k\ell}(x^0,\tilde{x}^1,\tilde{x}^2)+\int dk\psi_k^{(A)}(x^1,x^2)\psi_k^{(A)\dagger}(\tilde{x}^1,\tilde{x}^2)$$
$$=\delta(x^1-\tilde{x}^1)\delta(x^2-\tilde{x}^2)\,,\quad A=1,2 \tag{19}$$

which can be checked explicitly and which therefore implies that we have found all of the solutions.

Given the solutions of the Dirac equation in equations (14)–(18), the quantized Dirac field is

$$\psi^{(A)}(x)=\int dkdl\left\{\psi^{(A)}_{\sqrt{k^2+\ell^2}k\ell}(x)a^{(A)}(k,\ell)+\psi_{-\sqrt{k^2+\ell^2}-k,\ell}(x)b^{(A)\dagger}(k,\ell)\right\}+\int dk\psi_k^{(A)}(x)a^{(A)}(k) \tag{20}$$

Completeness of the solutions (19) implies that the second quantized field in Eq. (20) satisfies the canonical equal-time anti-commutation relation for the Dirac field

$$\left\{\psi_a^{(A)}(x),\psi_b^{(B)\dagger}(y)\right\}_{x^0=y^0}=\delta_{ab}\Delta^{AB}\delta^2(x-y) \tag{21}$$

$$\left\{\psi_a^{(A)}(x),\psi_b^{(B)}(y)\right\}_{x^0=y^0}=0\,,\quad \left\{\psi_a^{(A)\dagger}(x),\psi_b^{(B)\dagger}(y)\right\}_{x^0=y^0}=0 \tag{22}$$

when $a^{(A)}(k,\ell)$, $b^{(A)}(k,\ell)$, $a^{(A)}(k)$ and their Hermitian conjugates have the usual anti-commutation relations of fermion and anti-fermion annihilation and creation operators, respectively.

In particular, the creation and annihilation operators for bulk and edge degrees of freedom anti-commute with each other and we can find a basis of the set of states on which they operate that is composed of direct products of bulk and edge states. We are interested in the states of this system near the Dirac vacuum where all of the negative energy bulk states are occupied and all of the positive bulk energy states are empty,

$$a^{(A)}(k,\ell)|0>_{\rm bulk}=0\,,\quad b^{(A)}(k,\ell)|0>_{\rm bulk}=0$$

This state is unambiguous for the bulk states where there is a clear distinction between positive and negative energy states. The edge states must carry a representation of the anti-commutator algebra of the edge state creation and annihilation operators,

$$\{a^{(A)}(k), a^{(B)\dagger}(k')\} = \delta^{AB}\delta(k-k') \tag{23}$$

$$\{a^{(A)}(k), a^{(B)}(k')\} = 0\,,\ \{a^{(A)\dagger}(k), a^{(B)\dagger}(k')\} = 0 \tag{24}$$

We will denote a generic state in a representation of the algebra in equations (23) and (24) by $|\text{edge} >$ and we will call then edge states. We will be interested in states of the form

$$|0>_{\text{bulk}} \otimes |\text{edge}>$$

In the absence of interactions of the fermions, there is an enormous degeneracy of possible edge states in that any distribution of fermions in the edge states has the same energy as any other distribution. It is a priori unclear whether there are some edge states which are distinguished from others. Understanding possible answers to this question will be our goal in the following.

6 Scale Invariant States

It is interesting to ask what the possible scale invariant states of this scale invariant theory might be. Since the Dirac field has classical dimension one, a scale transformation, $x^\mu \to \Lambda x^\mu$ acts on the Dirac field by the replacement

$$\psi^{(A)}(x) \to \Lambda\psi^{(A)}(\Lambda x) \tag{25}$$

The wave-functions for the positive, negative and edge states then transform as

$$\psi^{(A)}_{\omega k\ell}(x) \;\to\; \Lambda\psi^{(A)}_{\omega k\ell}(\Lambda x) \;=\; \Lambda\psi^{(A)}_{\Lambda\omega\Lambda k\Lambda\ell}(x) \tag{26}$$

$$\psi^{(A)}_{k}(x) \;\to\; \Lambda\psi^{(A)}_{k}(\Lambda x) \;=\; \Lambda^{\frac{1}{2}}\psi^{(A)}_{\Lambda k}(x) \tag{27}$$

In order to construct scale invariant states, we are interested in complete orbits of the set of all possible scale transformations. An orbit of the scaling transformations acting on a function is the set of all functions that are generated from it by scale transformation with $0 < \Lambda < \infty$. An orbit of a particular bulk state under scaling is thus the infinite set of bulk states with all magnitudes but the same signs of k, ℓ and ω,

$$\text{Orbit of scaling} \;=\; \{\Lambda\psi^{(A)}_{\Lambda\omega\Lambda k\Lambda\ell}(x), \forall\Lambda \in (0,\infty)\}$$

The set of all bulk states thus divides into four orbits of the scaling transformations according to the signs of k and ω, each subset being an orbit of scaling. (Remember that for bulk states ℓ is always positive.) The Dirac ground state of the bulk, with all negative ω and all positive and negative values of k therefore has complete orbits of the scale transformation and it is scale invariant. Another way to put this is to say that the operator equations which define the Dirac vacuum for the bulk states

$$a(k,\ell)|0>_{\text{bulk}}=0\,,\quad b(k,\ell)|0>_{\text{bulk}}=0\,,\quad \forall k,\ell \tag{28}$$

is compatible with the existence of a unitary operator, $\Omega(\Lambda)$, acting on the bulk sector which implements

$$\Omega(\Lambda)a^{(A)}(k,\ell)\Omega^{\dagger}(\Lambda)=\Lambda a^{(A)}(k/\Lambda,\ell/\Lambda)\,,\quad \Omega(\Lambda)a^{(A)\dagger}(k,\ell)\Omega^{\dagger}(\Lambda)=\Lambda a^{(A)\dagger}(k/\Lambda,\ell/\Lambda) \tag{29}$$

$$\Omega(\Lambda)b^{(A)}(k,\ell)\Omega^{\dagger}(\Lambda)=\Lambda b^{(A)}(k/\Lambda,\ell/\Lambda)\,,\quad \Omega(\Lambda)b^{(A)\dagger}(k,\ell)\Omega^{\dagger}(\Lambda)=\Lambda b^{(A)\dagger}(k/\Lambda,\ell/\Lambda) \tag{30}$$

and which leaves the bulk vacuum invariant,

$$\Omega(\Lambda)|0>_{\text{bulk}}=|0>_{\text{bulk}} \tag{31}$$

On the other hand, as can be seen by inspecting equations (17) and (18), the set of all of the edge states in a given valley comprise a single orbit of scaling,

$$\text{Orbit of scaling} \;=\; \{\Lambda^{\frac{1}{2}}\psi^{(A)}_{\Lambda k}(x),\,\forall\Lambda\in(0,\infty)\} \tag{32}$$

The only possible scale invariant states of the edge in a particular valley must then have either all of the edge states completely filled or all of the edge states completely empty. Amongst all of the possible fillings of the edge states for each valley, it is only these two which are scale invariant. There are thus four possibilities for a scale-invariant edge state:

$$a^{(1)}(k)|ff>=0\;\forall k>0\,,\quad a^{(2)}(k)|ff>=0\;\forall k<0 \tag{33}$$

$$a^{(1)\dagger}(k)|ef>=0\;\forall k>0\,,\quad a^{(2)}(k)|ef>=0\;\forall k<0 \tag{34}$$

$$a^{(1)}(k)|fe>=0\;\forall k>0\,,\quad a^{(2)\dagger}(k)|fe>=0\;\forall k<0 \tag{35}$$

$$a^{(1)\dagger}(k)|ee>=0\;\forall k>0\,,\quad a^{(2)\dagger}(k)|ee>=0\;\forall k<0 \tag{36}$$

The scale invariant states of the entire system are then

$$|0>_{\text{bulk}}\otimes|ff>\;,\quad |0>_{\text{bulk}}\otimes|ee>\;,\quad |0>_{\text{bulk}}\otimes|fe>\;,\quad |0>_{\text{bulk}}\otimes|ef> \tag{37}$$

7 Momentum Density at the Edge

Let us examine the possibility that these states could have differing physical properties. Each state with wave-function in the bulk, $\psi^{(A)}_{\omega k\ell}(x)$, or at the edge, $\psi^{(A)}_{k}(x)$, carries a linear momentum k. This momentum is directed parallel to the boundary. Moreover, as we have already observed, all of the edge states in a given valley have the same sign of k. The contribution of a given occupied bulk state to the total momentum is $k\psi^{(A)\dagger}_{\omega k\ell}(x)\psi^{(A)}_{\omega k\ell}(x)$ and the contribution of an occupied edge state is $k\psi^{(A)\dagger}_{k}(x)\psi^{(A)}_{k}(x)$.

The linear momentum is the spatial volume integral of the stress tensor

$$\mathbf{T}^{\mu\nu}(x) = \frac{i}{4}\sum_A \bar{\psi}^{(A)}(x)(\gamma^\mu \overrightarrow{\partial}^\nu - \gamma^\mu \overleftarrow{\partial}^\nu + \gamma^\nu \overrightarrow{\partial}^\mu - \gamma^\nu \overleftarrow{\partial}^\mu)\psi^{(A)}(x) \tag{38}$$

We expect that parity symmetry of the bulk states would lead to cancellation of the linear momentum. It is instructive to confirm this more explicitly. If we add up the contribution of the occupied bulk states (after some algebra) we find

$$< \mathbf{T}^{02}(x) >_{\text{bulk}} = \lim_{x'\to x} \frac{1}{i}\frac{d}{dx^2}\int_{-\infty}^{\infty} dk \int_0^\infty d\ell \psi^{(A)\dagger}_{-\sqrt{k^2+\ell^2}k\ell}(x')\psi^{(A)}_{-\sqrt{k^2+\ell^2}k\ell}(x) \tag{39}$$

We can use a particle-hole-symmetry (CP), which is a good symmetry of each valley separately, to re-write the right-hand-side of the above equations as

$$< \mathbf{T}^{02}(x) >_{\text{bulk}} =$$
$$\sum_A \lim_{x'\to x} \frac{1}{2i}\frac{d}{dx^2}\int_{-\infty}^{\infty} d\ell \int_0^\infty dk \left[\psi^{(A)\dagger}_{\sqrt{k^2+\ell^2}k\ell}(x')\psi^{(A)}_{\sqrt{k^2+\ell^2}k\ell}(x) + \psi^{(A)\dagger}_{-\sqrt{k^2+\ell^2}k\ell}(x')\psi^{(A)}_{-\sqrt{k^2+\ell^2}k\ell}(x)\right] \tag{40}$$

Then we can use completeness of the set of all wave-functions to rewrite the right-hand-side of the above equation as

$$< \mathbf{T}^{02}(x) >_{\text{bulk}} =$$
$$\sum_A \lim_{x'\to x} \frac{1}{i}\frac{d}{dx^2}\frac{1}{2}\left\{\delta^2(x'-x) - \int dk\psi^{(A)\dagger}_k(x')\psi^{(A)}_k(x)\right\} = -\frac{1}{2}\sum_A \int dk\, k\, \psi^{(A)\dagger}_k(x)\psi^{(A)}_k(x)\,, \tag{41}$$

where we have taken the $x' \to x$ limit and we have assumed that the delta function (or a parity symmetric regularization of the delta function) obeys $\lim_{x'\to x}\frac{d}{dx_2}\delta^2(x-x') = 0$. Now, we note that

$$\int dk\, k\, \psi^{(1)\dagger}_k(x)\psi^{(1)}_k(x) = -\int dk\, k\, \psi^{(2)\dagger}_k(x)\psi^{(2)}_k(x) \tag{42}$$

Our conclusion is that the bulk Dirac sea contributes to the expectation value of the stress tensor cancel

$$< \mathbf{T}^{02}(x) >_{\text{bulk}} = 0 \tag{43}$$

Here, we note that this is due to an intricate cancellation between the momenta in each of the valleys. If there were only one valley, the Dirac sea of the bulk fermions would carry momentum density.

Now we see that momentum density distinguishes the scale invariant states of the edge. The states $|ff>$ and $|ee>$ where all of the edge states in both valleys are completely filled, or all of the edge states in both valleys are completely empty will carry no linear momentum, in that it cancels between the valley contributions. The other two states $|ef>$ and $|fe>$ will carry a linear momentum density which is given by

$$\int_0^\infty dk\, k\, \psi_k^{(1)\dagger}(x)\psi_k^{(1)}(x) = \frac{1}{4\pi}\frac{1}{[x^1]^3} \tag{44}$$

We conclude that the scale invariant states have the properties

$$< ef|\otimes_{\text{bulk}} < 0|\mathbf{T}^{02}(x)|0>_{\text{bulk}} \otimes|ef>= -\frac{1}{4\pi}\frac{1}{[x^1]^3} \tag{45}$$

$$< fe|\otimes_{\text{bulk}} < 0|\mathbf{T}^{02}(x)|0>_{\text{bulk}} \otimes|fe>= \frac{1}{4\pi}\frac{1}{[x^1]^3} \tag{46}$$

$$< ff|\otimes_{\text{bulk}} < 0|\mathbf{T}^{02}(x)|0>_{\text{bulk}} \otimes|ff>= 0 \tag{47}$$

$$< ee|\otimes_{\text{bulk}} < 0|\mathbf{T}^{02}(x)|0>_{\text{bulk}} \otimes|ee>= 0 \tag{48}$$

Another way to see that this should be the case is to note that P parity symmetry and translation invariance would make the above one-point functions vanish and it is only the states $|0>_{\text{bulk}} \otimes|0>_{ee}$ and $|0>_{\text{bulk}} \otimes|0>_{ff}$ which have both of these symmetries. The states $|0>_{\text{bulk}} \otimes|0>_{fe}$ and $|0>_{\text{bulk}} \otimes|0>_{ef}$ are a doublet which transform into each other under P.

8 U(1) Charge Density at the Edge

By an argument similar to the previous section where we studied the momentum density, we could examine U(1) currents. We note that each valley has a U(1) symmetry and each should have a conserved U(1) current density. Here we will only consider the total $U(1)$ current density which is the sum of the two. We define the operator current density using point splitting and we make it transform correctly under C and P by using the Dirac commutator operator ordering,

$$J^\mu(x) \equiv \sum_A \lim_{x'\to x} [\gamma^0\gamma^\mu]_{ab}\frac{1}{2}\left[\psi_a^{(A)\dagger}(x'), \psi_b^{(A)}(x)\right] \tag{49}$$

Since, under the C transformation

$$C:\ J^{\mu}(x)\to -J^{\mu}(x) \tag{50}$$

and since the bulk states transform covariantly under C, the contribution of the bulk states to the expectation value

$$<\text{edge}|\otimes_{\text{bulk}} <0|\ J^{\mu}(x)|0>_{\text{bulk}} \otimes|\text{edge}>$$

must vanish and the entire contribution comes from the edge states. The edge state wave-functions obey

$$\int_0^{\infty} dk\psi_k^{(1)\dagger}(x)\psi_k^{(1)}(x)=\int_{-\infty}^{0} dk\psi_k^{(2)\dagger}(x)\psi_k^{(2)}(x)=\frac{1}{4\pi}\frac{1}{[x^1]^2} \tag{51}$$

Then, adding the edge state contribution to the vanishing bulk contribution yields the U(1) current density, for example, for the completely filled edge states

$$<ff|\otimes_{\text{bulk}} <0|\ J^0(x)|0>_{\text{bulk}}\otimes|ff>=\frac{1}{2}\left[\int_0^{\infty} dk\psi^{(1)\dagger}(x)\psi^{(1)}(x)+\int_{-\infty}^{0} dk\psi^{(2)\dagger}(x)\psi^{(2)}(x)\right] \tag{52}$$

The factor of $\frac{1}{2}$ on the right-hand-side comes from the Dirac commutator. This leads to the one-point functions for the U(1) current in each of the scale invariant states,

$$<ff|\otimes_{\text{bulk}} <0|\ J^{\mu}(x)|0>_{\text{bulk}}\otimes|ff>=\frac{\delta_0^{\mu}}{4\pi}\frac{1}{[x^1]^2} \tag{53}$$

$$<ee|\otimes_{\text{bulk}} <0|\ J^{\mu}(x)|0>_{\text{bulk}}\otimes|ee>=-\frac{\delta_0^{\mu}}{4\pi}\frac{1}{[x^1]^2} \tag{54}$$

$$<ef|\otimes_{\text{bulk}} <0|\ J^{\mu}(x)|0>_{\text{bulk}}\otimes|ef>=0 \tag{55}$$

$$<fe|\otimes_{\text{bulk}} <0|\ J^{\mu}(x)|0>_{\text{bulk}}\otimes|fe>=0 \tag{56}$$

In the previous section we concluded that the P invariant states $|0>_{\text{bulk}}\otimes|ff>$ and $|0>_{\text{bulk}}\otimes|ee>$ had vanishing momenta. However, these states are not C invariant. They form a doublet under C and, as a consequence, they have a nonzero U(1) charge densities, with opposite signs.

9 Epilogue

In conclusion, we have outlined some properties of the scale invariant states of a simple scale invariant but not Lorentz or conformal invariant boundary field theory. The existence of edge states leads to the interesting anomalous effects that we find.

For our model with two valleys, there are four scale invariant states which split into two pairs. One pair of scale invariant edge states are seperately P parity invariant but they form a doublet of C charge conjugation symmetry. The other pair are separately C invariant but form a doublet of P. The states in the C doublet have oppositely signed nonzero charge densities whereas the states in the P doublet have nonzero oppositely signed linear momentum densities. The latter linear momentum is directed parallel to the edge and the charge and momentum densities in both cases decay by a scale coraviant power law with distance from the edge and with an oveall coefficient which is independent of the details of the system.

Edge states of the kind that we have studied here has long been known to occur at the lattice level in, for example, the tight binding model of graphene electrons and the have long been conjectured to have interesting properties and potentially important physical applications [6, 7]. For example it is known that states very similar to those which we have considered here but in a theory with an additional spin degree of freedom have anti-ferromagnetic states and there is a proof that they are the ground states of a sufficiently weakly coupled class of two-body interactions that could be added to the free Dirac theory are indeed ferromagnetic [8] and the ferromagnetic states are direct analogs of the scale invariant states that we have discussed here. It would be very interesting to study interacting field theories with the boundary and boundary conditions that we have discussed.

This work is supported in part by the Natural Sciences and Engineering Research Council of Canada (NSERC).

References

1. G.W. Semenoff, Condensed matter simulation of a three-dimensional anomaly. Phys. Rev. Lett. **53**, 2449 (1984)
2. A.H. Castro Neto, F. Guinea, N.M.R. Peres, K.S. Novoselov, A.K. Geim, The electronic properties of graphene. Rev. Mod. Phys. **81**, 109–162 (2009) arXiv:0709.1163 [cond-mat.other]
3. A.R. Akhmerov, C. Beenakker, Boundary conditions for dirac fermions on a terminated honeycomb lattice. Phys. Rev. B **77**, 085423 (2008)
4. J.A.M van Ostaay, A.R. Akhmerov, C.W.J. Beenakker, M. Wimmer, Dirac boundary condition at the reconstructed zigzag edge of graphene. Phys. Rev. **B 84**, 195434 (2011)
5. S. Biswas, G.W. Semenoff, Massless fermions on a half-space: the curious case of 2+1-dimensions. JHEP **10**, 045 (2022) arXiv:2208.06374 [hep-th]
6. M. Fujita, K. Wakabayashi, K. Nakada, K. Kusakabe, Peculiar localized state at zigzag graphite edge. J. Phys. Soc. Jpn. **65**, 1920v (1996)
7. K. Nakada, M. Fujita, G. Dresselhaus, M.S. Dresselhaus, Edge state in graphene ribbons: nanometer size effect and edge shape dependence. Phys. Rev. B **54**, 17954 (1996)
8. G.W. Semenoff, Boundary ferromagnetism in zigzag edged graphene. J. Math. Phys. **64**(7), 071902 (2023) arXiv:2211.09282 [cond-mat.mes-hall]

The Lattice Schwinger Model and Its Quantum Simulation

João C. Pinto Barros, Pierpaolo Fontana, Pasquale Sodano, and Andrea Trombettoni

Abstract In this chapter we review results on the lattice Schwinger model. In particular, we show how the effect of the anomaly is reproduced on the lattice. We connect these results to recent developments in the field of quantum simulation of interacting field theories. Schemes for the quantum simulation of (approximations of) Schwinger models are discussed.

Keywords Schwinger model · Quantum simulation

1 Introduction

It is a pleasure to contribute to a volume celebrating the many and relevant contributions given by Professor Gordon W. Semenoff to modern theoretical physics. One of the authors (PS) had the chance to meet Gordon when he and Gordon were very young graduate students in the theoretical physics department of the University of Alberta. We started to discuss physics very soon helped in that by the fact that we shared the same Ph.D. supervisor, Prof. Hiroomi Umezawa. Since then most of my career benefited from the stimulating collaboration with Gordon and my life in general enjoyed our friendship. Many anecdotes come to my mind, but I hope that there will be a chance to share them around a dinner table.

J. C. P. Barros
Institut für Theoretische Physik, ETH Zürich, Zürich, Switzerland

P. Fontana
Departament de Física, Universitat Autònoma de Barcelona, Bellaterra, Spain

P. Sodano (✉)
INFN, Sezione di Perugia, Perugia, Italy
e-mail: pasquale.sodano01@gmail.com

A. Trombettoni
Department of Physics, University of Trieste and INFN, Sezione di Trieste, Trieste, Italy

M. Paranjape et al. (eds.), *Proceedings of Gravity, Strings and Fields*,
CRM Series in Mathematical Physics,
https://doi.org/10.1007/978-3-031-91266-5_18

The contribution to this volume is centered on the lattice Schwinger model with staggered fermions. The Schwinger model is the one-flavor quantum electrodynamics (QED) in one spatial dimension. Historically, Schwinger considered this model [1] as an example of a gauge vector field that can have a non-zero mass [2]. Despite its simplicity as an Abelian gauge theory in in one spatial dimension, it has many interesting physical features like, for instance, anomalies [3], confinement [4] and chiral symmetry breaking [5]. This made the model very attractive to test analytical and numerical methods relevant for quantum field theory. Recently, analyses of the Schwinger model were motivated by a growing interest in its connections with quantum simulators [6], machine learning [7], matrix product states [8] and quantum computing [9].

The Schwinger model is an ideal laboratory to study the occurrence of anomalies in a field theory. In general, anomalies occur when all the symmetries of a classical field theory can not be realized simultaneously in the corresponding quantum field theory. The classic example is the Adler-Bell-Jackiw anomaly in a vector-like gauge theory, where gauge invariance and global axial symmetries are incompatible. In computations, this phenomenon can be seen as the impossibility of finding a regularization of the ultraviolet divergences which preserves both axial and gauge invariance. If the latter is preserved, the observable manifestation of the anomaly is the failure of conservation laws for axial currents and the resulting absence of their consequences in the spectrum of the quantized theory. The Schwinger model is the prototypical example of a solvable model where all of this occurs.

In the following we shall mainly focus on the one- and two-flavor Schwinger models, reviewing their lattice formulations using staggered fermions and the Hamiltonian approach to lattice gauge theories [10]. Even though the continuous axial symmetry is broken explicitly by the staggered fermions, the discrete axial symmetry remains and appears in the lattice theory as a translation by one site. An interesting issue which arises in the lattice regularization of gauge theories is how the lattice theory produces the effects of the axial anomaly.

Lattice field theories are manifestly gauge invariant by construction. Therefore, to properly recover the corresponding continuum theories, they must include a mechanism to violate axial current conservation. Typically, in lattice gauge theories axial anomalies are either cancelled by fermion doubling or the lattice regularization breaks the axial symmetry explicitly. However, in the lattice Schwinger model formulated in the Hamiltonian formalism with one species of staggered fermions, neither of these scenarios occurs. Indeed, the effects of the anomaly are not cancelled by doubling since the continuum limit of $(1+1)$-dimensional staggered fermions produces exactly a Dirac fermion, which is the matter content of the model. Furthermore, even if the continuum axial symmetry is explicitly broken, the remaining discrete axial symmetry must be spontaneously broken. The mass operator $\bar{\psi}\psi$ is odd under discrete axial transformations; if this symmetry is unbroken, then $\langle\bar{\psi}\psi\rangle = 0$. However, in the continuum theory it is known that [11] $\langle\bar{\psi}(x)\psi(x)\rangle = -e^{\gamma}e_c/2\pi\sqrt{\pi}$, where $\gamma = 0.577...$ is the Euler constant and e_c is the electric charge. This expectation value is called the chiral condensate. The chiral condensate is analogous to the magnetization in a spin model, with the mass m playing the role of the magnetic field. Therefore, even when studying a massless model, determining the sign of the chiral condensate requires the presence of a small mass term m which is subsequently sent to zero. The minus sign appearing in $\langle\bar{\psi}(x)\psi(x)\rangle$ arises when taking the limit $m \longrightarrow 0^+$ in the Schwinger model Lagrangian. The chiral condensate is non-zero even on the lattice, signaling the spontaneous breaking of discrete axial symmetry. In addition to the analysis of the chiral symmetry breaking pattern on the lattice, we show results rapidly converging to the continuum limit by performing a strong coupling expansion around a gauge invariant ground state [10]. Recent results on the Kogut-Susskind Hamiltonian approach with staggered fermions [12] include the study of the discrete chiral symmetry and mass shift [13] and of the real time dynamics and confinement in presence of an Abelian $\mathbb{Z}_n$ gauge group [14]. The two-flavor Schwinger model has been also thouroughfully investigated [15, 16], with new findings for the phase diagram presented in [17].

The one-flavor Schwinger model has also been studied as an example of how quantum link models [6, 18] can reproduce the physics of Hamiltonian lattice gauge theories. This versatility made the Schwinger model and its variations as a benchmark for the developments of schemes of quantum simulations of field theories [6]. The general goal is to physically implement, e.g., in ultracold atoms setups, an interacting system whose low energy description coincides with a target field theory. From one side one can study the effect of the space discretization—since in typical quantum simulations schemes the used non-relativistic atoms are in optical lattices—and as well of truncation of Hilbert space, as it is the case when one implements a discretization of a continuous symmetry. From the other side, one can use the remarkable amount of results from interacting field theories to test and improve the quantum simulation scheme at hand. It is clear that the Schwinger model is extremely useful in both directions and its properties have been widely revisited in the realm of quantum simulations.

We comment on these recent developments in Sect. 4, while Sects. 2 and 3 are devoted to reviews of relevant features of the Schwinger model on the lattice with one and two flavors. Our presentation partially follow the lines of the discussion presented in the two PhD Theses [19, 20], the former by Federico Berruto, who worked during his PhD project with Gordon W. Semenoff within the collaboration between University of Perugia and University of British Columbia.

2 The One-Flavor Schwinger Model

The continuum Schwinger model [1, 2] stands as the prototypical example of a solvable model where the anomaly occurs. This model is defined as $1+1$-dimensional QED with a single charged, massless Dirac spinor field. The action is

$$S = \int d^2x \left(\bar{\psi} \left(i\gamma_\mu \partial^\mu + \gamma_\mu A^\mu \right) \psi - \frac{1}{4e_c^2} F_{\mu\nu} F^{\mu\nu} \right) . \tag{1}$$

It is invariant under the gauge transformations $\psi(x) \to e^{i\theta(x)}\psi(x)$, $\psi^\dagger(x) \to \psi^\dagger(x)e^{-i\theta(x)}$, $A_\mu \to A_\mu + \partial_\mu\theta(x)$, and, formally, under the axial phase rotation $\psi(x) \to e^{i\gamma_5\alpha}\psi(x)$, $\psi^\dagger(x) \to \psi^\dagger(x)e^{-i\gamma_5\alpha}$, where the fifth gamma matrix $\gamma^5 = i\gamma^0\gamma^1$ has been introduced.

At the classical level the above symmetries lead to conservation laws for the vector and axial currents, defined respectively as $j^\mu(x) = \bar{\psi}(x)\gamma^\mu\psi(x)$ and $j_5^\mu(x) = \bar{\psi}(x)\gamma^5\gamma^\mu\psi(x)$. At the quantum level, both currents cannot be simultaneously conserved. If the chosen regularization is gauge invariant, ensuring the conservation of the vector current, then the axial current obeys the anomaly Eq. [5]

$$\partial_\mu j_5^\mu(x) = \frac{e_c^2}{2\pi} \epsilon^{\mu\nu} F_{\mu\nu}(x) , \tag{2}$$

where $\epsilon^{\mu\nu}$ is the two-dimensional Levi-Civita tensor. Furthermore, the correlation functions of the model do not exhibit axial symmetry.

In the spectrum of the Schwinger model, only neutral particles exist. Notably, the bosonic bound state created by the axial current operator, i.e., $\partial^\mu\phi(x) = \sqrt{\pi}\, j_5^\mu(x)$, appears as a free pseudo-scalar with mass $m_\phi = e_c/\sqrt{\pi}$.

Using the bosonic field ϕ, the action reads as

$$S = \int d^2x \left(\frac{1}{2} \partial_\mu\phi\partial^\mu\phi + \frac{e_c^2}{2\pi}\phi^2 \right) , \tag{3}$$

and the correlation functions for the vector and axial currents are obtained from the correlation functions of derivatives of the corresponding bosonic fields. The coupling constant e_c has the dimension of a mass, and the model is super-renormalizable.

The Schwinger model has often been used as a field theory where methods for lattice gauge theories, including the strong coupling expansion, can be tested and compared with exact results of the continuum model [21].

2.1 Lattice Schwinger Model

We now revisit the lattice quantization of the Schwinger model to show how the effects of the anomaly manifest through spontaneous symmetry breaking. A lattice Hamiltonian and constraint reproducing the corresponding ones in the continuum limit are:

$$H_S = \frac{e_L^2 a}{2} \sum_x E_x^2 - \frac{it}{2a} \sum_x (\psi_{x+1}^\dagger e^{iA_x}\psi_x - \psi_x^\dagger e^{-iA_x}\psi_{x+1}) , \tag{4}$$

$$E_x - E_{x-1} + \psi_x^\dagger \psi_x - \frac{1}{2} = 0\,. \tag{5}$$

Here, the fermionic fields are defined on the sites $x = -\frac{N}{2}, -\frac{N}{2}+1, ..., \frac{N}{2}$ of a one-dimensional lattice of length N and lattice spacing a. The gauge and the electric fields, A_x and E_x, respectively, are defined on the links joining two adjacent sites x and $x+1$. The length N is taken as an even integer and, when N is finite, periodic boundary conditions are considered. In the latter case, the continuum limit corresponds to the Schwinger model on a circle [11].

The coefficient t in the hopping term of Eq. (4) plays the role of the speed of light on the lattice. In the naive continuum limit, $e_L = e_c$ and $t = 1$. However, it is conveniennt to leave e_L and t as parameters that can be adjusted to match the lattice values of quantities such as the mass gap and the chiral condensate with those known in the continuum.

Both the Hamiltonian and gauge constraint exhibit the following discrete symmetries: parity P, discrete axial symmetry Γ and charge conjugation C, see, e.g., [10]. When $A_x = 0$, the spectrum of the hopping Hamiltonian in momentum space is $\epsilon(pa) = t \sin pa$, where $p \in [0, 2\pi/a]$. In the low-energy limit, this represents a massless relativistic spectrum for excitations near the Fermi level. The two intersections of the energy band with the Fermi level provide the continuum right and left moving massless fermions. The electron ground state is invariant under charge conjugation only when the Fermi level is at $\epsilon(p_F) = 0$, i.e., when exactly half of the fermion states are filled and $\sum_x \langle \rho(x) \rangle = 0$. In the remainder of this Section, we focus solely on this scenario, known as the half-filling case.

The lattice Schwinger model is equivalent to a one-dimensional quantum Coulomb gas on the lattice. To discuss this equivalence, it is convenient to employ the Coulomb gauge $A_x = A$. By eliminating the non-constant electric field and using the gauge constraint, the following effective Hamiltonian

$$H_S = H_u + H_p \equiv \left[\frac{e_L^2}{2N} E^2 + \frac{e_L^2 a}{2} \sum_{x,y} \rho(x) V(x-y) \rho(y) \right] + \left[-\frac{it}{2a} \sum_x (\psi_{x+1}^\dagger e^{iA} \psi_x - \psi_x^\dagger e^{-iA} \psi_{x+1}) \right] \tag{6}$$

is obtained, where the charge density is defined as $\rho(x) \equiv \psi_x^\dagger \psi_x - \frac{1}{2}$, and the potential

$$V(x-y) \equiv \frac{1}{N} \sum_{n=1}^{N-1} e^{i 2\pi n (x-y)/N} \frac{1}{4 \sin^2 \frac{\pi n}{N}} \tag{7}$$

represents the Fourier transform of the inverse Laplacian on the lattice for non-zero momentum. Here, the constant electric field is normalized so that $[A, E] = i$. The constant modes of the gauge field decouple in the thermodynamic limit $N \to \infty$.

The gauge fixed Hamiltonian (6) can also be represented as a quantum spin model using the Jordan-Wigner transformation [22] mapping the operator ψ's to spin S's. Consequently, the Hamiltonian is

$$H = \frac{t}{2a} \sum_{x} \left[e^{iA} S^+(x+1) S^-(x) + e^{-iA} S^-(x+1) S^+(x) \right] + \frac{e_L^2 a}{2} \sum_{x,y} S^3(x) V(x-y) S^3(y) + \frac{e_L^2 E^2}{2N} . \tag{8}$$

For now, we disregard the constant modes E and A. The first term in the Hamiltonian is the quantum XY model, known to have a disordered ground state (the exact solution of this model can be obtained by the Jordan-Wigner transformation). The second term is a long-range Ising interaction [23, 24] resulting from the infinite range of the Coulomb interaction. When both terms are present in the Hamiltonian, they compete with each other: the first one favors disorder, while the second one promotes order.

3 The Two-Flavor Schwinger Model

In this Section we discuss the $SU(2)$-flavor lattice Schwinger model in the Hamiltonian formalism using staggered fermions. The presence of the continuum internal isospin symmetry makes the model considerably more interesting than the one-flavor case; the spectrum is richer, exhibiting also massless excitations. Furthermore, the chiral symmetry breaking pattern differs from the one-flavor case. One is able to show [25, 26] that the strong coupling limit of the two-flavor lattice Schwinger model is mapped onto the one-dimensional spin-1/2 quantum Heisenberg antiferromagnet. The ground state of the antiferromagnetic chain and the complete spectrum have been known since many years [22].

The action of the $1+1$-dimensional electrodynamics with two charged Dirac spinor fields is

$$S = \int d^2x \left[\sum_{a=1}^{2} \overline{\psi}_a (i\gamma_\mu \partial^\mu + \gamma_\mu A^\mu) \psi_a - \frac{1}{4e_c^2} F_{\mu\nu} F^{\mu\nu} \right] . \tag{9}$$

The theory has an internal $SU_L(2) \otimes SU_R(2)$-flavor isospin symmetry (L and R standing for left and right); the Dirac fields are an isodoublet whereas the electromagnetic field is an isosinglet.

A lattice Hamiltonian and constraint are

$$H_S = \frac{e^2 a}{2} \sum_{x=1}^{N} E_x^2 - \frac{it}{2a} \sum_{x=1}^{N} \sum_{a=1}^{2} \left(\psi_{a,x+1}^\dagger e^{iA_x} \psi_{a,x} - \psi_{a,x}^\dagger e^{-iA_x} \psi_{a,x+1} \right) , \tag{10}$$

$$E_x - E_{x-1} + \psi^\dagger_{1,x}\psi_{1,x} + \psi^\dagger_{2,x}\psi_{2,x} - 1 = 0\,. \tag{11}$$

The fermion fields are defined on the sites, $x = 1, ..., N$, the gauge and electric fields, A_x and E_x, on the links $[x; x+1]$, N is an even integer and, when N is finite it is convenient to impose periodic boundary conditions. When N is finite, the continuum limit is the two-flavor Schwinger model on a circle. The coefficient t of the hopping term in (10) plays the role of the lattice light speed. In the naive continuum limit, $e_L = e_c$ and $t = 1$.

The Hamiltonian and gauge constraint exhibit the discrete symmetries parity P, axial symmetry Γ, charge conjugation C and as well well G-parity: $A_x \to -A_{x+1}$, $E_x \to -E_{x+1}$, $\psi_{1,x} \to \psi^\dagger_{2,x+1}$, $\psi^\dagger_{1,x} \to \psi_{2,x+1}$, $\psi_{2,x} \to -\psi^\dagger_{1,x+1}$ $\psi^\dagger_{2,x} \to -\psi_{1,x+1}$.

The lattice two-flavor Schwinger model is equivalent to a one dimensional quantum Coulomb gas on the lattice with two kinds of particles. To see this one can fix the gauge, $A_x = A$ (Coulomb gauge). Eliminating the non-constant electric field and using the gauge constraint, one obtains the effective Hamiltonian

$$H_S = H_u + H_p \equiv \left[\frac{e_L^2}{2N}E^2 + \frac{e_L^2 a}{2}\sum_{x,y}\rho(x)V(x-y)\rho(y)\right]$$
$$+\left[-\frac{it}{2a}\sum_x\sum_{a=1}^{2}(\psi^\dagger_{a,x+1}e^{iA}\psi_{a,x} - \psi^\dagger_{a,x}e^{-iA}\psi_{a,x+1})\right], \tag{12}$$

where the charge density is $\rho(x) = \psi^\dagger_{1,x}\psi_{1,x} + \psi^\dagger_{2,x}\psi_{2,x} - 1$, and the potential is again given by

$$V(x-y) = \frac{1}{N}\sum_{n=1}^{N-1} e^{i2\pi n(x-y)/N}\frac{1}{4\sin^2\frac{\pi n}{N}}\,. \tag{13}$$

The constant modes of the gauge field decouple in the thermodynamic limit $N \longrightarrow \infty$.

4 Quantum Simulation of the Schwinger Model

A significant amount of work surged in the last decade on the quantum simulation of interacting field theories [6, 27, 28]. Encoding gauge symmetry in quantum simulation experiments is the focal issue in quantum simulation of gauge theories. The Schwinger model, or variations of it, has been one of the prime targets of the community and has been largely used as a case study. This activity went along with the remarkable progress in the last years associated with several experiments that have realized a variety of modified versions of the model, with different degrees of complexity.

There are several strategies designed to achieve gauge symmetry or to encode an effective description of the model. We refer to [6, 27, 28] for useful material and reviews on quantum simulations of the Schwinger model and beyond. Here for the sake of completeness, we will remind one of the main approaches as the literature has investigated so far: energy penalty. We start with a description of the one-flavor case and then we briefly discuss how these can be generalized for the two-flavor case.

In the case of energy penalty, gauge symmetry emerges in the low-energy sector of the theory. While the Hilbert space dimension at any matter site is 2, the Hilbert space dimension of a single gauge link is infinite. Quantum link models (QLM) [18, 29] give a way of making this Hilbert space finite while preserving gauge invariance. We will start with a brief review of QLM and show how these models can emerge in the low-energy sector of a theory of one fermionic species coupled to two bosonic degrees of freedom, see [28, 30, 31] for more references.

In the second case, gauge fields are integrated out. One advantage of this approach is that no type of truncation is necessary. After removing the gauge degrees of freedom only an effective theory of one fermionic (or spin) degree of freedom per site is left out. The main price to pay is that the effective Hamiltonian is no longer local.

In the rest of the Section we shall denote respectively by c's and b's the fermionic and bosonic operators associated to the (non-relativistic) fermions and bosons with which the quantum simulation is performed. When two species of bosons are needed, they will be denoted by $b^{(1)}, b^{(2)}$. Moreover, for the one-flavor Schwinger model the abbreviation $G_x = E_x - E_{x-1} - c_x^\dagger c_x + \frac{1-(-1)^x}{2}$ will be used. The operators G_x are the generators of gauge transformations.

QLMs [18, 29] ensure that gauge invariance is preserved by maintaining the commutation relations and sacrificing unitarity through the identification with spins

$$U_x \to S_x^+, \quad U_x^\dagger \to S_x^-, \quad E_x \to S_x^z. \tag{14}$$

In this framework, the local Hilbert space becomes finite-dimensional, more precisely it has dimension $(2S+1)$ for a quantum link of spin-S. The operators U_x, previously unitary operators, no longer commute: $[U_x, U_y^\dagger] = 2\delta_{xy}E_x$.

We now discuss the effective low-energy gauge theory using the energy penalty approach. The idea behind this approach consists on building a Hamiltonian which does not prohibit the symmetry violation from occurring, but instead penalizes it with a large energy. More concretely, let suppose one wants to implement a set of symmetries that has a respective set of generators $\{G_x\}$ commuting with each other $\left[G_x, G_y\right] = 0$. Let H_0 be a Hamiltonian which does not respect these symmetries. Then one constructs the following Hamiltonian:

$$H = H_0 + \Gamma \sum_x G_x^2, \tag{15}$$

where Γ is a large energy scale (much larger than the energy scales involved in H_0). Since G_x are Hermitian G_x^2 have non-negative eigenvalues. One can choose the lowest eigenvalue to be zero by an appropriate definition of G_x. Then, at low energy ($\ll \Gamma$), the states will respect approximately the condition $G_x |\psi\rangle \simeq 0$. It is then possible to construct an effective Hamiltonian, valid in low energy, which will respect the symmetries generated by $\{G_x\}$.

Let G be the projector operator on the subspace of the total Hilbert space obeying $G_x |\psi\rangle = 0$ and let $P = 1 - G$. Then the low-energy Hamiltonian can be written as:

$$H_{eff} = G H_0 G - \frac{1}{\Gamma} G H_0 P \frac{1}{\sum_x G_x^2} P H_0 G + \mathcal{O}\left(\Gamma^{-2}\right) , \tag{16}$$

which respect the symmetries. Within this framework, an effective Abelian gauge theory can be constructed.

We now address how this can be used to simulate the Schwinger model. The fermionic matter fields will be described by fermionic species. For gauge fields one can use two species of bosons to write

$$U_x = b_x^{(2)\,\dagger} b_x^{(1)}, \; E_x = \frac{1}{2}\left(b_x^{(2)\,\dagger} b_x^{(2)} - b_x^{(1)\,\dagger} b_x^{(1)}\right) . \tag{17}$$

Each link is loaded with a total of $2S$ bosons where S is a half-integer. Then one has the desired representation for the quantum links in terms of atomic variables. In the case of atomic quantum simulations one can build a one-dimensional optical lattice where fermions are allowed to hop among lattice points and in each link there are a total of $2S$ bosons. The target Hamiltonian with Schwinger bosons can be written as

$$\begin{aligned} H = -t \sum_x \left(c_x^\dagger b_x^{(\bar{\sigma})\,\dagger} b_x^{(\sigma)} c_{x+1} + \text{h.c.}\right) + m \sum_x (-1)^x c_x^\dagger c_x \\ + \frac{g^2}{8} \sum_x \left(b_x^{(2)\,\dagger} b_x^{(2)} - b_x^{(1)\,\dagger} b_x^{(1)}\right)^2 , \end{aligned} \tag{18}$$

(using the convention $\sigma = 1$ and $\bar{\sigma} = 2$).

The last two terms can be, in principle, implemented directly properly tuning the interactions between the bosons and the potential for the fermions. The first term, instead, is a correlated hopping between bosons and fermions, less easy to implement. Furthermore the terms like $b_x^{(\bar{\sigma})\,\dagger} b_x^{(\sigma)}$ and $c_x^\dagger c_{x+1}$, which are not gauge invariant, must be suppressed. This is solved by the energy penalty approach. In general the non-gauge invariant Hamiltonian with the features described has then the form:

$$\begin{aligned} H_0 = &-\sum_x \left[t_F \left(c_x^\dagger c_{x+1} + \text{h.c.}\right) - t_B \left(b_x^{(2)\,\dagger} b_x^{(1)} + \text{h.c.}\right)\right] \\ &+ \sum_x \left(v_x^F c_x^\dagger c_x + \sum_\sigma v_x^{B\sigma} b_x^{(\sigma)\,\dagger} b_x^{(\sigma)}\right) + U \sum_x \left(b_x^{(2)\,\dagger} b_x^{(2)} - b_x^{(1)\,\dagger} b_x^{(1)} .\right)^2 \end{aligned} \tag{19}$$

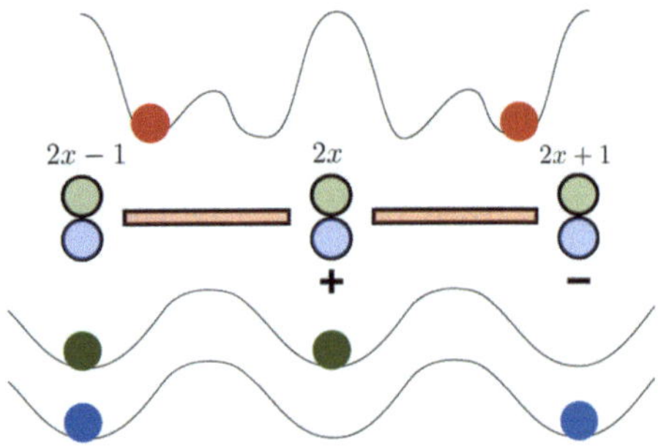

Fig. 1 Superlattice configurations for the bosonic and fermionic species towards the quantum simulation of the two-flavor Schwinger model. When compared to the one-flavor case, another fermionic species must be added that should also respect the correlated hopping with the bosons on the links

Using the generators for the $U(1)$ to penalize states out of the physical sector, we can write

$$H = H_0 + \Gamma \sum_x \left(E_x - E_{x-1} - c_x^\dagger c_x + \frac{1 - (-1)^x}{2} \right)^2 , \tag{20}$$

showing how the Schwinger model may emerge from the non-relativistic setup. It is crucial that one has access to the interactions that are introduced on the last term corresponding to the energy penalty.

The inclusion of extra flavors can be incorporated in both descriptions above. In the case of energy penalty, this entails introducing a new fermionic species. This implies further fine-tuning concerning fermion-fermion interactions (across species) and fermion-boson interactions with the new species. For the case of integration of gauge fields, one will have an effective spin ladder with long-range interactions. For the energy penalty scheme, there would be an extra fermionic species, which should be added to (19). The charge density at a point x is given by

$$\rho_x = c_{1x}^\dagger c_{1x} + c_{2x}^\dagger c_{2x} - 1 + (-1)^x . \tag{21}$$

As a consequence, (20) should be replaced by

$$H = H_0 + \Gamma \sum_x (E_x - E_{x-1} - \rho_x)^2 . \tag{22}$$

This will give rise to new types of interaction to the Hamiltonian but the principles are analogous. A representation of this case is presented in Fig. 1.

Acknowledgements The authors acknowledge discussions with F. Berruto, M. Burrello, A. Celi, M. Dalmonte, N. Defenu, M. C. Diamatini, D. Giuliano, G. Grignani, E. Langmann, L. Lepori, M. K. Marinković, G. Mussardo, A. Papa and U.-J. Wiese for discussions and collaborations on the different topics discussed in this chapter. Special thanks go of course to Gordon W. Semenoff for a lot of inspiring discussions on lattice gauge theories and many other topics.

References

1. J. Schwinger, Gauge invariance and mass. Phys. Rev. **125**, 397–398 (1962)
2. J. Schwinger, Gauge invariance and mass. ii. Phys. Rev. **128**, 2425–2429 (1962)
3. R. Arouca, A. Cappelli, T.H. Hansson, Quantum field theory anomalies in condensed matter physics. SciPost Phys. Lect. Notes, 62 (2022)
4. Y. Frishman, J. Sonnenschein. *Non-Perturbative Field Theory*. Cambridge University Press (2009)
5. S. Coleman, R. Jackiw, L. Susskind, Charge shielding and quark confinement in the massive schwinger model. Annals Phys. **93**(1), 267–275 (1975)
6. U.-J. Wiese, Ultracold quantum gases and lattice systems: quantum simulation of lattice gauge theories. Annalen der Physik **525**(10–11), 777–796 (2013)
7. J. Lin, D. Luo, X. Yao, P.E. Shanahan, Real-time dynamics of the schwinger model as an open quantum system with neural density operators (2024)
8. T. Angelides, L. Funcke, K. Jansen, S. Kühn, Computing the mass shift of wilson and staggered fermions in the lattice schwinger model with matrix product states. Phys. Rev. D **108**, 014516 (2023)
9. L. Funcke, T. Hartung, K. Jansen, S. Kühn, Review on quantum computing for lattice field theory. PoS, LATTICE2022:228 (2023)
10. F. Berruto, G. Grignani, G.W. Semenoff, P. Sodano, Chiral symmetry breaking on the lattice: A study of the strongly coupled lattice schwinger model. Phys. Rev. D **57**, 5070–5083 (1998)
11. J.E. Hetrick, Y. Hosotani, Qed on a circle. Phys. Rev. D **38**, 2621–2624 (1988)
12. J. Kogut, L. Susskind, Hamiltonian formulation of wilson's lattice gauge theories. Phys. Rev. D **11**(2), 395 (1975)
13. R. Dempsey, I.R. Klebanov, S.S. Pufu, B. Zan, Discrete chiral symmetry and mass shift in the lattice hamiltonian approach to the schwinger model. Phys. Rev. Res. **4**, 043133 (2022)
14. G. Magnifico, M. Dalmonte, P. Facchi, S. Pascazio, F.V. Pepe, E. Ercolessi, Real time dynamics and confinement in the $\mathbb{Z}_n$ schwinger-Weyl lattice model for 1+1 QED. Quantum **4**, 281 (2020)
15. S. Coleman, More about the massive schwinger model. Annals Phys. **101**(1), 239–267 (1976)
16. H. Georgi, Automatic fine-tuning in the two-flavor schwinger model. Phys. Rev. Lett. **125**, 181601 (2020)
17. R. Dempsey, I.R. Klebanov, S.S. Pufu, B.T. Søgaard, B. Zan, Phase diagram of the two-flavor schwinger model at zero temperature. Phys. Rev. Lett. **132**, 031603 (2024)
18. D. Horn, Finite matrix models with continuous local gauge invariance. Phys. Lett. B **100**(2), 149–151 (1981)
19. F. Berruto, *Strongly Coupled Lattice Gauge Theories and Antiferromagnetic Spin Systems*. PhD thesis, University of Perugia, available at arXiv:hep-th/9902036 (1998)
20. J.C. Pinto Barros, *Field and Gauge Theories with Ultracold Gauge Potentials and Fields*. Ph.D. thesis, SISSA (available at https://iris.sissa.it/handle/20.500.11767/57731#.XcQtPk6YWh4) (2017)
21. T. Banks, L. Susskind, J. Kogut, Strong-coupling calculations of lattice gauge theories: (1 + 1)-dimensional exercises. Phys. Rev. D **13**, 1043–1053 (1976)
22. V.E. Korepin, N.M. Bogoliubov, A.G. Izergin, Quantum inverse scattering method and correlation functions, pp. xv–xx. Cambridge University Press
23. N. Defenu, A. Codello, S. Ruffo, A. Trombettoni, Criticality of spin systems with weak long-range interactions. J. Phys. A Mathemat. Theoret. **53**(14), 143001 (2020)
24. N. Defenu, T. Donner, T. Macrì, G. Pagano, S. Ruffo, A. Trombettoni, Long-range interacting quantum systems. Rev. Mod. Phys. **95**, 035002 (2023)
25. F. Berruto, G. Grignani, G.W. Semenoff, P. Sodano, Spectrum of the two-flavor schwinger model from the heisenberg spin chain. Phys. Rev. D **59**, 034504 (1999)
26. F. Berruto, G. Grignani, G.W. Semenoff, P. Sodano, On the correspondence between the strongly coupled 2-flavor lattice schwinger model and the heisenberg antiferromagnetic chain. Annals Phys. **275**(2), 254–296 (1999)

27. *Strongly Coupled Field Theories for Condensed Matter and Quantum Information Theory: Proceedings, International Institute of Physics, Natal, Rn, Brazil, 2–21 August 2015*. Springer International Publishing (2020)
28. C. Jad, M. Aidelsburger, P. Hauke, F. Grusdt, B. Yang, Cold-atom quantum simulator of gauge theories. Nat. Comm. **21**, 25 (2025)
29. P. Orland, D. Rohrlich, Lattice gauge magnets: local isospin from spin. Nuclear Phys. B **338**(3), 647–672 (1990)
30. P. Fontana, J.C. Pinto Barros, A. Trombettoni, Quantum simulator of link models using spinor dipolar ultracold atoms. Phys. Rev. A **107**, 043312 (2023)
31. P. Fontana, *Quantum simulations of gauge theories and topological phases*. Ph.D. thesis, SISSA (available at https://iris.sissa.it/handle/20.500.11767/129710) (2022)

Two Toy Spin Chain Models of Decoherence

P. C. E. Stamp and Zhen Zhu

Abstract We solve for the decoherence dynamics of two models in which a simple qubit or 'Central Spin' couples to a bath of spins; the bath is made from a chain of spins. In Model 1, the bath spins are Ising spins; in Model 2, they are coupled by transverse spin-spin interactions, and the chain supports spin waves. We look at (i) the case where the Hamiltonian is static, with a constant system/bath coupling, and (ii) where this coupling varies in time.

1 Introduction

The dynamics of 'decoherence', ie., the gradual entanglement of a quantum system with its environment—is of considerable subtlety. In the simplest picture, one imagines some sort of exponential decay in time of off-diagonal matrix elements in a reduced density matrix for the central system. For a single central spin or 'qubit', this would occur on a timescale T_2, distinct from the time T_1 for the decay of the diagonal matrix elements.

This simple picture is only correct under very restrictive conditions - it is correct, eg., for a central spin coupled to random noise. Solutions for the dynamics of the 'spin-boson' model [1, 2] (where the central qubit couples to an oscillator bath) or the 'central spin' model [3] (where the central spin couples to a bath of spins) show much more complicated behaviour. In the central spin model, one can see quite complex non-monotonic behaviour in time of the density matrix elements. If one then looks at the correlations between the central system and the environmental variables, extremely complex features emerge, in which entanglement correlators at different levels exchange information in interesting ways [4, 5].

P. C. E. Stamp (✉)
Department of Physics and Astronomy, University of British Columbia, Vancouver, BC, Canada
e-mail: stamp@phas.ubc.ca

Z. Zhu
Pacific Institute of Theoretical Physics, University of British Columbia, Vancouver, BC, Canada

M. Paranjape et al. (eds.), *Proceedings of Gravity, Strings and Fields*,
CRM Series in Mathematical Physics,
https://doi.org/10.1007/978-3-031-91266-5_19

Even more complex features can emerge when the central system goes beyond a simple two-level 'qubit' system, to some object hopping on some lattice. This is a topic of considerable interest to those interested in the decoherence dynamics of quantum information processing systems (a huge topic), in the dynamics of quantum walks [6–8], in the well known problem of motion of holes in insulating magnetic systems [8–12], and anyone interested in general problems in quantum diffusion.

Quite generally the models that are employed fall into 2 classes, viz.,

(i) *Oscillator Bath Models*: In such models the environment is modelled as a set of independent oscillators [2]; these are supposed to represent extended modes (like phonons, photons, electron-hole pairs, gravitons, etc.). For a system containing N such oscillators, the coupling of the system to each oscillator is $\propto N^{-1/2}$, so that for large N (or for $N \to \infty$) the effect of the bath on the central system is independent of N, and one can assume an equilibrium bath at some temperature T.
(ii) *Spin Bath Models*: The environment is now modelled as a set of 'pseudospin' degrees of freedom [3], each possessing a finite-dimensional Hilbert space (often a 2-dimensional space, so that the environment is a set of 2-level systems). These represent localized modes (like defects, nuclear or paramagnetic spins, or dynamic impurities in the system). In this case the coupling between the central system and each bath 'spin' may not depend on N at all, and certainly there is no requirement that it be $\propto N^{-1/2}$. One cannot assume that this bath is in equilibrium.

For an elementary discussion of the difference between these two baths, see, eg., Refs. [13, 14]. A key difference is that for oscillator baths, decoherence is closely tied to dissipation: one expects a fluctuation/dissipation connection to exist. However for spin baths, no such necessary connection exists—in fact, one can have very large decoherence with almost no dissipation. This makes spin baths much more dangerous for, eg., quantum computation (Fig. 1).

In this paper we wish to show some very basic features of decoherence dynamics, by choosing two simple 'toy' models. The models are of some pedagogical interest, just because they are so simple—indeed, they are sufficiently simple that in one case we easily find an exact solution, and in the other case a very accurate solution in

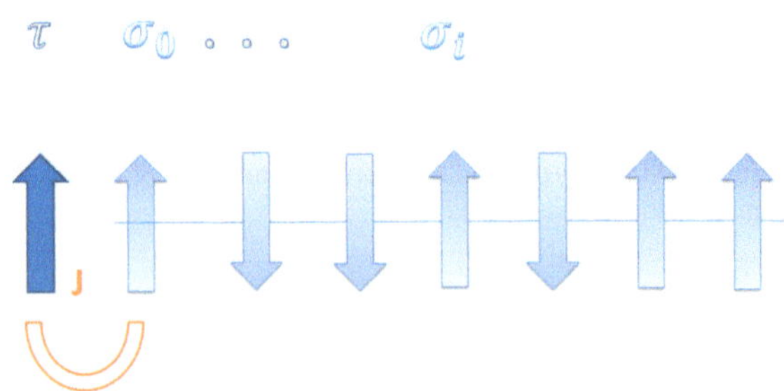

Fig. 1 Illustration of our spin chain model. A central spin couples to a spin chain

the regime of interest. In this short contribution we first introduce the models, and give the details of their solutions, and then make some observations based on these solutions. It will be seen that even in these simple models (about as simple as one can get) there are several interesting features, which persist in more complicated models.

It will be clear that much more can be said, which we have no space for in this short communication. We will also have no space to discuss the application of models of decoherence that are similar to the ones discussed here. For recent discussions of both anomalous diffusion and localization, which can occur when the decoherence is non-Markovian, work on optical lattices is of some interest [15–18]; models rather close in spirit to the models discussed here are examined in Refs. [7, 8, 19]. More general work on entanglement and decoherence in condensed matter systems is reviewed by LaFlorencie [20].

In both models, the environment is modelled as a 1-D spin chain $\{\sigma_i\}$, and the central system is a single spin τ, which is coupled to the first spin in the chain. The models differ only in the form of the bath Hamiltonian.

2 Model 1

The first model has an environment or bath with 'Ising' Hamiltonian, coupled transversely to a qubit, via a single coupling J between the qubit and the first spin on the spin chain. The Hamiltonian is:

$$H = J\tau^y\sigma_0^y + V\sum_{i=0}^{N-1}\sigma_i^z\sigma_{i+1}^z, \tag{1}$$

where τ^y is the central spin operator and the $\{\sigma_i\}$ are operators for N bath spins; J is the coupling between the central qubit and the first bath spin, and NV is the band width of the bath spins. We do *not* use periodic boundary conditions here.

The central spin has the reduced density matrix

$$\begin{pmatrix} |a|^2 & a^*b\kappa(t) \\ ab^*\kappa^*(t) & |b|^2 \end{pmatrix}. \tag{2}$$

This model can be solved exactly. Since H commutes with τ^y, the diagonal component (represented in τ^y eigenstates) of the reduced density matrix does not change in time. We write the off-diagonal components $\rho_{12}(t)$ and $\rho_{21}(t)$ as

$$\rho_{12}(t) = \kappa(t)\rho_{12}(0), \tag{3}$$

with $\kappa(t)$ being the "decoherence factor"

$$\kappa = \mathrm{tr}_b\left(e^{-iH^+t}\rho_b e^{-iH^-t}\right), \tag{4}$$

where $H^{\pm}$ are the block Hamiltonians in the $\tau^y = \pm 1$ subspaces respectively:

$$H^{\pm} = \pm J\sigma_0^y + V\sum_{i=0}^{N-1} \sigma_i^z \sigma_{i+1}^z \tag{5}$$

The solution to this model is

$$\begin{aligned}\kappa(t) &= \frac{V^2 + J^2 \cos(2\sqrt{1 + J^2/V^2}t)}{V^2 + J^2} \\ &= 1 - \frac{2J^2 \sin^2(\sqrt{1 + J^2/V^2}t)}{V^2 + J^2}.\end{aligned} \tag{6}$$

which is clear enough—one is basically dealing with a 2-spin problem, because there is no coupling between the spins on the chain. Thus all correlations and information existing initialy in the qubit are confined to the qubit/bath spin pair.

It is also interesting to obtain an approximate solution to this problem. To do this we use a Jordan-Wigner transformation to diagonalize H, by writing

$$\sigma_i^z = -\prod_{j<i}(1 - 2c_j^\dagger c_j)(c_i^\dagger + c_i), \tag{7}$$

$$\sigma_i^x = 1 - 2c_i^\dagger c_i, \tag{8}$$

$$\sigma_i^y = i\prod_{j<i}(1 - 2c_j^\dagger c_j)(c_i^\dagger - c_i). \tag{9}$$

Substituting this into (1) gives

$$\begin{aligned}H &= iJ\tau^y(c_0^\dagger - c_0) \\ &\quad + V\sum_{i=0}^{N-1}(c_i^\dagger c_{i+1}^\dagger - c_i c_{i+1}^\dagger + c_i^\dagger c_{i+1} - c_i c_{i+1}).\end{aligned} \tag{10}$$

Introducing the Fourier transform $c_{k_n} = \frac{1}{\sqrt{N}}\sum_j c_j e^{ijk_n}$, with $k_n = \frac{2\pi n}{N}$, and the Bogoliubov transformation

$$\gamma_{k_n} = e^{-\frac{ik_n}{2}}\left(\cos\left(\frac{k_n}{2}\right)c_{k_n} - i\sin\left(\frac{k_n}{2}\right)c_{-k_n}^\dagger\right), \tag{11}$$

$$c_{k_n} = \cos\left(\frac{k_n}{2}\right)e^{\frac{ik_n}{2}}\gamma_{k_n} + i\sin\left(\frac{k_n}{2}\right)e^{\frac{ik_n}{2}}\gamma_{-k_n}^\dagger. \tag{12}$$

the Hamiltonian becomes

$$\begin{aligned} H &= \sum_n \left(V(2\gamma_{k_n}^\dagger \gamma_{k_n} - 1) + \frac{iJ}{\sqrt{N}} \tau^y (\gamma_{k_n}^\dagger - \gamma_{k_n}) \right) \\ &= \sum_n H_n, \end{aligned} \tag{13}$$

with $H_n = V(2\gamma_{k_n}^\dagger \gamma_{k_n} - 1) + \frac{iJ}{\sqrt{N}} \tau^y (\gamma_{k_n}^\dagger - \gamma_{k_n})$.

Now the individual Hamiltonians H_n do not commute with each other; we have

$$[H_n, H_{n'}] = -\frac{2J^2}{N} (\gamma_{k_n}^\dagger - \gamma_{k_n})(\gamma_{k_{n'}}^\dagger - \gamma_{k_{n'}}). \tag{14}$$

However, in the weak interaction limit $J/\sqrt{N}V \ll 1$, these commutators can be neglected (see Appendix), and the Hamiltonian (13) becomes block diagonal. In each $\{\gamma_{k_n}^\dagger, \gamma_{k_n}\}$ subspace, we have

$$H_n(\tau^y = +1) = \begin{pmatrix} 1 & \frac{iJ}{\sqrt{N}V} \\ -\frac{iJ}{\sqrt{N}V} & -1 \end{pmatrix} \tag{15}$$

$$H_n(\tau^y = -1) = \begin{pmatrix} 1 & -\frac{iJ}{\sqrt{N}V} \\ \frac{iJ}{\sqrt{N}V} & -1 \end{pmatrix}. \tag{16}$$

Let's assume the bath is initially in a fully mixed state $\rho_B = \frac{1}{2}\mathbf{I}$. It is then straightforward to calculate $\kappa(t)$:

$$\begin{aligned} \kappa &= 1 - 2\sin^2\left(\sqrt{1 + \frac{J^2}{NV^2}}t\right) \frac{J^2}{NV^2 + J^2} \\ &\approx 1 - 2\frac{J^2}{NV^2} \sin^2 t. \end{aligned} \tag{17}$$

so that as $N \to \infty$, we have:

$$\kappa = \lim_{N\to\infty} \left(1 - 2\frac{J^2}{NV^2} \sin^2 t, \frac{J^2}{NV^2}\right)^N = e^{-2J^2 \sin^2 t / V^2}. \tag{18}$$

which gives the correct result up to order J^2/V^2.

We see that κ is periodic in time (see Fig. 2). This is because, as already noted, information cannot flow into the whole spin chain—it is *confined* to the first local bath spin and the central spin.

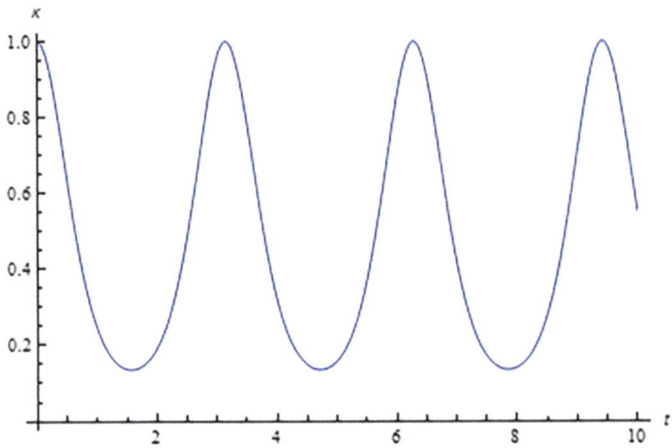

Fig. 2 Decoherence factor $\kappa(t)$ (18) of model 1 as a function of time. There is no real decoherence and κ goes back to 1 periodically

3 Model 2

We now consider another model in which the spin chain is the environment, but in which there are interactions between the bath spins (here, simple transverse spin-spin coupling). The model Hamiltonian is

$$H = J\tau^y\sigma_0^y + V\sum_{i=0}^{N-1}(\sigma_i^+\sigma_{i+1}^- + h.c.). \tag{19}$$

in which spin waves can propagate down the chain. Thus the environment is now a set of oscillators (spin waves); this model is a variant of the spin-boson model.

Again using the Jordan-Wigner transformation, we have

$$H = \sum_n \left(2V\cos k_n\left(\gamma_{k_n}^\dagger\gamma_{k_n} - \frac{1}{2}\right) - \frac{iJ}{\sqrt{N}}\tau^y(\gamma_{k_n}^\dagger - \gamma_{k_n})\right). \tag{20}$$

which has the same structure as (13), except now with bath spectrum $\cos k_n V$ instead of a constant. We now follow the same approximate procedure used for model 1, to get the decoherence factor up to order J^2:

$$
\begin{aligned}
\frac{1}{2}\mathrm{tr}\left(e^{-iH_n^- t}e^{-iH_n^\dagger t}\right) &= 1 - \frac{2\sin^2(\sqrt{V^2\cos^2 k_n + \frac{J^2}{N}}t)}{\cos^2 k_n + \frac{J^2}{NV^2}} \cdot \frac{J^2}{NV^2} \\
&\approx 1 - \frac{2J^2\sin^2(Vt\cos k_n)}{NV^2\cos^2 k_n}; \qquad (N\to\infty) \\
&\approx e^{-\frac{2J^2\sin^2(Vt\cos k_n)}{NV^2\cos^2 k_n}}.
\end{aligned}
\tag{21}
$$

After including all the H_ns, we finally get

$$
\kappa(t) = \exp\left(-\sum_n \frac{2J^2\sin^2(Vt\cos k_n)}{NV^2\cos^2 k_n}\right) \tag{22}
$$

If we now take the limit $N\to\infty$, this reduces to

$$
\begin{aligned}
\kappa(t) &\approx \exp\left(-\frac{J^2}{\pi V^2}\int_{-\pi}^{\pi} dk \frac{\sin^2(Vt\cos k)}{\cos^2 k}\right); \qquad (N\to\infty) \\
&= \exp\left(-2J^2t^2\,{}_1F_2\left(\frac{1}{2};\frac{3}{2},2;-V^2t^2\right)\right).
\end{aligned}
\tag{23}
$$

where ${}_1F_2$ is a hypergeometric function.

The resulting behaviour of $\kappa(t)$ is illustrated in Fig. 3. Its long time asymptotic behaviour can still be studied. When $t\to\infty$, we can expand the generalized hypergeometric function as

$$
(Vt)^2\,{}_1F_2\left(\frac{1}{2};\frac{3}{2},2;-V^2t^2\right) = Vt + \frac{1}{2}\sqrt{\frac{1}{\pi Vt}}\cos\left(2Vt - \frac{\pi}{4}\right) + O\left(\frac{1}{(Vt)^{\frac{3}{2}}}\right). \tag{24}
$$

and we see that $\kappa \to e^{-J^2t/V} \to 0$ as $t\to\infty$.

We see that in this model the system decoheres completely as $t\to\infty$; the quantum information is forever lost to the environment, into the spin wave bath. In fact, we are dealing here with a particular species of spin-boson model.

4 Time-Varying System-Bath Coupling

What if we are allowed to switch on or off the coupling between the central system and the bath? One can envisage several such situations - for example, we can simply switch the coupling off, or we can switch it off and then bring it back again, in an effort to reverse the decoherence.

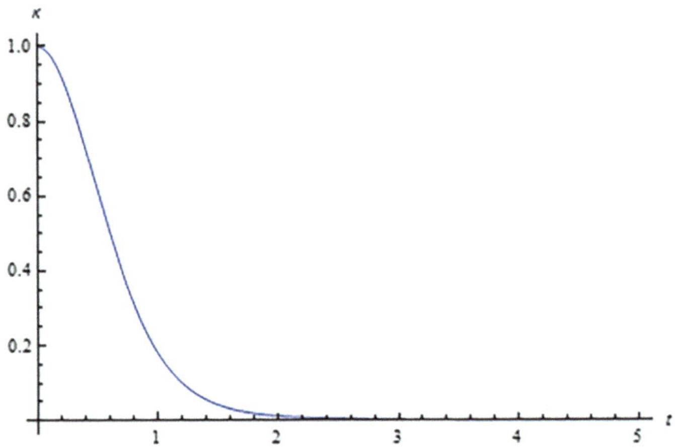

Fig. 3 This is the decoherence factor of model 2 as a function of time. The loss of coherence is clear over time

4.1 Switch-Off Process

Let us assume a time-varying coupling $J(t)$ of form:

$$H = J(t)\tau^y\sigma_0^y + H_{bath}, \tag{25}$$

where $J(t) \to J_0$ as $t \to -\infty$ and $J(t) \to 0$ as $t \to \infty$. We choose a convenient model form for $J(t)$ to be

$$J(t) = \frac{J_0}{e^{kt}+1}. \tag{26}$$

so that $\frac{d}{dt}J(t) = -J_0ke^{kt}/(e^{kt}+1)^2$.

We see that when $k \to 0$, $\frac{d}{dt}J(t)$ will go to zero; when $k \to \infty$, it will become a Heaviside step function with a jump at $t = 0$ (see Fig. 4). These 2 limits describe two complementary processes. The $k = 0$ limit simply adiabatically decouple the system from the bath, whereas the $k \to \infty$ limit suddenly switches off the coupling. For this sudden decoupling case, the result should be the same as what we get from previous sections since the central system stops evolving after the turn-off. But for the adiabatic decoupling case, things are more complicated—we need to solve the problem explicitly—we will do this in the long-time limit.

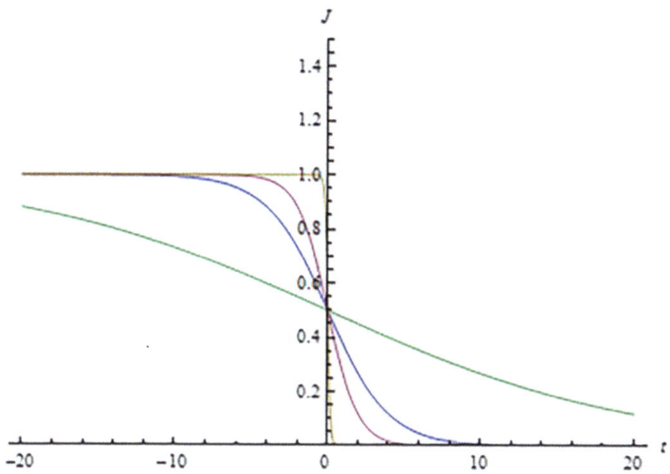

Fig. 4 Figure of the coupling $J(t)$ over time. These lines represent $k = 0.1, 0.5, 1, 10$ respectively

4.2 Time-Dependent Solutions

With the time-dependent $J(t)$ we can follow the same Jordan-Wigner/Bogoliubov procedure as before, and now transform the Hamiltonian to

$$H = \sum_n 2E_n(\gamma_n^\dagger \gamma_n - \frac{1}{2}) + \sum \frac{iJ(t)}{\sqrt{N}}(\gamma_n^\dagger - \gamma_n), \tag{27}$$

where $E_n = V$ for model 1, and $E_n = V \cos k_n$ for model 2. Then we have

$$H_n^\pm = \begin{pmatrix} 2E_n & \mp\frac{iJ(t)}{\sqrt{N}} \\ \pm\frac{iJ(t)}{\sqrt{N}} & 0 \end{pmatrix}. \tag{28}$$

Our goal is to solve this Hamiltonian in each $\{\gamma_n^\dagger|0\rangle, |0\rangle\}$ subspace. A vector $(a(t), b(t))$ in this subspace evolves as

$$-i\begin{pmatrix} \frac{d}{dt}a(t) \\ \frac{d}{dt}b(t) \end{pmatrix} = \begin{pmatrix} 2E_n & \mp\frac{iJ(t)}{\sqrt{N}} \\ \pm\frac{iJ(t)}{\sqrt{N}} & 0 \end{pmatrix}\begin{pmatrix} a(t) \\ b(t) \end{pmatrix}. \tag{29}$$

so that

$$a(t) = \mp\frac{\sqrt{N}}{J(t)}\dot{b}(t), \tag{30}$$

$$\ddot{b}(t) - \left(iE_n + \frac{\dot{J}(t)}{J(t)}\right)\dot{b}(t) + \frac{J^2(t)}{N}b(t) = 0. \tag{31}$$

After some derivation (see Appendix), we get a result for $\kappa(t)$ from this. The long time asymptotic limit $t \to \infty$ of $\kappa(t)$ is then:

$$\kappa(t) = \prod_n \left(1 - \frac{2J^2}{NE_n^2}\right). \tag{32}$$

For the two models we then have:

(i) *Model 1*: Here we have $H_{bath} = V \sum_{i=0}^{N-1} \sigma_i^z \sigma_{i+1}^z$, and $E_n = V$ for all subspaces. Then the decoherence factor is

$$\kappa_1(t) = \left(1 - \frac{2J^2}{NE_n^2}\right)^N = e^{-2J_0^2}. \tag{33}$$

(ii) *Model 2*: Now we have $H_{bath} = V \sum_{i=0}^{N-1} (\sigma_i^+ \sigma_{i+1}^- + h.c.)$, and $E_n = V \cos k_n$ for each subspaces. The decoherence factor is

$$\kappa_2(t) = e^{-\frac{1}{2\pi}\int_0^{2\pi} dk \frac{2J^2}{V^2 \cos^2 k}} \to 0. \tag{34}$$

Thus, in this adiabatically decoupled limit, we find "partial decoherence" for model 1 ($\kappa \to e^{-2J^2}$) and "complete decoherence" for model 2 ($\kappa \to 0$). Apparently, system 1 ends up in a mixed state and the coherence is partially lost.

However, appearances are deceptive in this case. To see this, we can imagine reversing the coupling back to its original value. We therefore consider what happens for a general slow-varying $J(t)$. In this case (31) still holds, and we can consider both models 1 and 2.

(i) *Model 1*: Here again we have $E_n = V$. If the coupling is varying slowly, $\dot{J}(t) \ll J(t)E_n$, we can safely neglect the dJ/dt term in (31) and we have

$$\begin{aligned} b(t) =& C_1 e^{-iVt/2 - iVt/2\sqrt{1+\frac{4J^2(t)}{NV^2}}} + C_2 e^{-iVt/2 + iVt/2\sqrt{1+\frac{4J^2(t)}{NV^2}}}, \\ a(t) =& -\frac{i2C_1\sqrt{N}}{J(t)}\left(V + V\sqrt{1 + \frac{4J^2(t)}{NV^2}}\right) e^{-iVt/2 - iVt/2\sqrt{1+\frac{4J^2(t)}{NV^2}}} \\ &- \frac{i2C_2\sqrt{N}}{J(t)}\left(V - V\sqrt{1 + \frac{4J^2(t)}{NV^2}}\right) e^{-iVt/2 + iVt/2\sqrt{1+\frac{4J^2(t)}{NV^2}}}. \end{aligned} \tag{35}$$

We can see that in this adiabatic limit, the result has no dependence on dJ/dt. The decoherence factor $\kappa_1(t)$ should therefore be solely dependent on the form of $J(t)$. This means that if one slowly changes the coupling strength back to its original value, $\kappa_1(t)$ would go back to its original value regardless of the apparent loss of decoherence we obtained in (33). One can thus recover the full coherence by decoupling the model 1 to its environment and then recoupling it back adiabatically.

The coherence is not truly lost into the environment, but simply encoded there in the entanglement between the central spin and the first environmental spin. It can be transferred back to the central spin simply by reversing the interaction.

(ii) *Model 2*: In model 2 we cannot neglect the $\dot{J}(t)$ term, since $E_n = V \cos k_n$ can be zero. The slow varying condition $dJ/dt \ll J(t)E_n$ breaks down for the modes in the center of the band. Actually it is easy to prove that $\kappa(t)$ must stay at 0 after adiabatically decoupling from the environment, and that never come back to its initial state even if one reverses the interaction. In this case, the coherence is truly lost through the modes near $k_n \approx \frac{\pi}{2}, \frac{3\pi}{2}$, once we take the limit $N \to \infty$.

This result is of course typical of a system coupled to an oscillator bath. One can easily generalize the above considerations to a finite-temperature bath.

5 Discussion

It is always helpful, in discussing some general physical phenomenon, to have simple toy models which illustrate basic features of the physics. Decoherence is of course very general—it has been studied intensively in contexts ranging from black hole physics [21] and infrared features of quantum field theory [22], as well as in a huge variety of condensed matter systems. Understanding decoherence is also key to doing quantum computation.

As noted in the introduction, there exist thorough studies of decoherence in several well-known models [1–3]. Even in a simple central spin model, there are several decoherence mechanisms in play [3]; these mechanisms generalize to more general central systems like particles hopping on rings [23] or moving on hyperlattices [7]. In all these models one finds no connection between decoherence and dissipation, quite unlike the case for oscillator bath models like the spin-boson model [1, 2].

The two models here, taken in conjunction with results for the central spin model [3], make it very clear why this is the case. Let us recall the form of the central spin model Hamiltonian: one has

$$H = H_{CS} + H_{SB}, \tag{36}$$

where the central spin part is

$$H_{CS} = \left[\Delta_o \hat{\tau}_x e^i \sum_k (\phi_k + \alpha_k \cdot \sigma_k) + H.c. \right] \tag{37}$$

$$+(\epsilon_o + \sum_k \lambda^k \cdot \sigma_k)\hat{\tau}_z \tag{38}$$

and the spin bath Hamiltonian is

$$H_{SB} = \sum_k \mathbf{h}_k \cdot \sigma_k + \sum_{k,k'} V_{kk'}^{\alpha\beta} \sigma_k^\alpha \sigma_{k'}^\beta. \tag{39}$$

in which σ_k is again the bath spin operator.

Some of the terms in this Hamiltonian are of no special interest here. The Berry phase coupling ϕ_k and the topological phase α_k (which leads to topological decoherence [3]) naturally arise in the truncation of the central system to a 2-level system, but are usually quite small in realistic systems.

The 'orthogonality blocking' vector coupling λ^k is not so easily dismissed—it leads to 'precessional decoherence', which is often the major source of decoherence for systems at low T. Nevertheless we shall drop it, noting that further work, in which it is included, will be of considerable interest when it comes to applications of these toy models.

Finally, we note that the interaction $V_{kk'}^{\alpha\beta}$ between bath spins is often very small (particularly if the spins are nuclear, or if the bath spins are widely separated in space). On longer time scales it does play an essential role (in creating 'fluctuational decoherence' [3]), but at shorter timescales it can be treated as negligible.

Let us therefore assume that we now drop all these interactions, so that $V_{kk'}^{\alpha\beta} = 0$, and that $\phi_k = \alpha_k = 0$. Then this reduced Central Spin (CS) model is simply a generalization of our 'model 1', in which the central spin now couples simultaneously to all of the bath spins instead of just the first one in the chain. It is known from the solution to this reduced CS model [3] that decoherence exists in it, provided one averages over the bath spin states. However, if we do not make such an average, then an initially pure state gradually distributes itself over the N spins in the bath [5]. The time evolution will be periodic, just as for 'model 1', but now the period is exponentially long (with timescale $\propto \exp N^{1/3}$); after this timescale, the information 'comes back' to reconstitute the original central spin state.

Thus model 1 is just the maximum simplification that one can make of the original CS model. Because model 1 is essentially just a 2-spin problem, we get the simple results given above. If we generalize model 1 to allow coupling of the central qubit to some reduced set of M bath spins, then the basic result of the calculation will now be clear - quantum correlations and entanglement will simply be shared over time between the central qubit and the M bath spins, with entanglement between different bath spins mediated by the central spin. Very accurate solutions for a problem of this kind (which is very relevant to experiment) can be found using entanglement correlator methods [4, 5].

Model 2 is not a simplification of any central spin model; instead, as already noted, it is just a special case of the spin-boson model. If we had wanted to, we could have simply derived the form of the Caldeira-Leggett [1] spectral function $J(\omega)$ for this system, and proceeded from there. Note however that if the number N of spins in the chain is finite, we will also get, in model 2, the quantum analogue of Poincaré recurrences—but in the calculations given above, we have let $N \to \infty$.

In model 2, the spin wave modes with $k_n \approx \frac{\pi}{2}, \frac{3\pi}{2}$ serve the purpose of carrying quantum information away from the central spin.

Obviously there is a lot more one can say about these 2 simple models. It will be of interest, for example, to analyse the higher-spin 'entanglement correlators' [4, 5] in model 2, and to see what happens as one adds fields $\{\mathbf{h}_k\}$ acting on the bath spins, and/or gradually 'switches on' the interaction between the bath spins. Many models are then possible, all of them generalizations of the 2 simple central spin models discussed here.

Finally, we note that one can also generalize these 2 models to deal with a central system in which a particle hops around some site Hamiltonian. This is nothing but a 1-dimensional 'polaron' model - with a spin bath background, it is a type of spin polaron system. It is also related to the original model discussed by Feynman for quantum computation [24]. The central system now has a Hamiltonian

$$H_o = -\sum_{<ij>} \left[t_{ij} c_i^\dagger c_j \, e^{iA_{ij}^0} + H.c. \right] + \sum_j \varepsilon_j c_j^\dagger c_j \tag{40}$$

in which we also admit a Berry phase A_{ij}^0 connected to hopping between nearest-neighbour sites.

The most general coupling to the spin bath then takes the form

$$\begin{aligned} V_{int} = \sum_k^N \Bigg[& \sum_j F_j^k(\sigma_k) \hat{c}_j^\dagger \hat{c}_j \\ & + \sum_{ij} (G_{ij}^k(\sigma_k) \hat{c}_i^\dagger \hat{c}_j + H.c.) \Bigg] \end{aligned} \tag{41}$$

with both diagonal coupling F_j^k and non-diagonal coupling G_{ij}^k of the hoping particle to the k-th bath spin.

We see immediately that if we choose a simple site-diagonal coupling $F^k(\sigma_k)$ on each site, making all these couplings the same for each site, and also let the site energy ε_j be the same for each site, then we have another toy model for which concrete calculations can be done. A related model was treated some time ago [23], in which the line was closed to form a ring.

This work was supported by the National Science and Engineering Council of Canada. We thank Tim Cox for discussions.

Dedication (from PCES): This paper is dedicated to my long-time colleague Gordon Semenoff, whom I first met while I was a postdoc, and he already a professor; at that time I was rather awed by him! Since then he has been a welcome source of advice, wisdom, and friendship, with similar views on many things; and we even once wrote a paper [25] together! I wish him many more years happily carving his own path, in the inimitable way he has done up to now.

6 Appendix

Here we derive two key results used in the text, viz., the effect of the non-commutativity of the $\{H_n\}$ in the discussion of model 1, and the derivation of the solution used for adiabatic switching for both models.

6.1 Neglecting Commutators

In (14), we noticed that the H_n's do not commute with each other; in fact

$$[H_n, H_{n'}] = -\frac{2J^2}{N}(\gamma^{\dagger}_{k_n} - \gamma_{k_n})(\gamma^{\dagger}_{k_{n'}} - \gamma_{k_{n'}}). \tag{42}$$

We can use the Zassenhaus formula [26] to get

$$\begin{aligned} e^{i\sum_n H_n t} &= \prod_n e^{iH_n t} \prod_{n_1>n_2} e^{\frac{t^2}{2}[H_{n_1},H_{n_2}]} \\ &\times \prod_{n_1\geq n_2\geq n_3} e^{\frac{-it^3}{6}([H_{n_1},[H_{n_2},H_{n_3}]]+[[H_{n_1},H_{n_2}],H_{n_3}])}\cdots. \end{aligned} \tag{43}$$

Although the Hamiltonian is not block diagonal in each $\{\gamma^{\dagger}_{k_n}|0\rangle, |0\rangle\}$ subspace, if we restrict our model to the weak interaction region, i.e. $J \ll 1$, we can omit this commutator since it always comes into the final expression in higher order. For example , for the $\prod_{n_1,n_2} e^{\frac{t^2}{2}[H_{n_1},H_{n_2}]}$ term, we could expand it in powers of $1/N$ and take expectation values

$$\begin{aligned} &\mathrm{tr}_b\left(\prod_{n_1>n_2} e^{\frac{t^2}{2}[H^+_{n_1},H^+_{n_2}]} \prod_{n_1,n_2} e^{\frac{t^2}{2}[H^-_{n_1},H^-_{n_2}]}\right) \\ &= \mathrm{tr}_b\left(e^{-\frac{J^2t^2}{N}\sum_{n_1>n_2}(\gamma^{\dagger}_{k_{n_1}}-\gamma_{k_{n_1}})(\gamma^{\dagger}_{k_{n_2}}-\gamma_{k_{n_2}})}\right) \\ &\approx 1 - \frac{J^4t^4}{N^2}N(N-1). \end{aligned} \tag{44}$$

As we can see in the main text this already happens in the J^2 order. Therefore in the weakly interacting region, our calculation can be a good approximation to the exact result.

6.2 Adiabatic Decoupling

The goal is to solve the asymptotic solution to the following equation

$$a(t) = \mp\tfrac{\sqrt{N}}{J(t)}\dot{b}(t) \tag{45}$$

$$\ddot{b}(t) - \left(iE_n + \tfrac{\dot{J}(t)}{J(t)}\right)\dot{b}(t) + \tfrac{J^2(t)}{N}b(t) = 0. \tag{46}$$

We first study the evolution of the state (0, 1). Actually, the state (1, 0) can be treated in the same way just with an opposite sign of $J(t)$ and E_n. Then we have the initial condition

$$b(t_0) = 1 \tag{47}$$

$$\dot{b}(t_0) = 0. \tag{48}$$

Define the following quantity for simplification, keeping in mind that $\frac{J_0^2}{N}$ is very small

$$\epsilon = \frac{J_0^2}{N}. \tag{49}$$

The general solution of the equation (31) is

$$b(t) = C_1 e^{-iE_n t/2 - iE_n\sqrt{1+4\epsilon/E_n^2}/2}(1+e^{kt})^{i\sqrt{\epsilon}/k} f_1(-e^{kt}) \tag{50}$$

$$+C_2 e^{-iE_n t/2 + iE_n\sqrt{1+4\epsilon/E_n^2}/2}(1+e^{kt})^{i\sqrt{\epsilon}/k} f_2(-e^{kt}) \tag{51}$$

Here we have

$$f_1(x) = {}_2F_1\left(1+\frac{iE_n}{2k}+\frac{i\sqrt{\epsilon}}{k}-\frac{iE_n}{2k}\sqrt{1+\frac{4\epsilon}{E_n^2}};\ -\frac{iE_n}{2k}+\frac{i\sqrt{\epsilon}}{k}-\frac{iE_n}{2k}\sqrt{1+\frac{4\epsilon}{E_n^2}};\ 1-\frac{iE_n}{2k}\sqrt{1+\frac{4\epsilon}{E_n^2}};\ x\right)$$

$$f_2(x) = {}_2F_1\left(1+\frac{iE_n}{2k}+\frac{i\sqrt{\epsilon}}{k}+\frac{iE_n}{2k}\sqrt{1+\frac{4\epsilon}{E_n^2}};\ -\frac{iE_n}{2k}+\frac{i\sqrt{\epsilon}}{k}+\frac{iE_n}{2k}\sqrt{1+\frac{4\epsilon}{E_n^2}};\ 1+\frac{iE_n}{2k}\sqrt{1+\frac{4\epsilon}{E_n^2}};\ x\right) \tag{52}$$

We set our initial time $t_0 \to -\infty$, then $e^{kt} \to 0$ and

$$f_{1,2}(-e^{kt}) \to 1. \tag{53}$$

Therefore

$$\begin{aligned} b(t_0) &= 1 \\ &= C_1 e^{-iE_n t_0/2 - iE_n t_0\sqrt{1+4\epsilon/E_n^2}/2} + C_2 e^{-iE_n t_0/2 + iE_n t_0\sqrt{1+4\epsilon/E_n^2}/2}; \end{aligned} \tag{54}$$

$$\begin{aligned}\dot{b}(t_0) &= 1 \\ &= C_1(-iE_n/2 - iE_n\sqrt{1+4\epsilon/E^2}/2)e^{-iE_nt_0/2-iE_nt_0\sqrt{1+4\epsilon/E_n^2}/2} \\ &\quad + C_2(-iE_n/2 + iE_n\sqrt{1+4\epsilon/E_n^2}/2)e^{-iE_nt_0/2+iE_nt_0\sqrt{1+4\epsilon/E_n^2}/2}\end{aligned} \tag{55}$$

Then we can get the constant C_1, C_2 as

$$C_1 = \frac{1}{2}e^{iE_nt_0/2+iE_nt_0\sqrt{1+4\epsilon/E_n^2}/2}(-1+(1+\frac{4\epsilon}{E_n^2})^{-\frac{1}{2}}) \tag{56}$$

$$C_2 = \frac{1}{2}e^{iE_nt_0/2-iE_nt_0\sqrt{1+4\epsilon/E_n^2}/2}(1+(1+\frac{4\epsilon}{E_n^2})^{-\frac{1}{2}}). \tag{57}$$

After we get the solution , we can study its behavior when $t \to \infty$. We use the fact that when $t \to \infty$

$$f_1(-e^{kt}) \to Ae^{iE_nt/2-i\sqrt{\epsilon}t+\frac{itE_n}{2}\sqrt{1+4\epsilon/E_n^2}/2} \tag{58}$$

$$f_2(-e^{kt}) \to Be^{iE_nt/2-i\sqrt{\epsilon}t-\frac{itE_n}{2}\sqrt{1+4\epsilon/E_n^2}/2} \tag{59}$$

with

$$A = \frac{\Gamma(1-\frac{iE_n}{k}\sqrt{1+4\epsilon/E_n^2})\Gamma(1+\frac{iE_n}{k})}{\Gamma(1+\frac{iE_n}{2k}-\frac{i\sqrt{\epsilon}}{k}-\frac{iE_n}{k}\sqrt{1+4\epsilon/E_n^2})} \times \frac{1}{\Gamma(1+\frac{iE_n}{2k}+\frac{i\sqrt{\epsilon}}{k}-\frac{iE_n}{k}\sqrt{1+4\epsilon/E_n^2})} \tag{60}$$

$$B = \frac{\Gamma(1+\frac{iE_n}{k}\sqrt{1+4\epsilon/E_n^2})\Gamma(1+\frac{iE_n}{k})}{\Gamma(1+\frac{iE_n}{2k}-\frac{i\sqrt{\epsilon}}{k}+\frac{iE_n}{k}\sqrt{1+4\epsilon/E_n^2})} \times \frac{1}{\Gamma(1+\frac{iE_n}{2k}+\frac{i\sqrt{\epsilon}}{k}+\frac{iE_n}{k}\sqrt{1+4\epsilon/E_n^2})}. \tag{61}$$

Recalling that $\epsilon = \frac{J_0^2}{N}$ is a very small number, we can expand the solution according in powers of ϵ. The results are as follows:

Zeroth order: if $\epsilon = 0$, then $C_1 \to 0$, $C_2 \to 1$, and

$$B \to \frac{\gamma(1+\frac{iE_n}{k})\gamma(1+\frac{iE_n}{k})}{\gamma(1+\frac{iE_n}{k})\gamma(1+\frac{iE_n}{k})} \to 1. \tag{62}$$

Thus we find that $b(t) \to 1$, which is exactly what we expected.

First order Since $\sqrt{1+\frac{4\epsilon}{E_n^2}} \approx 1 + \frac{2\epsilon}{E^2}$, we have

$$C_1 \to e^{iE_nt_0}(1+\frac{i\epsilon t}{E_n})(-\frac{\epsilon}{E^2}) = -\frac{\epsilon}{E_n^2}e^{iE_nt_0} \tag{63}$$

$$C_2 \to e^{-i\epsilon t_0/E_n^2}(1-\frac{\epsilon}{E_n}) = 1 - \frac{\epsilon}{E_n^2} - \frac{i\epsilon t}{E_n^2}. \tag{64}$$

As a result

$$A \rightarrow \Gamma(1-\frac{iE_n}{k})\Gamma(1+\frac{iE_n}{k}) \times \left[1-\frac{i\epsilon}{kE_n}\left(\frac{\Gamma'(1-iE_n/k)}{\Gamma(1-iE_n/k)}+i\Gamma'(1)\right) + \frac{\epsilon}{E_n^2}(\Gamma'(1)-\Gamma''^2(1))\right]$$
$$B \rightarrow 1-\frac{\Gamma'^2(1+\frac{iE_n}{k})}{\Gamma(1+\frac{iE_n}{k})^2}\frac{\epsilon}{k^2}+\frac{\Gamma''(1+\frac{iE}{k})}{\Gamma(1+i\frac{E_n}{k})}\frac{\epsilon}{k^2}; \tag{65}$$

so that for $b(t)$ we get the behaviour

$$b(t) \rightarrow 1-\frac{\epsilon}{k^2}e^{iE_nt_0}\Gamma(1-\frac{iE_n}{k})\Gamma(1+\frac{iE_n}{k})-\frac{\epsilon}{E_n^2}-\frac{i\epsilon t}{E_n^2}-\frac{\epsilon}{k^2}\frac{\Gamma'^2(1+\frac{iE_n}{k})}{\Gamma(1+\frac{iE_n}{k})^2}+\frac{\epsilon}{k^2}\frac{\Gamma''(1+\frac{iE}{k})}{\Gamma(1+i\frac{E_n}{k})}+O(\epsilon^{\frac{3}{2}}) \tag{66}$$

Then the trace of $\mathrm{tr}(e^{iH_+t}e^{iH_-t})$ is $|b(t)|^2-|a(t)|^2$. Since $|b(t)|^2-|a(t)|^2=1$, $\mathrm{tr}(e^{iH_+t}e^{iH_-t})=2|b(t)|^2-1$. For the adiabatic decoupling we mentioned in the main text, we take $k\rightarrow 0$, then $\frac{d}{dt}J(t)\rightarrow 0$. The system is slowly decoupled from the bath. We use the fact that when $|z|\rightarrow\infty$

$$(\ln\Gamma(z))' = \ln z + \frac{z-1/2}{z} + O(z^{-2}) \tag{67}$$

$$(\ln\Gamma(z))'' = \frac{\Gamma''(z)}{\Gamma(z)} - \left(\frac{\Gamma'(z)}{\Gamma(z)}\right)^2 \tag{68}$$

$$(\ln\Gamma(z))'' = \frac{1}{z} + \frac{1}{2z^2} + O(z^{-3}) \tag{69}$$

We can get the factor $b(t)$ as well as the decoherence factor κ_n for this subspace; one gets

$$|b(t)|^2 \rightarrow 2\left(1-\frac{\epsilon}{E_n^2}\right) \tag{70}$$

and also that

$$\kappa_n(t) = \mathrm{tr}(e^{iH_n^+t}e^{iH_n^-t}) \rightarrow 1-\frac{2J_0^2}{NE_n^2} \tag{71}$$

This is the result used in the main text to get $\kappa(t)$ for models 1 and 2.

References

1. A.J. Leggett et al., Rev. Mod. Phys. **59**, 1 (1987)
2. U. Weiss, *Quantum Dissipative Systems*. World scientific, 4th edn (2012)
3. N.V. Prokof'ev, P. Stamp, Rep. Prog. Phys. **63**, 669 (2000)
4. T. Cox, P. Stamp, Phys. Rev. A **98**, 062110 (2018)
5. T. Cox, Ph.D. Thesis, University of British Columbia (2019); see https://open.library.ubc.ca/soa/cIRcle/collections/ubctheses/24/items/1.0378509
6. S.E. Venegas-Andraca, Quantum Inf. Process. **11**, 1015 (2012)

7. N.V. Prokof'ev, P. Stamp, Phys. Rev. A **74**, 020102 (2006)
8. J. Carlstrom, N.V. Prokof'ev, B. Svistunov, Phys. Rev. Lett. **116**, 247202 (2016)
9. Y. Nagaoka, Phys. Rev. **147**, 392 (1966)
10. W.F. Brinkman, T.M. Rice, Phys. Rev. B **2**, 1324 (1970)
11. G. Ji et al., Phys. Rev. X **11**, 021022 (2021)
12. K. Knakkergaard Nielsen, Phys. Rev. B **106**, 115144 (2022)
13. P. Stamp, A. Gaita-Arino, J. Mat. Chem. **19**, 1718 (2009)
14. P. Stamp, Stud. Hist. Phil. Mod. Phys. **37**, 467 (2006)
15. P.M. Preiss, R. Ma, M.E. Tai, A. Lukin, M. Rispoli, P. Zupancic, Y. Lahini, R. Islam, M. Greiner, Science **347**, 1229 (2015)
16. W. Zheng, N.R. Cooper, Phys. Rev. A **97**, 021601(R) (2018)
17. J. Major, G. Morigi, J. Zakrzewski, Phys. Rev. A **98**, 053633 (2018)
18. G. Afek, N. Davidson, D.A. Kessler, E. Barkai, Rev. Mod. Phys. **95**, 031003 (2023)
19. B. Danacı, I. Yalçınkaya, B. Çakmak, G. Karpat, S.P. Kelly, A.L. Subası, Phys. Rev. A **103**, 022416 (2021)
20. N. LaFlorencie, Phys. Rep. **646**, 1 (2016)
21. D. Danielson, G. Satishchandran, R.M. Wald, Phys. Rev. D **108**, 025007 (2023); Int. J. Mod. Phys. D **31**, 2241003 (2022); and refs therein
22. For a general reference on infrared divergences in both quantum gravity and quantum electrodynamics, see A. Strominger, *Lectures on the Infrared Structure of Gravity and Gauge Theory*. Princeton University Press (2018). For our recent point of view on the subject, see C. DeLisle, J. Wilson-Gerow, P.C.E. Stamp, JHEP **03**, 290 (2021)
23. Z. Zhu, A. Aharony, O. Entin-Wohlman, P. Stamp, Phys. Rev. A **81**, 062127 (2010)
24. R.P. Feynman, Opt. News **11**(2), 11 (1985)
25. M. Hasselfield, Taejin Lee, G.W. Semenoff, P.C.E. Stamp, Ann. Phys. (NY) **321**, 2849 (2006)
26. W. Magnus, Commun. Pure Appl. Mathem. **7**(4), 649 (1954)

Gravity Versus Noncommutative Gauge Theory: A Double Copy Perspective

Richard J. Szabo

Abstract We discuss how Moyal deformations of gauge theories, which arise naturally from open string theory, fit into the paradigm of colour-kinematics duality and the double copy of gauge theory to gravity. Along the way we encounter novel noncommutative scalar field theories with rigid colour symmetry that have no interacting commutative counterparts. These scalar theories offer new perspectives on old ideas that rank one noncommutative gauge theories are gravitational theories. This is rendered explicit in four dimensions where they and their double copy images yield deformations of integrable theories describing the self-dual sectors of Yang-Mills theory and gravity.

1 Introduction

It is an endearing pleasure for me to write this contribution on the occasion of Gordon Semenoff's 70th birthday. Gordon was my MSc and then Ph.D. supervisor at the University of British Columbia from 1990–1995. From him I learned how to become a careful and independent researcher. The present article describes interactions between quantum field theory, string theory, gravity and mathematical physics, which are themes that have permeated Gordon's research interests.

The basic premise of this contribution is the double copy of noncommutative Yang-Mills theory, as it appears in low-energy effective field theories on D-branes in constant B-fields, to gravity, which we describe in Sect. 2. It is based on the lengthy recent paper [1], which is engulfed in the modern language of the homotopy algebraic approach to quantum field theory. In the remaining sections we give a concise summary of some of the main points of that work, in a manner which we hope is more palatable to a broader audience. We skip most technical details, while referring to [1] for all mathematical constructions as well as more thorough and complete citations to the relevant literature.

R. J. Szabo (✉)
Department of Mathematics, Heriot-Watt University Maxwell Institute for Mathematical Sciences, Edinburgh, Scotland
e-mail: R.J.Szabo@hw.ac.uk

M. Paranjape et al. (eds.), *Proceedings of Gravity, Strings and Fields*,
CRM Series in Mathematical Physics,
https://doi.org/10.1007/978-3-031-91266-5_20

2 Double Copying Noncommutative Yang-Mills Theory

KLT Double Copy Relations. The double copy of gauge theory to gravity has its origins in string theory, going back to the mid-1980s with the famous relations between tree-level scattering amplitudes of closed and open strings discovered by Kawai, Lewellen and Tye (KLT) [2]. These relations reflect the topological property that a sphere, representing a closed string worldsheet, can be realised by gluing together two discs, representing open string worldsheets:

In the field theory limit $\alpha' \longrightarrow 0$, the left-hand side represents a tree-level amplitude for graviton scattering while the right-hand side consists of amplitudes of gluons. This results in the famous double copy slogan:

$$\text{Gravity} = (\text{Gauge Theory})^2 . \tag{1}$$

At a more quantitative level, the KLT double copy relations between n-point amplitudes read

$$M_n^{L\otimes R} = \sum_{w,w'} A_n^L(w)\, \boldsymbol{S}_n(w|w')\, A_n^R(w') , \tag{2}$$

where the sums run over orderings of n external gluons, and A_n are colour-stripped amplitudes appearing in the full gluon amplitudes

$$\mathcal{A}_n = \sum_{S_n/\mathbb{Z}_n} \mathrm{Tr}\,(T^{a_1}\cdots T^{a_n})\, A_n(1,\dots,n) \tag{3}$$

expressed as a sum over cyclic orderings of external particles.

The quantity $\boldsymbol{S}_n(w|w')$ which glues left and right gauge theory amplitudes together is called the *KLT momentum kernel*. It was realised almost 30 years after the advent of the KLT relations that they can be expressed as the inverses

$$\boldsymbol{S}_n(w|w') = \boldsymbol{m}_n(w|w')^{-1} \tag{4}$$

of bi-coloured scalar amplitudes $\boldsymbol{m}_n(1,\dots,n|1,\dots,n)$ in a field theory known as the biadjoint scalar theory [3], which we shall discuss in more detail later on.

Noncommutative Field Theory. In the late 1990s it was realised that, on $\mathbb{R}^d$ with a constant NS–NS B-field and flat background metric g, the low-energy interactions of open strings ending on stacks of N D-branes are described by a non-local deformation of $U(N)$ Yang-Mills theory to a *noncommutative gauge theory* [4–10]. Open strings see an effective metric G and a bivector θ related to the closed string background (g, B), such that $G = g$ and $\theta = 0$ when $B = 0$.

Colour-stripped open string amplitudes then have a simple θ-dependence given by phase factors

$$A_n(1, \dots, n)_{G,\theta} = \mathrm{e}^{-\mathrm{i} \sum_{i<j} p_i \times p_j} A_n(1, \dots, n)_{G,\theta=0} , \tag{5}$$

where $k \times q = \frac{1}{2} k_\mu \theta^{\mu\nu} q_\nu$ and p_i are the external momenta. The right-hand side involves the open string amplitude in the absence of the B-field but computed with the open string metric G. In the Seiberg-Witten scaling limit where $\alpha' \longrightarrow 0$ with $g \sim (\alpha')^2 \longrightarrow 0$, these phase factors are captured by replacing the commutative pointwise product $\mu(\phi_1 \otimes \phi_2) := \phi_1 \phi_2$ of fields $\phi_1, \phi_2 \in \Omega^0(\mathbb{R}^d, \mathfrak{u}(N))$ with their associative Moyal product defined by

$$\phi_1 \star \phi_2 = \mu \circ \exp\big(\tfrac{\mathrm{i}}{2} \theta^{\mu\nu} \partial_\mu \otimes \partial_\nu\big)(\phi_1 \otimes \phi_2) . \tag{6}$$

KLT Relations with *B*-Fields. The B-field modifications of the KLT relations are trivial: we simply rewrite the original KLT formula (2) by substituting (5) to get

$$M_n^{L\otimes R} = \sum_{w,w'} A_n^L(w)_{G,\theta} \, \boldsymbol{S}_n(w|w')_{G;\theta,\bar\theta} \, A_n^R(w')_{G,\bar\theta} , \tag{7}$$

where the gravitational amplitudes on the left-hand side are now computed with the background metric G. The new momentum kernel is

$$\boldsymbol{S}_n(w|w')_{G;\theta,\bar\theta} = \Theta_n(w)^{-1} \, \boldsymbol{S}_n(w|w')_{G;\theta=0,\bar\theta=0} \, \bar\Theta_n(w')^{-1} , \tag{8}$$

where

$$\Theta_n(1, \dots, n) = \mathrm{e}^{-\mathrm{i} \sum_{i<j} p_i \times p_j} . \tag{9}$$

We have allowed for different bivectors θ and $\bar\theta$ in the left and right gauge theory amplitudes, which appear when distinct gauge fluxes on the worldvolumes of different stacks of D-branes are turned on.

So far this is a boring result: it simply modifies the slogan (1) to

$$\text{Gravity} = (\text{Noncommutative Gauge Theory})^2 . \tag{10}$$

That is, the double copy of noncommutative Yang-Mills theory is *ordinary* general relativity, rather than some deformation of gravity as one may have anticipated. However, this result raises several theoretical questions; for example:

Question 1: What is the scalar theory sourcing the relation $\boldsymbol{S}_n = \boldsymbol{m}_n^{-1}$ for the B-field modified KLT momentum kernel?

The answers to this and other questions lead to new perspectives on the double copy relations, which require careful considerations of non-locality, with interesting answers and applications. These are the topic of the remaining sections.

3 Colour and Kinematics in Moyal-Deformed Theories

Colour-Kinematics Duality. Fastforwarding ahead 20 years from the KLT paper, Bern, Carrasco and Johansson (BCJ) [11] posit a stripping of colour from kinematic factors in numerators of tree-level gauge theory amplitudes in the form

$$\mathcal{A}_n^{L/R} = \sum_I \frac{c_I \, n_I^{L/R}}{D_I} \, , \tag{11}$$

where the sum runs over all planar trivalent Feynman graphs with n external edges, c_I are colour factors constructed from structure constants f^{abc} of the gauge algebra $\mathfrak{g}$, n_I depend only on purely kinematic data (momentum, spin, flavour, etc.), and D_I are propagator factors associated to internal edges.

The colour-stripping in (11) depends on a choice of basis, which is dictated by demanding that it be compatible with *colour-kinematics duality*: any relation of the schematic form $c_I + c_J + c_K + \cdots = 0$ involving only colour factors, such as the Jacobi identity among f^{abc} for the Lie algebra $\mathfrak{g}$, is mirrored by a relation $n_I + n_J + n_K + \cdots = 0$ among kinematic factors. These are called generalized Jacobi identities. The duality predicts the existence of an infinite-dimensional kinematic algebra underlying scattering amplitudes, whose general structure is currently one of the main open problems in the double copy programme.

When the colour-stripping is compatible with colour-kinematics duality, then the double copy relations replace colour by kinematics $c_I \longmapsto n_I^{R/L}$ in (11) to give amplitudes of a theory with linearized diffeomorphism symmetry via two copies of gauge theory kinematic factors:

$$M_n^{L\otimes R} = \sum_I \frac{n_I^L \, n_I^R}{D_I} \, . \tag{12}$$

The BCJ relations ensure that the amplitudes are independent of the chosen basis for the decomposition (11).

This operation also allows one to replace kinematic factors $n_I^{L/R} \longmapsto \bar{c}_I$ in (11) by colour factors constructed from structure constants $\bar{f}^{\bar{a}\bar{b}\bar{c}}$ of another gauge algebra $\bar{\mathfrak{g}}$, regarded as a kinematic Lie algebra since the resulting amplitudes trivially satisfy colour-kinematics duality. This gives the amplitudes of a *biadjoint scalar theory*, called the "zeroth copy" of the gauge theory because it acts as the identity theory in the space of copyable theories: the double copy of any field theory with the biadjoint scalar theory simply returns the original theory.

These considerations now bring us to our second question:
Question 2: What is the meaning of colour-kinematics duality in noncommutative gauge theories?

To understand why this is a question, we review the well-known mixing between gauge and kinematic degrees of freedom in Moyal-deformed field theories.

Colour-Kinematics Mixing. Consider the noncommutative associative algebra $\Omega^0(\mathbb{R}^d, \mathfrak{u}(N))$ with the Moyal product. The star-commutator

$$[\phi_1 \stackrel{\star}{,} \phi_2] = \phi_1 \star \phi_2 - \phi_2 \star \phi_1 \tag{13}$$

makes it an infinite-dimensional Lie algebra, which we denote as $\mathfrak{u}_\star(N)$. As a vector space, $\Omega^0(\mathbb{R}^d, \mathfrak{u}(N)) \simeq \mathfrak{u}(N) \otimes \Omega^0(\mathbb{R}^d)$, but not as the Lie algebra $\mathfrak{u}_\star(N)$ unless $\theta = 0$, because expanding the fields in a basis $\{T^a\}$ of $\mathfrak{u}(N)$ shows

$$[\phi_1 \stackrel{\star}{,} \phi_2] = \tfrac{1}{2}\left(f^{abc}\,\{\phi_1^a \stackrel{\star}{,} \phi_2^b\} + \mathrm{i}\,d^{abc}\,[\phi_1^a \stackrel{\star}{,} \phi_2^b]\right) T^c\,, \tag{14}$$

where $[T^a, T^b] = f^{abc}\,T^c$ and $\{T^a, T^b\} = \mathrm{i}\,d^{abc}\,T^c$. This loss of factorization spoils standard colour-stripping and colour-kinematics duality.

Kinematic Lie Algebras. On the other hand, this mixing provides a novel realisation of the kinematic algebras for some theories. The momentum space basis $e_k^a = \mathrm{e}^{\,\mathrm{i}\,k\cdot x}\,T^a$ of $\Omega^0(\mathbb{R}^d, \mathfrak{u}(N))$ obeys the star-commutation relations

$$\left[e_k^a \stackrel{\star}{,} e_p^b\right] = F^{abc}(k, p, k+p)\,e_{k+p}^c\,, \tag{15}$$

with the momentum dependent structure constants

$$F^{abc}(k, p, k+p) = f^{abc}\cos(k \times p) + \mathrm{i}\,d^{abc}\sin(k \times p)\,. \tag{16}$$

For $N = 1$ this makes $\Omega^0(\mathbb{R}^d)$ into the *kinematic Lie algebra* $\mathfrak{u}_\star(1)$ with

$$[e_k \stackrel{\star}{,} e_p] = F(k, p, k+p)\,e_{k+p} \quad , \quad F(k, p, k+p) = 2\,\mathrm{i}\,\sin(k \times p)\,. \tag{17}$$

In the semi-classical limit, i.e. to first order in the deformation bivector θ, the star-commutator truncates to the Poisson bracket $\{\phi_1, \phi_2\}_\theta = \theta^{\mu\nu}\,\partial_\mu\phi_1\,\partial_\nu\phi_2$, and $\mathfrak{u}_\star(1)$ reduces to the Lie algebra $\mathfrak{sdiff}(\mathbb{R}^d)$ of *symplectic diffeomorphisms*. The full deformation $\mathfrak{u}_\star(1)$ of $\mathfrak{sdiff}(\mathbb{R}^d)$ is related to the infinite unitary algebra $\mathfrak{u}(\infty)$ as well as a certain $W_{1+\infty}$-algebra known as the symplecton algebra, see e.g. [12–15]. For a description of the $N > 1$ kinematic Lie algebras $\mathfrak{u}_\star(N)$, see [16].

Twisted Colour-Stripping. The way to handle the mixing issue comes from a new perspective on colour-stripping [1], and ultimately a new notion of colour-kinematics duality, tailored to the momentum dependent structure constants of these inherently non-local field theories. It is based on the beautiful formalism of twisted tensor products of homotopy algebras developed by Borsten et al. [17], which has the additional power to provide off-shell formulations of colour-kinematics duality and the double copy relations.

Skipping the formal definitions and details, the upshot is that the Lie algebra

$$\mathfrak{u}_\star(N) \;=\; \mathfrak{u}(N)\otimes_\tau \Omega^0(\mathbb{R}^d) \tag{18}$$

can be factorized as a *twisted tensor product* of the Lie algebra $\mathfrak{u}(N)$ with the commutative algebra $(\Omega^0(\mathbb{R}^d),\,\mu)$ of functions, where

$$\tau : \mathfrak{u}(N)\otimes\mathfrak{u}(N) \longrightarrow \mathfrak{u}(N)\otimes \mathrm{End}\big(\Omega^0(\mathbb{R}^d)\big)\otimes\mathrm{End}\big(\Omega^0(\mathbb{R}^d)\big) \tag{19}$$

is the *twist datum* defined by

$$\begin{aligned}\tau(T^a,T^b) \;=\;& [T^a,T^b]\otimes\cos\big(\tfrac12\,\theta^{\mu\nu}\,\partial_\mu\otimes\partial_\nu\big)\\ &+\mathrm{i}\,\{T^a,T^b\}\otimes\sin\big(\tfrac12\,\theta^{\mu\nu}\,\partial_\mu\otimes\partial_\nu\big)\,.\end{aligned} \tag{20}$$

Then in Sweedler notation

$$\begin{aligned}[T^a\otimes\phi_1^a \overset{\star}{,} T^b\otimes\phi_2^b] \;=\;& \tau^{(1)}(T^a,T^b)\\ &\otimes\mu\big(\tau^{(2)}(T^a,T^b)(\phi_1^a)\,,\;\tau^{(3)}(T^a,T^b)(\phi_2^b)\big)\,.\end{aligned} \tag{21}$$

This strips colour from kinematics, with each factor of the twisted tensor product a conventional algebraic object: noncommutativity is now completely absorbed into the twist datum τ. For a given theory, this twisted factorization must be compatible with a form of colour-kinematics duality in order to be able to apply the double copy relation. This answers Question 2 above.

Noncommutative Biadjoint Scalar Theory. Let us illustrate this operation by constructing a new non-local scalar field theory of relevance to our ensuing discussions. The *biadjoint scalar theory* is a massless cubic scalar field theory with rigid symmetry under a pair of quadratic Lie algebras $\mathfrak{g}$ and $\bar{\mathfrak{g}}$. Its fields

$$\phi \;=\; T^a\otimes\bar T^{\bar a}\otimes\phi^{a\bar a} \;\in\; \Omega^0(\mathbb{R}^d,\mathfrak{g}\otimes\bar{\mathfrak{g}}) \;\simeq\; \mathfrak{g}\otimes\big(\bar{\mathfrak{g}}\otimes\Omega^0(\mathbb{R}^d)\big) \tag{22}$$

propagate according to the action functional

$$S_{\rm bi}[\phi] \;=\; \int \mathrm{d}^dx\;\frac12\,\phi^{a\bar a}\,\Box\,\phi_{a\bar a} - \frac{g}{3!}\,f^{abc}\,\bar f^{\bar a\bar b\bar c}\,\phi^{a\bar a}\,\phi^{b\bar b}\,\phi^{c\bar c}\,. \tag{23}$$

Deform the biadjoint scalar theory by replacing the Lie algebra $\bar{\mathfrak{g}}\otimes\Omega^0(\mathbb{R}^d)$ with the Lie algebra $\mathfrak{u}_\star(\bar N)\simeq\mathfrak{u}(\bar N)\otimes_{\bar\tau}\Omega^0(\mathbb{R}^d)$. The new action functional is

$$S_{\rm bi}^\star[\phi] \;=\; \int \mathrm{d}^dx\;\frac12\,\phi^{a\bar a}\star\Box\,\phi_{a\bar a} - \frac{g}{3!}\,f^{abc}\,\bar\lambda^{\bar a\bar b\bar c}\,\phi^{a\bar a}\star\phi^{b\bar b}\star\phi^{c\bar c}\,, \tag{24}$$

where $\bar{\lambda}^{\bar{a}\bar{b}\bar{c}} := \bar{f}^{\bar{a}\bar{b}\bar{c}} + \mathrm{i}\,\bar{d}^{\bar{a}\bar{b}\bar{c}}$. This theory has the standard scalar propagator, but now the 3-point Feynman vertex

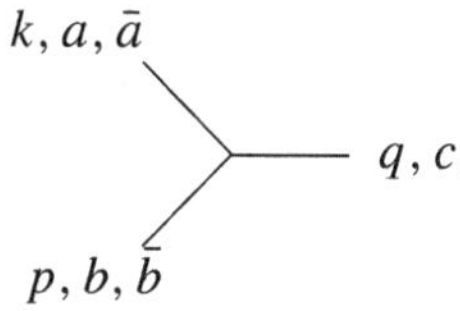

$$= -\mathrm{i}\, g\, f^{abc}\, \bar{F}^{\bar{a}\bar{b}\bar{c}}(k, p, q)$$

comes from replacing the structure constants $\bar{f}^{\bar{a}\bar{b}\bar{c}}$ of $\bar{\mathfrak{g}}$ with the momentum dependent structure constants $\bar{F}^{\bar{a}\bar{b}\bar{c}}(k, p, q)$ of $\mathfrak{u}_\star(\bar{N})$ in the usual biadjoint scalar vertex.

4 Noncommutative Chern-Simons Theory

The simplest example in which to illustrate the new double copy prescription is the Moyal deformation of Chern-Simons gauge theory in three dimensions. The gluon fields are $\mathfrak{u}(N)$-valued 1-forms $A \in \Omega^1(\mathbb{R}^3, \mathfrak{u}(N))$ with action functional

$$S^\star_{\rm CS}[A] = \int \mathrm{Tr}\Big(\frac{1}{2}\, A \wedge_\star \mathrm{d}A + \frac{g}{3!}\, A \wedge_\star [A \stackrel{\star}{,} A]\Big)\,, \tag{25}$$

where the Moyal-deformed exterior product $\wedge_\star$ of forms is defined analogously to (6) by replacing partial derivatives with Lie derivatives along the coordinate directions. The propagator is the same as in ordinary Chern-Simons theory, but now the 3-gluon Feynman vertex is

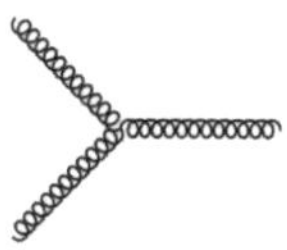

$$= -\mathrm{i}\, g\, \epsilon_{\mu\nu\rho}\, F^{abc}(k, p, q)$$

Twisted Colour-Kinematics Duality. Like its commutative version, noncommutative Chern-Simons theory has no propagating degrees of freedom and so all gluon scattering amplitudes vanish. However, one can go slightly off-shell and consider amplitudes of 'harmonic' gluon states [1]. Then numerators of these amplitudes in Lorenz gauge are built from the Lie brackets of the twisted tensor product $\mathfrak{u}(N) \otimes_\tau \mathfrak{vdiff}(\mathbb{R}^3)$ with the infinite-dimensional Lie algebra $\mathfrak{vdiff}(\mathbb{R}^3)$ of volume-preserving diffeomorphisms of $\mathbb{R}^3$. The generators are

$$\varepsilon^a_\mu(p) = e^a_p\, \Pi_{\mu\nu}(p)\, \partial^\nu\,, \tag{26}$$

where $\Pi_{\mu\nu}$ are transverse projection operators with $\Pi_{\mu\nu}(p)\, p^\mu = 0$. The commutation relations are

$$[\varepsilon^a_\mu(k) \,\overset{\star}{,}\, \varepsilon^b_\nu(p)] \;=\; \mathrm{F}^{abc}_{\mu\nu\rho}(k,p,k+p)\,\varepsilon^c_\rho(k+p)\,, \tag{27}$$

where the momentum dependent structure constants are given by

$$\mathrm{F}^{abc}_{\mu\nu\rho}(k,p,k+p) \;=\; F^{abc}(k,p,k+p)\,\Pi_{\mu\mu'}(k)\,\epsilon_{\mu'\nu'\rho}\,\Pi_{\nu'\nu}(p)\,. \tag{28}$$

Double Copy. As before, using Lie derivatives we factorize the (differential graded) Lie algebra $\big(\Omega^\bullet(\mathbb{R}^3,\mathfrak{u}(N)),[-\overset{\star}{,}-]\big)$ as the twisted tensor product $\mathfrak{u}(N)\otimes_\tau\Omega^\bullet(\mathbb{R}^3)$ with the (differential graded) commutative algebra $\Omega^\bullet(\mathbb{R}^3)$. This factorization is compatible with twisted colour-kinematics duality.

Replacing now the full *twisted* colour-stripping by kinematic factors leads to the same double copy theory as that of ordinary Chern-Simons theory with itself [18]. This is a non-local theory of fields $H_{\bar\mu\mu}=A_{\bar\mu}\otimes A_\mu$, with action functional

$$\widehat{S}_{\mathrm{CS}}[H] \;=\; \int \mathrm{d}^3x\;\epsilon^{\bar\mu\bar\nu\bar\rho}\,\epsilon^{\mu\nu\rho}\Big(H_{\bar\mu\mu}\,\frac{1}{\Box}\,\partial_{\bar\nu}\,\partial_\nu H_{\bar\rho\rho}+\frac{g}{3}\,H_{\bar\mu\mu}\,H_{\bar\nu\nu}\,H_{\bar\rho\rho}\Big)\,. \tag{29}$$

This example illustrates a general result, which in particular explains what we found in Sect. 2: Double copies of Moyal-deformed gauge theories coincide with the ordinary double copy theories of their commutative counterparts.

5 Coloured Moyal Deformations of Cubic Scalar Theory

Among other things, in this section we will answer Question 1 from Sect. 2.

Adjoint Scalar Theory. Setting $\bar N=1$ in the noncommutative biadjoint scalar theory from Sect. 3 leads to a non-local theory of a massless scalar field $\phi\in\Omega^0(\mathbb{R}^d,\mathfrak{g})$ with rigid colour symmetry and action functional

$$S_{\mathrm{ad}}[\phi] \;=\; \int \mathrm{d}^dx\;\frac{1}{2}\,\phi^a\star\Box\,\phi_a-\frac{\mathrm{i}\,g}{3}\,f^{abc}\,\phi^a\star\phi^b\star\phi^c\,. \tag{30}$$

This theory has a standard scalar propagator and the 3-point Feynman vertex

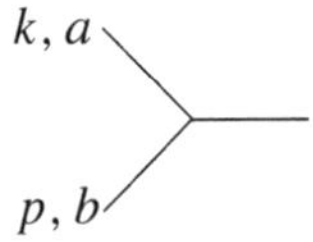

$q,c \;=\; -\mathrm{i}\,g\,f^{abc}\,F(k,p,q)$

An unusual feature of the adjoint scalar theory is that it becomes free when $\theta=0$ (hence the omission of the adjective 'noncommutative'). This is not what normally happens in Moyal-deformed scalar theories, where the effect of noncommutativity is to replace local interactions of an ordinary scalar field theory with non-local interactions induced by the Moyal product.

The adjoint scalar theory possesses manifest colour-kinematics duality: numerators of its amplitudes factorize according to the untwisted tensor product $\mathfrak{g}\otimes\mathfrak{u}_\star(1)$. The kinematic algebra is the Lie algebra $\mathfrak{u}_\star(1)$.

Following the relations of the twist formalism, the zeroth copy of the adjoint scalar theory replaces the Lie algebra $\mathfrak{u}_\star(1)=\mathfrak{u}(1)\otimes_\tau\Omega^0(\mathbb{R}^d)$ with $\bar{\mathfrak{g}}\otimes\Omega^0(\mathbb{R}^d)$. This results in the commutative biadjoint scalar theory from Sect. 3.

On the other hand, the double copy relation applied to the adjoint scalar theory replaces $\mathfrak{g}\otimes\mathfrak{u}_\star(1)$ with $\mathfrak{u}(1)\otimes_{\bar\tau}\big(\mathfrak{u}(1)\otimes_\tau\Omega^0(\mathbb{R}^d)\big)$. This results in a complicated cubic theory of a scalar field $\phi\in\Omega^0(\mathbb{R}^d)$ deformed by two bivectors θ and $\bar\theta$, with action functional

$$\widehat{S}_{\rm ad}[\phi] = \int \mathrm{d}^d x\ \frac{1}{2}\,\phi\star\Box\phi - \frac{\mathrm{i}\,g}{3}\sum_{n=0}^{\infty}\frac{(-1)^n}{(2n+1)!}\,\bar\theta^{\mu_1\nu_1}\cdots\bar\theta^{\mu_{2n+1}\nu_{2n+1}} \times \phi\,[\partial_{\mu_1}\cdots\partial_{\mu_{2n+1}}\phi\stackrel{\star}{,}\partial_{\nu_1}\cdots\partial_{\nu_{2n+1}}\phi]\ . \tag{31}$$

Binoncommutative Biadjoint Scalar Theory. Taking things one step further, one can modify the double copy of the adjoint scalar theory to include colour: the scalar fields now live in the double twisted tensor product

$$\mathfrak{u}(N)\otimes_\tau\big(\mathfrak{u}(\bar N)\otimes_{\bar\tau}\Omega^0(\mathbb{R}^d)\big)\ . \tag{32}$$

The interactions in this theory involve a binoncommutative product which is a binonabelian extension of the product used to construct the double copy theory for the adjoint scalar. This theory is the *zeroth copy* of the noncommutative Chern-Simons theory from Sect. 4.

This scalar field theory is quite complicated when written down in position space in terms of an action functional, but it has a simple representation in momentum space as the standard scalar propagator with the Feynman vertex

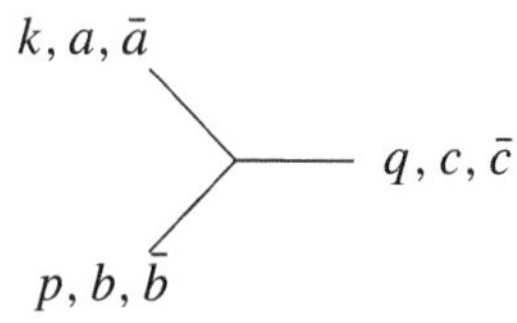

$$= -\mathrm{i}\,g\,F^{abc}(k,p,q)\,\bar F^{\bar a\bar b\bar c}(k,p,q)$$

The n-point colour-stripped tree-level binoncommutative biadjoint scalar amplitudes $\boldsymbol{m}_n^{\star\bar\star}$ can be obtained by dressing the corresponding amplitudes $\boldsymbol{m}_n$ of the usual biadjoint scalar theory from Sect. 3 with the phase factors (9) as [1]

$$\boldsymbol{m}_n^{\star\bar{\star}}(w|w') = \Theta_n(w)\,\boldsymbol{m}_n(w|w')\,\bar{\Theta}_n(w')\,. \tag{33}$$

Noncommutative Yang-Mills Theory. We finally return to our story from Sect. 2 and reconsider the Moyal deformation of Yang-Mills theory on $\mathbb{R}^d$. It is defined by the action functional

$$S^\star_{\rm YM}[A] = \frac{1}{2}\int {\rm Tr}\left(F^\star_A \wedge_\star *_G F^\star_A\right)\,, \tag{34}$$

where $F^\star_A = {\rm d}A + \frac{g}{2}\,[A \overset{\star}{,} A]$ is the noncommutative field strength of a gauge field $A \in \Omega^1(\mathbb{R}^d, \mathfrak{u}(N))$, and $*_G$ denotes the Hodge duality operator with respect to the open string metric G.

The detailed analysis of this theory is rather involved and was discussed extensively in [1]. The double copy relations require a *strictification* of the theory, which amounts to replacing the quartic interactions of gluons by cubic interactions through the introduction of suitable auxiliary fields. This is possible to do in a way that twisted colour-stripping is compatible with colour-kinematics duality at each multiplicity order. The corresponding tree-level scattering amplitudes obey the appropriate versions of the BCJ relations.

From (4), (8) and (33) it follows that the B-field modified KLT momentum kernel $\boldsymbol{S}_n(w|w')_{G;\theta,\bar{\theta}}$ of the double copy relations for noncommutative Yang-Mills theory is sourced by the binoncommutative biadjoint scalar theory as

$$\boldsymbol{S}_n(w|w')_{G;\theta,\bar{\theta}} = \boldsymbol{m}_n^{\star\bar{\star}}(w|w')_G^{-1}\,. \tag{35}$$

This answers Question 1 from Sect. 2. The novel scalar theories introduced in this section have further interesting applications which we now describe.

6 $\mathfrak{u}_\star(1)$ Gauge Theories are Double Copy Theories

The rank $N = 1$ limits of noncommutative gauge theories are interesting interacting theories of photons. Like the adjoint scalar theory, they become free theories in the commutative limit $\theta = 0$. They have no colour symmetry, but they are invariant under the kinematic Lie algebra $\mathfrak{u}_\star(1)$ of noncommutative gauge transformations. These theories have long been believed to contain gravitation. Evidence is provided for instance through the interplay between gauge transformations and spacetime diffeomorphisms discussed in Sect. 3, as well as emergent gravity phenomena; see e.g. [19, 20] for reviews.

The double copy relations offer a new perspective on the emergence of gravitation: a rank one noncommutative gauge theory can be realised as the double copy of the adjoint scalar theory from Sect. 5 with the ordinary non-abelian version of the gauge theory in question, for any gauge algebra $\mathfrak{g}$. Symbolically

$$\mathfrak{u}_\star(1)\ \text{Gauge Theory} \;=\; \text{Adjoint Scalar Theory} \otimes \text{Gauge Theory}\ . \tag{36}$$

For example, in the case of Chern-Simons theory, the left-hand side of this relation realises a theory with local symmetry under a subalgebra of spacetime diffeomorphisms, along the lines discussed in Sect. 4.

Self-Dual Yang-Mills Theory and Gravity. The general relation (36) encompasses recent discussions of the best understood example of the double copy. Set $d = 4$ and work in coordinates $(x^+, x^-, z, \bar{z})$ on $\mathbb{R}^4 \simeq \mathbb{R}^2 \times \mathbb{C}$. Turn on bivectors θ and $\bar{\theta}$ whose only non-zero components are equal to ± 1 along the null (z, x^-)-plane.

Then the semi-classical limit in $\bar{\theta}_{-z}$ of the $\bar{N} = 1$ binoncommutative biadjoint scalar theory from Sect. 5 is the usual integrable deformation of self-dual Yang-Mills theory given by the noncommutative instanton equations [21]

$$\Box\phi - \mathrm{i}\, g\, [\partial_z\phi \overset{\star}{,} \partial_-\phi] \;=\; 0\ , \tag{37}$$

where $\phi \in \Omega^0(\mathbb{R}^4, \mathfrak{u}(N))$ is the Leznov prepotential representing the single polarization state of a gluon that remains after projection to the self-dual sector. The self-dual sector of noncommutative Yang-Mills theory has a non-trivial commutative limit to ordinary self-dual Yang-Mills theory and arises (in Leznov gauge) from quantizing open $\mathcal{N} = 2$ strings in a constant B-field background [22].

For $N = 1$, this theory is the double copy of the adjoint scalar theory constructed in Sect. 5 and (37) coincides with the deformed Plebański equation for noncommutative gravity [23, 24]. In the semi-classical limit in θ_{-z} it reduces to Plebański's equation $\Box\phi = g\,\{\partial_z\phi, \partial_-\phi\}_\theta$ for self-dual gravity, and the theory with action functional (31) is related to the closed string field theory of $\mathcal{N} = 2$ strings [25]. Thus self-dual $\mathfrak{u}_\star(1)$ Yang-Mills theory is noncommutative self-dual gravity.

The semi-classical limit of the adjoint scalar theory itself is self-dual Yang-Mills theory in Leznov gauge [26], with equation of motion $\Box\phi = g\,[\partial_z\phi, \partial_-\phi]$. It reproduces the well-known double copy relation in the self-dual sector

$$\text{Self-Dual Gravity} \;=\; (\text{Self-Dual Yang-Mills Theory})^2\ , \tag{38}$$

where the kinematic algebra is the infinite-dimensional Lie algebra $w_{1+\infty}$ of area-preserving diffeomorphisms of the null (z, x^-)-plane [27]. Thus the adjoint scalar theory is a deformation of self-dual Yang-Mills theory consistent with colour-kinematics duality; however, it is not integrable [15]. On the other hand, the double copy relation (36) now reads

$$\begin{matrix}\text{Noncommutative} \\ \text{Self-Dual Gravity}\end{matrix} \;=\; \begin{matrix}\text{Adjoint Scalar} \\ \text{Theory}\end{matrix} \;\otimes\; \begin{matrix}\text{Self-Dual} \\ \text{Yang-Mills Theory}\end{matrix}\ . \tag{39}$$

In both instances (38) and (39) the integrability of gravity is "inherited" from the gauge theory [15, 28].

7 Outlook: A Braided Double Copy?

In this contribution we have described recently developed twisted versions of colour-stripping and colour-kinematics duality that can handle the non-locality of Moyal-deformed gauge theories which induce momentum dependent structure constants in amplitudes [1]. The result of the double copy relations is then classical, in that ordinary gravitational theories are always produced. It is natural to wonder whether there is a way in which a noncommutative theory of gravity is related to noncommutative gauge theory. We assert that the answer is affirmative, through a suitable variant of the double copy.

Moyal-deformed theories of gravity have been developed with "twisted" diffeomormphism symmetry, though it is unclear whether these models can arise in string theory; see e.g. [19] for a review. These theories lie in the broad class of 'braided quantum field theories', which are theories with braided symmetries that can be defined and treated in the new language of *braided homotopy algebras* [29, 30]. The key point is that if one works with braided homotopy algebras, then *untwisted* colour-stripping and factorization are possible, with the kinematic factors now retaining information about noncommutativity in the form of (differential graded) braided commutative algebras.

Hence it should be possible to realise twisted noncommutative gravity theories through

Conjecture: There is a braided version of the double copy construction in terms of factorizations into braided kinematic algebras and braided noncommutative field theories such that

$$\text{Twisted Noncommutative Gravity} \;=\; (\text{Braided Gauge Theory})^2 \,. \tag{40}$$

Acknowledgements The author thanks Guillaume Trojani for the collaboration on [1]. This article is based upon work from COST Actions CaLISTA CA21109 and THEORY-CHALLENGES CA22113 supported by COST (European Cooperation in Science and Technology).

References

1. R.J. Szabo, G. Trojani, Homotopy double copy of noncommutative gauge theories. Symmetry **15**(8), 1543 (2023). arXiv:2306.12175 [hep-th]
2. H. Kawai, D.C. Lewellen, S. Tye, A relation between tree amplitudes of closed and open strings. Nucl. Phys. B **269**, 1–23 (1986)
3. F. Cachazo, S. He, E.Y. Yuan, Scattering of massless particles: scalars, gluons and gravitons. JHEP **07**, 033 (2014). (arXiv:1309.0885 [hep-th])
4. M.R. Douglas, C.M. Hull, D-branes and the noncommutative torus. JHEP **02**, 008 (1998). (arXiv:hep-th/9711165)

5. F. Ardalan, H. Arfaei, M.M. Sheikh-Jabbari, Noncommutative geometry from strings and branes. JHEP **02**, 016 (1999). (arXiv:hep-th/9810072)
6. C.S. Chu, P.M. Ho, Noncommutative open string and D-brane. Nucl. Phys. B **550**, 151–168 (1999). (arXiv:hep-th/9812219)
7. V. Schomerus, D-branes and deformation quantization. JHEP **06**, 030 (1999). (arXiv:hep-th/9903205)
8. N. Seiberg, E. Witten, String theory and noncommutative geometry. JHEP **09**, 032 (1999). (arXiv:hep-th/9908142)
9. M.R. Douglas, N.A. Nekrasov, Noncommutative field theory. Rev. Mod. Phys. **73**, 977–1029 (2001). (arXiv:hep-th/0106048)
10. R.J. Szabo, Quantum field theory on noncommutative spaces. Phys. Rept. **378**, 207–299 (2003). (arXiv:hep-th/0109162)
11. Z. Bern, J. Carrasco, H. Johansson, New relations for gauge theory amplitudes. Phys. Rev. D **78**, 085011 (2008). (arXiv:0805.3993 [hep-ph])
12. J. Hoppe, Diffeomorphism groups, quantization and $SU(\infty)$. Int. J. Mod. Phys. A **4**, 5235–5248 (1989)
13. C.N. Pope, L.J. Romans, X. Shen, W_∞ and the Racah-Wigner Algebra. Nucl. Phys. B **339**, 191–221 (1990)
14. F. Lizzi, R.J. Szabo, A. Zampini, Geometry of the gauge algebra in noncommutative Yang-Mills theory. JHEP **08**, 032 (2001). (arXiv:hep-th/0107115)
15. E. Chacón, H. García-Compeán, A. Luna, R. Monteiro, C.D. White, New heavenly double copies. JHEP **03**, 247 (2021). (arXiv:2008.09603 [hep-th])
16. R. Monteiro, From Moyal deformations to chiral higher-spin theories and to celestial algebras. JHEP **03**, 062 (2023). (arXiv:2212.11266 [hep-th])
17. L. Borsten, B. Jurčo, H. Kim, T. Macrelli, C. Sämann, M. Wolf, Double copy from homotopy algebras. Fortsch. Phys. **69**(8–9), 2100075 (2021). arXiv:2102.11390 [hep-th]
18. M. Ben-Shahar, H. Johansson, Off-shell color-kinematics duality for Chern-Simons. JHEP **08**, 035 (2022). (arXiv:2112.11452 [hep-th])
19. R.J. Szabo, Symmetry, gravity and noncommutativity. Class. Quant. Grav. **23**, R199–R242 (2006). (arXiv:hep-th/0606233)
20. H. Steinacker, Emergent geometry and gravity from matrix models: an introduction. Class. Quant. Grav. **27**, 133001 (2010). (arXiv:1003.4134 [hep-th])
21. K. Takasaki, Anti-self-dual Yang-Mills equations on noncommutative spacetime. J. Geom. Phys. **37**, 291–306 (2001). (arXiv:hep-th/0005194)
22. O. Lechtenfeld, A.D. Popov, B. Spendig, Open $\mathcal{N}=2$ strings in a B-field background and non-commutative self-dual Yang-Mills. Phys. Lett. B **507**, 317–326 (2001). (arXiv:hep-th/0012200)
23. I. Strachan, The Moyal algebra and integrable deformations of the self-dual Einstein equations. Phys. Lett. B **283**, 63–66 (1992)
24. K. Takasaki, Dressing operator approach to Moyal algebraic deformation of self-dual gravity. J. Geom. Phys. **14**, 111–120 (1994). (arXiv:hep-th/9212103)
25. H. Ooguri, C. Vafa, Geometry of $\mathcal{N}=2$ strings. Nucl. Phys. B **361**, 469–518 (1991)
26. A. Parkes, A cubic action for self-dual Yang-Mills. Phys. Lett. B **286**, 265–270 (1992). (arXiv:hep-th/9203074)
27. R. Monteiro, D. O'Connell, The kinematic algebra from the self-dual sector. JHEP **07**, 007 (2011). (arXiv:1105.2565 [hep-th])
28. R. Monteiro, Celestial chiral algebras, colour-kinematics duality and integrability. JHEP **01**, 092 (2023). (arXiv:2208.11179 [hep-th])

29. M. Dimitrijević Ćirić, G. Giotopoulos, V. Radovanović, R.J. Szabo, Braided L_∞-algebras, braided field theory and noncommutative gravity. Lett. Math. Phys. **111**(6), 148 (2021). arXiv:2103.08939 [hep-th]
30. G. Giotopoulos, R.J. Szabo, Braided symmetries in noncommutative field theory. J. Phys. A **55**(35), 353001 (2022). arXiv:2112.00541 [hep-th]

Combinatorial Quantum Gravity: Q-it From Bit?

Carlo A. Trugenberger

Abstract I shall review the main points of combinatorial quantum gravity, a model for the emergence of dynamical geometric manifolds from random graphs, driven by the combinatorial Ollivier-Ricci curvature. In the simplest version, the manifold is a negative-curvature surface with two scales, an ultraviolet Planck length and an infrared radius of curvature. At scales larger than the latter, there appears an holographic Lorentzian picture with one additional dimension, in which geodesic observers move away from each other and see the residual slow dynamics of particles as 3D quantum mechanics.

Keywords Quantum gravity · Quantum mechanics · Graph theory

1 Introduction

General relativity (GR) and quantum mechanics (QM) are incompatible. On one side, the background independence of GR is at odds with the absolute time of QM [1]; on the other side, GR is perturbatively non-renormalizable. If the rules of QM are taken as granted, this means that GR is either an effective field theory, with new physics emerging on short scales [2, 3] or it is dominated by a non-perturbative ultraviolet fixed point [4–6].

An alternative, point of view is to formulate a microscopic theory in terms of completely different variables and recover GR at large scales. This is the path first proposed by Wheeler [7] in his "it from bit" hypothesis, which posits that the fundamental theory of the universe should be formulated exclusively in terms of information-theoretic binary degrees of freedom, with no reference to either time or space, which are emergent concepts at large scales. This bottom-up approach has then been adopted in group field theory [8] and the tensor model [9–11].

C. A. Trugenberger (✉)
SwissScientific Technologies SA, Geneva, Switzerland
e-mail: ca.trugenberger@bluewin.ch

M. Paranjape et al. (eds.), *Proceedings of Gravity, Strings and Fields*,
CRM Series in Mathematical Physics,
https://doi.org/10.1007/978-3-031-91266-5_21

The prevailing idea today is that QM is more fundamental than GR and has to be taken as the correct theory governing any microscopic degree of freedom. Correspondingly, Wheeler's original idea has been promoted to a "it from qubit" hypothesis, in which space-time is expected to emerge from quantum entanglement (for a review see [12]), which is at odds with the original proposal [7]. Recently, I proposed a model of quantum gravity [13] in which dynamical geometry and QM seem to emerge simultaneously from information-theoretic bits, encoded in the edges of random graphs [14], exactly as proposed by Wheeler, with the added bonus of a first-principles derivation of the dimension 3 of the space in which we live [15, 16]. In this talk I shall review the main points of this model.

2 Combinatorial Quantum Gravity

Combinatorial quantum gravity [13, 17, 18] is formulated as a statistical partition function on the space of all incompressible 2D (even)-regular graphs (IERG) with N vertices,

$$\mathcal{Z} = \sum_{IERG} \mathrm{e}^{-\frac{1}{g}H} \,, \tag{1}$$

where g is the dimensionless coupling. Incompressible graphs are those for which triangles, squares and pentagons do not share more than one edge. This condition is the loop equivalent of the hard-core condition for bosonic point particles. As the hard-core condition prevents the infinite compressibility of Bose gases, the graph incompressibility condition prevents graph crumpling.

The Hamiltonian is the total Ollivier-Ricci curvature [19, 20], a discrete graph equivalent of the Riemannian Einstein-Hilbert action which converges to the continuum Ricci curvature for random geometric graphs [21, 22],

$$H = -2D \sum_{i \in G} \kappa(i) = -2D \sum_{i \in G} \sum_{j \sim i} \kappa(ij) \,, \tag{2}$$

where we denote by $j \sim i$ the neighbour vertices j to vertex i in graph G, i.e. those connected to i by one edge (ij). Ricci curvature on manifolds is a measure of how much (infinitesimal) spheres around a point contract (positive Ricci curvature) or expand (negative Ricci curvature) when they are transported along a geodesic with a given tangent vector at the point under consideration. The Ollivier curvature is a discrete version of the same measure. For two vertices i and j it compares the Wasserstein (or earth-mover) distance $W\left(\mu_i, \mu_j\right)$ between the two uniform probability measures $\mu_{i,j}$ on the spheres around i and j to the distance $d(i, j)$ on the graph and is defined as $\kappa(i, j) = 1 - W\left(\mu_i, \mu_j\right)/d(i, j)$. The Wasserstein distance between two probability measures μ_i and μ_j on the graph is defined as $W\left(\mu_i, \mu_j\right) = \inf \sum_{i,j} \xi(i, j) d(i, j)$, where the infimum has to be taken over all couplings (or transference plans) $\xi(i, j)$.

While the Ollivier-Ricci curvature is cumbersome to compute in the generic case, it simplifies substantially on 2D-regular incompressible graphs [17],

$$\kappa(ij) = \frac{T_{ij}}{2D} - \left[1 - \frac{2 + T_{ij} + S_{ij}}{2D}\right]_+ - \left[1 - \frac{2 + T_{ij} + S_{ij} + P_{ij}}{2D}\right]_+ . \quad (3)$$

where the subscript "+" is defined as $[\alpha]_+ = \max(0, \alpha)$ and T_{ij}, S_{ij} and P_{ij} denote the number of triangles, squares and pentagons supported on egde (ij). This gives

$$\begin{aligned} H &= H_{\text{global}} + H_{\text{local}} , \\ H_{\text{global}} &= 16\left(\frac{D(D-1)}{2}N - \frac{9}{8}T - S - \frac{5}{8}P\right) , \\ H_{\text{local}} &= \sum_{(ij)\in E_1} \left((T_{ij} + S_{ij}) - (2D-2)\right) \\ &+ \sum_{(ij)\in E_2} \left((T_{ij} + S_{ij} + P_{ij}) - (2D-2)\right) , \end{aligned} \quad (4)$$

where T, S and P are the total numbers of triangles, squares and pentagons on the graph and E_1 and E_2 are the ensembles of edges for which the respective summands are strictly positive.

3 Loop Condensation, the Emergence of Hyperbolic Manifolds

The model undergoes a continuous phase transition from a random graph phase to a geometric phase in which graphs become quadrangulations of manifolds [13, 17], a phase transition due to a condensation of square loops on the graph, while triangles and pentagons are suppressed and survive at best as isolated defects [17]. The minimum energy is reached for a regular lattice configuration, with $S = D(D-1)N/2$ and (2D-2) squares per edge, with N the number of vertices and D the topological dimension of the emerged manifold. In the case $D = 2$, on which we will focus from now on, the ground state is thus a torus graph. Since random graphs can be embedded without edge crossings on a Riemann surface of genus N, decreasing the coupling, in this phase, entails decreasing the genus of the emerged manifold until the torus graph at $g = 0$ is reached. In Fig. 1 we show the example of a two-torus emerged at a small but finite value of g in the geometric phase.

Random regular graphs are locally tree-like, with sparse cycles, governed by a Poisson distribution with mean $(2D-1)^k/k$ for cycles of length k [23]. The order parameter for the transition is the relative number of squares S/N where N is the number of squares for a torus graph, shown in Fig. 2.

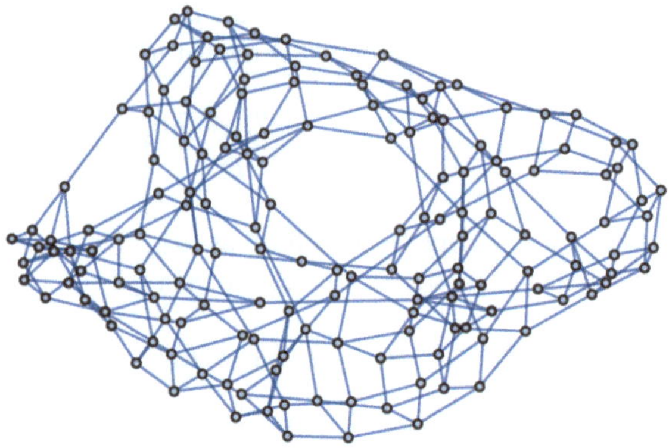

Fig. 1 A two-torus at a small but finite value of the coupling g in the geometric phase

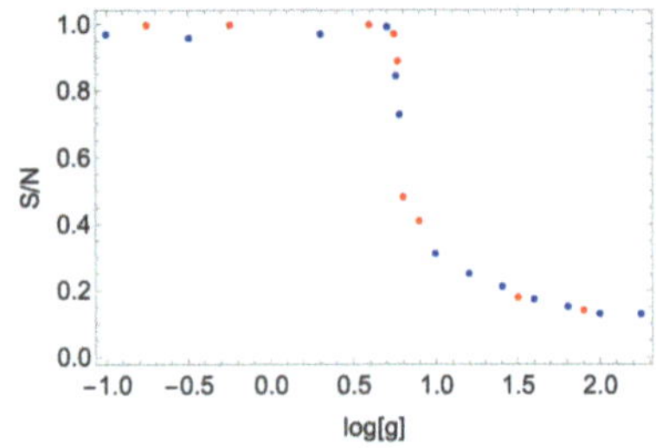

Fig. 2 The order parameter S/N for the random-to-geometric phase transition for $D = 2$

The red dots are obtained by increasing the coupling g, the blue dots by decreasing it. Note the absence of hysteresis, as expected for a continuous phase transition. In the random phase, S/N approaches $(3^4/4)/160 = 0.126\,(N = 160$ in this example), as expected from the sparse cycle distribution in this phase. Unfortunately, the Monte Carlo algorithm does not always manage to reach an exact ground state with $S/N = 1$ deep in the geometric phase when starting from a random graph.

As always in the case of continuous phase transitions, there is a length scale $\xi(g)$ diverging at the critical point, encoded in the connected correlation function

$$c(d, g) = \frac{\langle \left(S_i - \bar{S}\right) \left(S_j - \bar{S}\right) \rangle}{\sigma^2(S)} , \qquad (5)$$

between the number of squares S_i and S_j based on vertices i and j at fixed graph distance d, with $\sigma^2(S)$ denoting the variance (over the vertices) of the number of squares. The typical behaviour of this correlation in the geometric phase, as a function of d (at fixed g) is shown in Fig. 3.

There is an intermediate domain in which correlations turn negative, because there is no correlation between squares on these distances but the average square number $\bar{S}$ is large. On these intermediate distance the graph behaves like a random graph. Correlations then suddenly pick up at the distance $\xi(g) = 10$, before decaying again to zero on even larger distances. In this case, the approach to zero is not due to

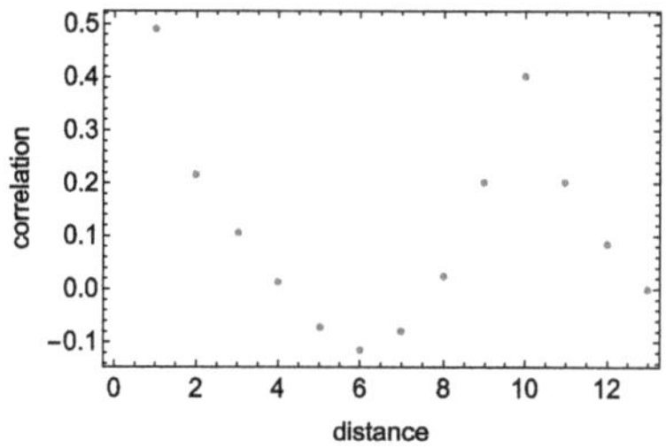

Fig. 3 The typical behaviour of the connected correlation function between the numbers of squares at distance d on the graph in the geometric phase

absence of correlations but, rather, to the fact that each vertex has the same number of squares approaching $\bar{S}$. This is the geometric manifold phase on large scales.

The graphs in the geometric phase behave thus as geometric manifolds on distances larger than $\xi(g)$ and as random graphs below this scale. In random graphs, distances scale as $\log N_\xi$ [14], where $N_\xi = N^{\alpha(g)}$ is the their number of vertices, with $0 \leq \alpha(g) \leq 1$, whereas they scale linearly with $\sqrt{N}$ in the geometric phase. The Hausdorff dimension is thus a scale-dependent quantity that diverges on the scale $\xi(g)$ and decreases to 2 (D in general) on larger scales. We can now take the continuum limit by introducing a fixed length ℓ for the graph edges and letting simultaneously $\ell \to 0$ and $N \to \infty$ so that $\ell_P = \alpha(g)\ell \log N$ is fixed. We obtain so an infinite surface (manifold in general), sprinkled on the scale ℓ_P by bubbles of random phase (the disordered phase).

To establish the character of the emerging surfaces on large scales, let us temporarily forget the microscopic structure and consider only infinite graphs with a fixed number of squares per vertex, say 3. In order to "geometrize" the graphs, we must turn these purely combinatorial objects into 1-skeletons of topological spaces by assigning a fixed length ℓ to the edges. When this is done, the 2-cell embeddings become 1-skeletons of tilings of the sought after continuum surface. In general this is not easy. In 2D, however, it can be done [24]. The surface defined by this tiling is the hyperbolic plane. On large scales, the emerged manifolds are surfaces of negative curvature, so called Cartan-Hadamard manifolds (for a review see [25]) in general or hyperbolic manifolds (for a review see [26]) when no defects are present and the negative curvature is constant. The spectral dimension fits confirm that the geometric phase manifolds on intermediate scales, just above ℓ_P are indeed hyperbolic surfaces [15].

To conclude this section, let us focus on the small-scale bubbles of randomness. Since the Hamiltonian of the model is the negative Ollivier curvature and random graphs have the lowest possible Ollivier curvature, these bubbles have an excitation energy corresponding to minus the excess negative curvature with respect to the geometric manifold at large scales. Note also that the UV scale ℓ_P is inversely related to the IR scale given by the radius of curvature of the geometric phase. When ℓ_P vanishes for $g \to 0$, the radius of curvature diverges and the surface becomes flat. For $g \to g_{\text{crit}}^-$, the radius of curvature diverges, while ℓ_P increases until the whole space fits into a dot of size ℓ_P within which distances scale logarithmically.

4 Emergence of Time and Effective de Sitter Space

As in any statistical mechanics model, there will be fluctuations about the static free energy minima. This presupposes, however the existence of a universal time t. This has nothing to do with the coordinate time of GR, however, we need only an ordinal quantity that defines "before vs. after". But such a quantity is already present in the model: it is the gravitational coupling g itself. We can thus take $\beta_g = 1/g$ as universal time t, which, of course, immediately leads to a cosmological model, as we will see below.

As a consequence of fluctuations, the randomness bubbles will undergo small changes in shape, size and position. A full discussion of these fluctuations is beyond the scope of the present review. Here we shall consider the bubbles as point-like objects and focus only on their positions. In this limit, the fluctuations amount to Brownian motion, which is very sensitive to the geometry of the fluctuating background: in the case of negative curvature it is anomalous and becomes asymptotically ballistic and deterministic.

Let us consider a constant negative curvature surface as the upper sheet of a two-sheeted hyperboloid embedded in 3D Minkwoski space with metric $(+1, +1, -1)$ as

$$\mathbf{x} = \left(\frac{1}{H} \sinh Hz \cos\theta, \frac{1}{H} \sinh Hz \sin\theta, \frac{1}{H} \cosh Hz \right) , \tag{6}$$

The metric is $ds^2 = dz^2 + G^2(z) d\theta^2$, with $G(z) = \sinh Hz/H$ and $-H^2$ the curvature.

Brownian motion on this manifold decomposes in two independent radial and angular Brownian motions (z_t, θ_t) [27–29] (for a review see [30–32]). For $Hz \gg 1$ the radial equation reduces to

$$z_t = \frac{H}{2} t + R_t , \tag{7}$$

up to exponentially small corrections. Here, R_t is a sub-dominant random component such that $R_t/t = O\left(t^{-1/2}\right)$ for large t. Note that, for simplicity of presentation, a diffusion coefficient $\mathcal{D}$ is understood as absorbed in the universal time t, which has thus dimension $[length]^2$. The parameter $\tilde{t} = t/\mathcal{D}$ is thus measured in seconds and $v = H\mathcal{D}$ is a velocity. This equation means that, asymptotically, the manifold coordinate z can be identified with the absolute time [15, 16], which represents the dynamic emergence of coordinate time in this region of the manifold. We shall discuss the sub-dominant random component below.

The angular Brownian motion converges to a limiting angle θ_∞. Near this, its quadratic variation is given by [30, 32]

$$\langle \Delta \left(\theta_\infty - \theta_t\right)^2 \rangle = \int_t^\infty ds \frac{1}{G\left(z_s\right)^2} . \tag{8}$$

If we consider large values of t we can substitute z_s with its limiting value (7). Then, using the above value of $G(z)$ we obtain the dominant contribution to the angular Brownian motion as

$$\langle \Delta \left(\theta_\infty - \theta_z \right)^2 \rangle = 4\, \mathrm{e}^{-2Hz} \, . \tag{9}$$

This is the exact inverse of the usual Brownian motion, whose quadratic variation is quadratic in t (and thus ballistic) at short times, below the typical scattering time and linear in t (and thus diffusive) at large times, when scatterings become frequent. The convergence to a limiting angle has the consequence that scatterings become less and less frequent asymptotically until they stop altogether and motion becomes ballistic and deterministic in the time parameter $\tau = 2 \exp(-Hz)$, with the distribution probability $u(z, \theta)$ satisfying $\left(\partial^2/\partial\tau^2 - \partial^2/\partial\theta^2\right) u = 0$, which is equivalent to

$$\left(\frac{\partial^2}{\partial z^2} + H \frac{\partial}{\partial z} - 4H^2 \mathrm{e}^{-2Hz} \frac{\partial^2}{\partial \theta^2} \right) u(z, \theta) = 0 \, . \tag{10}$$

This is the wave equation on Lorentzian de Sitter space-time [33] of positive curvature $+H^2$, obtained from (6) by the substitution $\cosh(Hz) \leftrightarrow \sinh(Hz)$. Up to exponentially small correction and the sub-dominant random component to be discussed below, Brownian motion in negative curvature is equivalent to null geodesic motion on de Sitter space-time. The same picture applies to the flat slicing of both manifolds [15, 16]. Lorentzian de Sitter space-time appears thus as an effective, rather than fundamental description of the fluctuating manifold in the region $Hz \gg 1$. The prescription for going from the fundamental to the effective description of the manifold is a simultaneous change of sign of the time component of the metric and the curvature. For flat space operators, this amounts to the usual Wick rotation which, however, here is induced dynamically.

In the effective description, the randomness bubbles have a natural interpretation as matter particles of the Planck size ℓ_{P}, with an intrinsic rest energy proportional to the excess (positive) curvature with respect to the de Sitter background. In this model, thus space-time and matter are just two different phases of the bits ensemble. In the course of absolute time, i.e. when the coupling g decreases, the whole space-time emerges from a random graph dot of the Planck size and its curvature decreases when more and more matter is transformed into space until a flat, empty manifold is reached. In the course of this evolution, the Lorentzian effective region is pushed out more and more until it vanishes completely.

5 Spectral Dimension 3 and Quantum Behaviour

In general, diffusion processes probe the intrinsic geometry of a manifold via the return probability kernel $K(t) = \mathrm{Tr} \exp(t\Delta)$. The quantity $d_{\mathrm{s}} = -2d\ln K(t)/d\ln t$ is called the spectral dimension and measures the effective number of dimensions available to a random walker on the scale reached by time t, essentially on small scales for $t \to 0$ and on very large scales for $t \to \infty$. In the mathematical literature the latter is called "geometry at infinity". However, this does not work in the present case, because the Laplacian on a constant negative curvature manifold has a spectral gap $\lambda_0 = -\lim_{t\to\infty} \ln K(t)/t = (D-1)^2 H^2/4$, representing the bottom of the spectrum of the positive operator $-\Delta$. As a consequence, the return probabilities $K(t)$ are dominated by an exponential behaviour at large t [34, 35],

$$K(t) \asymp \frac{\left(1 + H^2 t\right)^{(D-3)/2}}{(H^2 t)^{D/2}} \mathrm{e}^{-\frac{(D-1)^2}{4} H^2 t} , \tag{11}$$

which gives a spectral dimension linearly diverging in time.

The large-scale geometry, however, can be probed by the sub-dominant random component of Brownian motion [36]. This is called the infinite Brownian loop [36], and is the infinite-Θ limit of the Brownian bridge $BB(t, \Theta)$, which is the Brownian motion $B(t)$ constrained to come back to the origin $x = 0$ for $t = \Theta$. Let us decompose the Brownian process $B(t)$ asymptotically for $t \to \infty$ as

$$B(t) \sim \lim_{\Theta\to\infty} \left(\frac{t}{\Theta} B(\Theta) + BB(t, \Theta) \right) . \tag{12}$$

When Brownian motion is not asymptotically ballistic, the infinite Brownian loop is the Brownian process itself. Otherwise, it represents the subdominant random component after the dominant deterministic component has been subtracted. The corresponding return probabilities are given by (11) without the dominant exponential factor and give the spectral dimension function

$$d_{\mathrm{s}}(t) = D - (D-3) \left(\frac{H^2 t}{1 + H^2 t} \right) . \tag{13}$$

Thus, the spectral dimension on scales smaller than $1/H$ is D while the large-scale spectral dimension, above distances $1/H$, becomes 3, independently of D. In the mathematical literature this is called the "pseudo-dimension", or "dimension at infinity" of the constant negative curvature manifold [36]. The large-scale spectral dimension 3 is actually a consequence of the local limit theorem in negative curvature [37].

The uniform continuous estimate of the kernel of the infinite Brownian loop is known [34]:

$$K(t, \rho) \asymp t^{-\frac{3}{2}} \mathrm{e}^{-\frac{\rho^2}{4t}} , \tag{14}$$

where ρ is the hyperbolic distance. This is the isotropic heat kernel on a 3D Euclidean manifold with distance norm ρ "inherited" from the 2D hyperbolic distance [36]. The corresponding kernel in the effective Lorentzian description is

$$K_{\mathrm{eff}}(\tilde{t}, \rho) \asymp \left(\frac{m}{i\hbar\tilde{t}} \right)^{\frac{3}{2}} \mathrm{e}^{-\frac{m\rho^2}{2i\hbar\tilde{t}}} , \tag{15}$$

where we have used the previously introduced time $\tilde{t}$ measured in seconds. This is the 3D Schrödinger propagator for a particle of mass $m = \hbar/2\mathcal{D}$. At large scales, the effective description of slow processes, as seen by co-moving observers, is 3D quantum mechanics with Riemannian distances inherited from the fundamental 2D hyperbolic surface, i.e. quantum mechanics in a 3D Newton-Cartan manifold.

References

1. K.V. Kuchar, Time and interpretations of quantum gravity. Int. J. Mod. Phys. D **20**, 3 (2011)
2. J. Polchinski, *String Theory* (Cambridge University Press, Cambridge, 1998)
3. P. Horava, Quantum gravity at a Lifshitz point. Phys. Rev. D **79**, 084008 (2009)
4. Eichhorn, A., Asymptotically safe gravity, in *Proceedings of the 57th Course of the Erice International School of Subnuclear Physics*, "In search for the unexpected", arXiv: 2003.00044 (2019)
5. J. Ambjorn, J. Görlich, J. Jurkiewicz, R. Loll, Nonperturbative quantum gravity. Phys. Rep. **519**, 127–210 (2012)
6. R. Loll, Quantum gravity from causal dynamical triangulations: a review. Classical Quantum Gravity **37**, 013002 (2019)
7. J.A. Wheeler, Information, physics, quantum: the search for links, in *Proceedings of the III International Symposium on the Foundations of Quantum Mechanics*, pp. 354–358, Tokyo (1989)
8. D. Oriti, The microscopic dynamics of quantum space as a group field theory, in *Foundations of Space and Time: Reflections on Quantum Gravity*; Aug 10–14 2009, Cape Town, South Africa; Cambridge University Press 257 (2011)
9. J. Ambjorn, B. Durhuus, T. Jonsson, Three-dimensional simplicial quantum grvaity and generalized matrix models. Mod. Phys. Lett. A **06**, 1133–1146 (1991)
10. N. Sasakura, Tensor model for gravity and orientability of manifolds. Mod. Phys. Lett. A **06**, 2613–2624 (1991)
11. N. Sasakura, Phase profile of the wave function of canonical tensor model and emergence of large spacetimes. Int. J. Mod. Phys. **36**, 2150222 (2021)
12. X.-L. Qi, Does gravity come from quantum information? Nat. Phys. **4**, 984–987 (2018)
13. C.A. Trugenberger, Combinatorial quantum gravity: geometry from random bits. JHEP **09**, 045 (2017)
14. R. Albert, L. Barabasi, Statistical mechanics of complex networks. Rev. Mod. Phys. **74**, 47 (2002)
15. C.A. Trugenberger, Emergent time, cosmological constant and boundary dimension at infinity in combinatorial quantum gravity. JHEP **04**, 019 (2022)
16. C.A. Trugenberger, Effective de Sitter space, quantum behaviour and large-scale spectral dimension (3+1). JHEP **03**, 186 (2023)
17. C. Kelly, C.A. Trugenberger, F. Biancalana, Self-assembly of geometric space from random graphs. Classical and Quantum Gravity **36**, 125012 (2019)
18. C. Kelly, C.A. Trugenberger, F. Biancalana, Emergence of the circle in a statistical model of random cubic graphs. Classical and Quantum Gravity **38**, 075008 (2021)
19. Y. Ollivier, Ricci curvature of metric spaces. C. R. Math. Acad. Sci. Paris **345**, 643–646 (2007)
20. Y. Linn, L. Lu, S.T. Yau, Ricci curvature of graphs. Tohoku Math. J. **63**, 605–627 (2011)
21. P. van der Hoorn, W.J. Cunningham, G. Lippner, C.A. Trugenberger, D. Krioukov, Ollivier-Ricci curvature convergence in random geometric graphs. Phys. Rev. Res. **3**, 013211 (2021)
22. C. Kelly, C.A. Trugenberger, F. Biancalana, Convergence of combinatorial gravity. Phys. Rev. D **105**, 124002 (2022)
23. P.E. O'Neil, Asymptotics in random (0,1)-matrices. Bull. Am. Math. Soc. **75**, 1276 (1969)

24. B. Datta, S. Gupta, Semi-regular tilings of the hyperbolic plane. Discrete Comput. Geom. **65**, 531 (2019)
25. K. Shiga, Hadamard manifolds, in *Geometry of Geodesics and Related Topics, Advanced Studies in Pure Mathematics* 3, pp. 239–281 (1984)
26. J.W. Anderson, *Hyperbolic Geometry* (Springer, Berlin, 2005)
27. J.-J. Prat, Etude asymptotique et convergence angulaire du mouvement brownien sur une variété à courbure négative, C. R. Acad. Sci. Paris Sér. A-B 280 (22):Aiii, A1539-A1542 (1975)
28. W.S. Kendall, Brownian motion on 2-dimensional manifolds of negative curvature. Trans. Amer. Math. Soc. **275**, 751–760 (1983)
29. E.P. Hsu, W.S. Kendall, Limiting angle of Brownian motion in certain two-dimensional Cartan-Hadamard manifolds. Annales de la faculté des Sciences de Toulouse **1**, 169–186 (1982)
30. E.P. Hsu, A brief introduction to Brownian motion on a Riemann manifold, Summer School in Kyushu (2008)
31. E.P. Hsu, Stochastic analysis on manifolds, Graduate studies in mathematics 38, Providence (RI) (2002)
32. M. Arnaudon, A. Thalmeier, Brownian motion and negative curvature. Progress Prob. **64**, 145–163 (2011)
33. M. Spradlin, A. Strominger, A. Volovich, Les Houches lectures on de Sitter space, arXiv:hep-th/0110007
34. E.B. Davies, N. Mandouvalos, Heat Kernel bounds on hyperbolic space and Kleinian groups. Proc. London Math. Soc. **3**(52), 182–208 (1988)
35. A. Grigorýan, Estimates of heat kernels on Riemannian manifolds, in *Spectral Theory and Geometry*, Cambridge University Press, Cambridge (2010)
36. J.-P. Anker, P. Bougerol, T. Jeulin, The infinite Brownian loop on a symmetric space. Rev. Mat. Iberoamericana **18**, 41–97 (2002)
37. F. Ledrappier, S. Lim, Local limit theorem in negative curvature, arXiv:1503.04156

Cosmology Without Scalar Fields is Like Quantum Field Theory Without RG Flow

Mark Van Raamsdonk

Abstract Time-dependent scalar fields provide a candidate explanation for the dark energy. For these to vary on cosmological time scales, the derivative of the scalar potential in Planck units should have roughly the same magnitude as the potential itself. We emphasize that scalars with this property are present in any four-dimensional gravitational effective theory with a known UV completion via holography, provided that the dual CFT has scalar operators with dimensions of order one. Cosmological solutions without time-dependent scalars are analogous to solutions dual to the vacuum states of such CFTs while solutions with time-dependent scalars are analogous to solutions dual to the vacuum state of quantum field theories with RG flow where these CFTs provide the UV or IR fixed point. If time-dependent scalars do explain the dark energy, the gravitational effective theory describing our universe could be a $\Lambda < 0$ model associated to a holographic CFT.

1 Introduction

Cosmological observations are consistent with the idea that our universe is homogeneous and isotropic on large scales. Standard theoretical models take this homogeneity and isotropy of the background cosmology as a basic assumption. With this choice, the most general classical background has a metric of FLRW form, and a set of scalar fields that depend only on time. For example, in the spatially flat case, we have[1]

$$ds^2 = -dt^2 + a(t)dx_i dx_i \qquad \phi_i = \phi_i(t) \; , \tag{1}$$

The standard ΛCDM model of cosmology assumes that there are no time-dependent scalar fields, though many other models have considered them (See [1–3]

[1] We also have matter and radiation whose effects can be described by a stress-energy tensor of perfect fluid form.

M. Van Raamsdonk (✉)
Department of Physics and Astronomy, University of British Columbia, Vancouver, BC, Canada
e-mail: mav@phas.ubc.ca

M. Paranjape et al. (eds.), *Proceedings of Gravity, Strings and Fields*,
CRM Series in Mathematical Physics,
https://doi.org/10.1007/978-3-031-91266-5_22

for early discussions). Including the possibility of time-dependent scalars can only improve fits to observations, though at the cost of introducing more parameters. Current observations only constrain the dark energy equation of state parameter w (equal to $(\dot{\phi}^2/2 - V)/(\dot{\phi}^2/2 + V)$ for a scalar field model with potential V) to within around 10 percent of the value $w = -1$ for time-independent dark energy (see [4, 5] for recent discussions), so the possibility of time-varying scalars is apparently still viable.[2]

It is interesting to ask whether theoretical considerations suggest the presence or absence of time-dependent scalar fields. From the point of view of effective field theory, there isn't much guidance. Choosing the simplest model consistent with observations suggests the pure ΛCDM model without scalars.[3] On the other hand, including time-dependent scalars is a more generic possibility, which is furthermore already assumed to be required during the presumed inflationary phase of cosmology.

Having scalar fields whose scale of time variation is the same order of magnitude as the age of the universe requires that the derivative of the scalar potential is similar to the scale of the potential itself: if the scalar potential varies on cosmological times scales, we have $H = \dot{a}/a \sim \dot{V}/V = V'\dot{\phi}/V$. From the scalar evolution equation

$$\ddot{\phi} + 3H\dot{\phi} + V'(\phi) = 0 , \tag{2}$$

we have V'/H as a typical scale for $\dot{\phi}$ and from the Friedmann equation we have $H \sim \sqrt{V}$ when potential energy is important (we take Planck units where $8\pi G/3 = 1$). Combing these, we have

$$\frac{1}{V}\frac{dV}{d\phi} = \mathcal{O}(1) . \tag{3}$$

In other words, we want the potential to have an order one variation when the scalar field varies by order one in Planck units. It is not possible to say whether or not this is natural from the point of view of effective field theory alone.[4]

One might hope that the likely presence or absence of time-dependent scalar fields in cosmology would be informed by quantum gravity considerations. It is expected that only certain effective field theories and perhaps only certain solutions of these effective theories are allowed in a UV complete theory (such as string theory) [7]. With complete knowledge of these allowed theories and solutions, we would ideally like to ask whether among the space of possible homogeneous and isotropic cosmological solutions with curvature scale in Planck units similar to our

[2] If there are scalar fields that vary on cosmological scales, the fluctuations in these fields will correspond to nearly massless scalar particles, and these can mediate long-range forces. These particles or the corresponding forces have not been observed, but that could be because they are part of the dark sector that is belived to contain most of the matter in the universe.

[3] Note however that this model may be in tension with some observations [6].

[4] It is sometimes suggested that having both V and V' sufficiently small in Planck units represents two separate fine tunings in the effective theory, making the case with time-dependent scalars less natural. However, we will see that there are natural reasons why these would represent the same scale.

own universe, the presence of time-dependent scalars is generic or rare (e.g. requiring some fine tuning).

At present, we are far from having a complete understanding of cosmological solutions in quantum gravity, so we won't be able to make any direct conclusions about the likelihood of time-dependent scalars. However, we will argue in this note that the holographic approach to quantum gravity (the AdS/CFT correspondence [8])—probably our most powerful tool for understanding UV complete quantum gravitational physics - does have something to say about the question. We will suggest that for gravitational effective theories with UV completion via a dual field theory, it is natural to have scalar fields satisfying condition (3). While we don't have a microscopic understanding of cosmological solutions of these effective theories, a very similar class of the solutions (making t spacelike and one of the x directions timelike in (1)) are well understood and correspond to the vacuum physics of quantum field theories. Solutions without time-varying scalars correspond to the vacuum of a dual CFT associated with our effective gravity theory, while solutions with time-varying scalars correspond to quantum field theories with renormalization group flow such that the IR or UV is governed by this CFT. The latter are more generic provided that the CFTs in question have scalar operators with dimensions of order one.

The AdS/CFT correspondence The AdS/CFT correspondence allows us to make some general statements about a very large class of gravitational effective field theories without knowing explicitly what these are. According to the correspondence, any four-dimensional effective gravitational theory with a UV completion and a stable AdS solution should correspond to a dual three-dimensional CFT, whose correlation functions are defined by asymptotic gravitational observables.[5,6] The vacuum state of this conformal field theory corresponds to the solution of the effective gravitational theory with pure anti-de-Sitter geometry (supported by the negative scalar potential at the extremum) and vanishing scalars.

Such four-dimensional gravitational effective theories with a dual three-dimensional CFT are arguably the only ones which we can at present be sure exist as full-fledged quantum theories. Thus, it makes sense to ask our question about the genericity of scalars with potential satisfying condition (3) in the context of these theories.

In a consistent gravitational effective theory with a dual CFT, the scalar fields are in one-to-one correspondence with the scalar operators in the dual CFT [9]. The various scalar fields correspond to the various independent directions that we can move away from the extremum of the potential. The m^2 of each scalar field determines the scale associated with the derivatives of the potential as we move away from the extremum in the direction corresponding to that scalar. The mass of a scalar field (working in Planck units) is related to the dimension Δ of the corresponding scalar operator in the dual CFT by

---

[5] This class of theories has a scalar potential with a negative extremum $-V_0$ for $\phi_i = 0$ and obeys $d^2V/d\phi^2 \geq -9/4V_0$ for each scalar in order to ensure stability of the AdS solution.

[6] In known examples, the UV completion of the effective gravitational theory involves string theory or M theory, but we do not need to assume this.

$$m^2/|V_0| = (\Delta^2 - 3\Delta) . \tag{4}$$

From this we see the key point: for a potential $V = V_0 + \frac{1}{2}m^2\phi^2 + \ldots$ (ignoring higher order terms), V will change by an order one fraction when ϕ changes by order one in Planck units provided that the scalar ϕ is associated to a CFT operator whose dimension is of order one. Thus, from the CFT point of view, having scalars for which the scale associated with derivatives of the potential is the same as the cosmological scale (our condition (3)) is completely natural. For example, $\Delta = 3$ represents the boundary between relevant and irrelevant operators in a three-dimensional CFT. A complete absence of operators with order one dimension would mean that the theory has no possible relevant deformations and no irrelevant deformations except by very large dimension operators. We are not aware of any CFT with these properties, though the space of three-dimensional CFTs is not very well understood.

Of course, CFTs dual to realistic gravitational effective field theories are special and it may be that among this special class of theories, it is more common not to have any low-dimension scalar operators. Indeed, having relatively few operators with dimensions of order one (i.e. having a "gap" in the spectrum) is a hallmark of holographic CFTs required to have a dual gravitational effective theory with relatively few light fields. But the best understood examples still do typically have some low-dimension scalar operators. For example, the AdS vacua of string theory constructed in [10, 11] with realistically small (negative) cosmological constants and small extra dimensions have a variety of light scalars associated with possible deformations of the extra dimensions in the model.

A standard concern is that in theories without supersymmetry, the scalar masses would receive quantum corrections that bring them up to the scale of supersymmetry breaking. However, similar corrections would naively bring up the scale of vacuum energy also. The dual CFT picture relates these two scales, suggesting that if there is some bulk mechanism that keeps the cosmological constant very small, it may also keep the scalar potential flat.[7]

In summary, in the gravitational effective field theories whose UV completions are understood, the cosmological constant scale appears to be the natural scale for the derivatives $dV/d\phi$ for the lightest scalars unless the dual CFT has absolutely no scalar operators with dimensions of order one.

Solutions Even if an allowed effective field theory has scalars satisfying the condition (2), there will generally still be solutions with time-varying scalars and solutions where the scalars are fixed. Further, some solutions of the effective field theory may correspond to legitimate spacetimes in the fundamental theory, while others may not. What we would like to ask is: given the space of cosmological solutions of the effective field theory that are valid from a microscopic point of view, how generic is the presence of time-varying scalars?

[7] Of course, it may be that non-supersymmetric large N CFTs with gravity duals typically have no light scalar operators, but we are not aware of any direct evidence for this. It may also be that there are no such theories, as conjectured in [12]. In this case, the consistent gravitational effective theories we are discussing would necessarily be supersymmetric, but this supersymmetry could be broken by the time-dependent scalars in cosmological solutions.

Unfortunately, while the effective field theories dual to CFTs have many cosmological solutions, we do not generally have a microscopic interpretation for these,[8] so we can't really say anything about the space of valid cosmological solutions. On the other hand, there is a very similar class of solutions for which we do have a good microscopic understanding. These are solutions of the same form as (1) where we switch the signature so that the previous time direction is spatial and one of the spatial directions is timelike[9]

$$ds^2 = d\tau^2 + a(\tau)dx_\mu dx^\mu \qquad \phi_i = \phi_i(\tau) . \tag{5}$$

Within this space of solutions, there is a large class where a microscopic description is understood. These solutions are dual to the vacuum physics of quantum field theories for which the CFT associated with our effective field theory describes either the UV or the IR physics.[10]

For a given gravitational effective theory, we have a solution of this type with no scalar fields turned on and solutions with scalar fields. The solution without scalar fields corresponds to the vacuum state of the unperturbed conformal field theory while solutions with τ-dependent scalars correspond to the vacuum state of some field theory with an RG flow were our CFT governs the UV or IR physics.

Provided that the conformal field theory in question has more than one relevant operator or can arise in more than one way via RG flow from a UV theory, the solution without varying scalars represents a measure zero subset of the space of solutions.[11] In this case, the solutions with varying scalars are more generic.

Summary We have reviewed that in the best understood gravitational effective theories, corresponding to holographic CFTs, having scalars with a potential that permits variations on cosmological time scales corresponds to having scalar operators in the dual CFT with order one dimensions. To the extent that having scalar operators of dimension one is a generic expectation for such CFTs, the condition (3) for the corresponding effective field theories seems generic rather than finely tuned.

We do not have a microscopic understanding of cosmological solutions of these effective theories, but we noted that for a very similar class of solutions, those without varying scalars holographically describe the vacuum physics of the CFT associated with the effective theory, while those with varying scalars describe the vacuum physics of quantum field theories with RG flows for which the CFT governs the UV or IR physics.

---

[8] See [13, 14] for a possible microscopic description of such cosmological solutions.

[9] This analogy was also employed recently in [15] to gain insight into inflationary cosmology. A more direct relation between such solutions and $\Lambda > 0$ cosmologies was suggested in [16].

[10] In these solutions, $a(\tau)$ is exponentially growing or decaying for small or large τ, so that the solution locally looks like AdS there.

[11] With only one relevant operator, the coefficient of this relevant operator is a dimensionful parameter, and all values of this parameter are physically equivalent.

Thus, from the point of view of holographic models of quantum gravity, having scalars which vary on cosmological time scales would appear to be a generic feature rather than a finely tuned one, unless there is some special property of CFTs with realistic gravity duals (complete absence of scalar operators with order one dimensions) for which there is so far no evidence.

Ultimately, the existence or not of time-dependent scalars should be determined by observations. But since at present these observations still allow for models with time-dependent scalars, we suggest that such models should be taken as the most natural extension of the ΛCDM model to be considered in situations when ΛCDM appears not to align with observations (e.g. when confronting the Hubble tension).

Relevance of $\Lambda < 0$ *models* The reader may wonder whether it is really appropriate to attempt to gain intuition about our cosmology by thinking about gravitational theories with $\Lambda < 0$ when we live in an accelerating universe. Our motivation for considering the $\Lambda < 0$ theories was simply to find a context where can ask about the genericity of condition (2) and of cosmology-like solutions with rolling scalars with some theoretical control. If V sets the typical scale for the smallest derivatives of the potential in situations where we have a complete microscopic understanding, it is plausible that this continues to be true more generally.[12] A separate reason that it may be relevant to consider $\Lambda < 0$ models is that if our universe does have time dependent scalars, these might well be moving towards some extremum of the potential with a negative value even though the current value is positive.[13] That is, if we don't assume from the start that the acceleration is caused by a cosmological constant, and we accept that scalars varying on cosmological time scales may be a generic situation in cosmology, then an effective field theory with a $\Lambda < 0$ extremum becomes a reasonable possibility for describing our observed physics.

Acknowledgements We would like to thank Stefano Antonini, Daniel Green, Arjun Kar, Lampros Lamprou, Juan Maldacena, Petar Simidzija, Brian Swingle, and Chris Waddell for helpful discussions. This work was performed in part at Aspen Center for Physics, which is supported by National Science Foundation grant PHY-1607611. This work is supported in part by the National Science and Engineering Research Council of Canada (NSERC) and in part by the Simons foundation via a Simons Investigator Award and the "It From Qubit" collaboration grant.

[12] It is interesting that for $V > 0$, the de Sitter swampland conjecture in string theory [17] suggests that the scale of V gives the lowest *possible* scale for the derivatives of the scalar potential, though it does not say whether this scale is typical.

[13] According to the de Sitter swampland conjecture in string theory [17], there are no $\Lambda > 0$ vacua , so time-dependent scalars moving towards a negative or zero value of the potential is required.

References

1. P.J.E. Peebles, B. Ratra, Cosmology with a time variable cosmological constant. Astrophys. J. Lett. **325**(L17) (1988). https://doi.org/10.1086/185100
2. B. Ratra, P.J.E. Peebles, Cosmological consequences of a rolling homogeneous scalar field. Phys. Rev. D **37**, 3406 (1988). https://doi.org/10.1103/PhysRevD.37.3406
3. R.R. Caldwell, R. Dave, P.J. Steinhardt, Cosmological imprint of an energy component with general equation of state. Phys. Rev. Lett. **80**, 1582–1585 (1998), https://doi.org/10.1103/PhysRevLett.80.1582 [astro-ph/9708069]
4. D. Brout et al., The pantheon+ analysis: cosmological constraints. Astrophys. J.**938**, 110 (2022), https://doi.org/10.3847/1538-4357/ac8e04 [2202.04077]
5. P. Agrawal, G. Obied, P.J. Steinhardt, C. Vafa, On the cosmological implications of the string swampland. Phys. Lett. B **784**, 271–276 (2018)
6. E. Di Valentino, O. Mena, S. Pan, L. Visinelli, W. Yang, A. Melchiorri et al., In the realm of the hubble tension-a review of solutions. Classical and Quantum Gravity **38**, 153001 (2021)
7. E. Palti, The swampland: introduction and review. Fortschritte der Physik **67**, 1900037 (2019)
8. J.M. Maldacena, The Large N limit of superconformal field theories and supergravity. Int. J. Theor. Phys. **38**, 1113–1133 (1999), https://doi.org/10.1023/A:1026654312961 [hep-th/9711200]
9. O. Aharony, S.S. Gubser, J.M. Maldacena, H. Ooguri, Y. Oz, Large N field theories, string theory and gravity. Phys. Rept. **323**183–386 (2000), https://doi.org/10.1016/S0370-1573(99)00083-6 [hep-th/9905111]
10. M. Demirtas, M. Kim, L. McAllister, J. Moritz, A. Rios-Tascon, Exponentially small cosmological constant in string theory. Phys. Rev. Lett. **128**, 011602 (2022), https://doi.org/10.1103/PhysRevLett.128.011602 [2107.09065]
11. M. Demirtas, M. Kim, L. McAllister, J. Moritz, A. Rios-Tascon, Small cosmological constants in string theory. JHEP **12**, 136 (2021), https://doi.org/10.1007/JHEP12(2021)136 [2107.09064]
12. H. Ooguri, C. Vafa, Non-supersymmetric AdS and the Swampland. Adv. Theor. Math. Phys. **21**, 1787–1801 (2017), https://doi.org/10.4310/ATMP.2017.v21.n7.a8 [1610.01533]
13. J.M. Maldacena, L. Maoz, Wormholes in AdS. JHEP **02**, 053 (2004), https://doi.org/10.1088/1126-6708/2004/02/053[hep-th/0401024]
14. S. Antonini, P. Simidzija, B. Swingle, M. Van Raamsdonk, Cosmology from the vacuum, 11220 (2203)
15. D. Baumann, D. Green, T. Hartman, Dynamical constraints on RG flows and cosmology. JHEP **12**, 134 (2019), https://doi.org/10.1007/JHEP12(2019)134 [1906.10226]
16. P. McFadden, K. Skenderis, Holography for cosmology. Phys. Rev. D **81**, 021301 (2010), https://doi.org/10.1103/PhysRevD.81.021301[0907.5542]
17. G. Obied, H. Ooguri, L. Spodyneiko, C. Vafa, De sitter space and the swampland, 1806.08362

Entanglement of Free-Fermion Systems, Signal Processing and Algebraic Combinatorics

Pierre-Antoine Bernard, Nicolas Crampé, Rafael I. Nepomechie, Gilles Parez, and Luc Vinet

Abstract This paper offers a review of recent studies on the entanglement of free-fermion systems on graphs that take advantage of methods pertaining to signal processing and algebraic combinatorics. On the one hand, a parallel with time and band limiting problems is used to obtain a tridiagonal matrix commuting with the chopped correlation matrix in bispectral situations and on the other, the irreducible decomposition of the Terwilliger algebra arising in the context of P-polynomial association schemes is seen to yield a simplifying framework.

Keywords Free fermions · Graphs · Entanglement · Heun operators · Association scheme · Terwilliger algebra

---

Dedicated to Gordon Semenoff on the occasion of his 70th birthday.

P.-A. Bernard · G. Parez · L. Vinet (✉)
Centre de Recherches Mathématiques and Département de Physique, Université de Montréal, Montréal, Québec, Canada
e-mail: luc.vinet@umontreal.ca

P.-A. Bernard
e-mail: pierre-antoine.bernard@umontreal.ca

G. Parez
e-mail: gilles.parez@umontreal.ca

N. Crampé
Institut Denis-Poisson CNRS/UMR 7013, Université de Tours - Université d'Orléans, Tours, France
e-mail: crampe1977@gmail.com

R. I. Nepomechie
Physics Department, University of Miami, Coral Gables, FL, USA
e-mail: nepomechie@miami.edu

L. Vinet
IVADO, Montréal, Québec, Canada

M. Paranjape et al. (eds.), *Proceedings of Gravity, Strings and Fields*,
CRM Series in Mathematical Physics,
https://doi.org/10.1007/978-3-031-91266-5_23

1 Introduction

Quantifying the entanglement of quantum many-body problems is a meaningful issue and free-fermion systems provide fertile ground for such studies. Roughly speaking, the model is taken in its ground state, split in two parts and the entanglement of one part relative to the other is explored. This review offers a survey of some papers written on the subject by the authors and collaborators over the recent years [1–8].

2 Free Fermions on Weighted Paths

We first consider fermionic chains with dynamics described as follows.

2.1 *The Hamiltonian and its Eigenstates*

The system is an open chain of length $N+1$ described by the Hamiltonian

$$H = \sum_{n=0}^{N-1} J_n (c_{n+1}^\dagger c_n + c_n^\dagger c_{n+1}) - \sum_{n=0}^{N} B_n c_n^\dagger c_n, \quad J_{-1} = 0, \tag{1}$$

with the constants J_n and B_n real and positive and c_n, $c_n^\dagger$ the fermionic operators obeying the anticommutation relations:

$$\{c_m, c_n^\dagger\} = \delta_{m,n}, \qquad \{c_m, c_n\} = 0, \qquad m, n = 0, \dots N. \tag{2}$$

Let $\{|n\rangle,\ n = 0, \dots, N\}$ be the orthonormal *position basis* made out of the characteristic vectors $|n\rangle$ of $\mathbb{C}^{N+1}$. Introducing the matrix Λ defined by

$$\Lambda|n\rangle = J_{n-1}|n-1\rangle - B_n|n\rangle + J_n|n+1\rangle, \tag{3}$$

the Hamiltonian H can be written in the form: $H = \sum_{m,n=0}^{N} \Lambda_{mn} c_m^\dagger c_n$ with $\Lambda_{mn} = \langle m|\Lambda|n\rangle$. Note that Λ can be viewed as the adjacency matrix of a weighted path with self-loops.

The eigenstates of H are obtained by diagonalizing Λ. This brings the orthonormal *energy basis* of $\mathbb{C}^{N+1}$:

$$\{|\omega_k\rangle,\ k = 0, \dots, N \ \mid\ \Lambda|\omega_k\rangle = \omega_k|\omega_k\rangle\}. \tag{4}$$

We order the energies as $\omega_k \leqslant \omega_{k+1}$. Owing to (3) and (4), the wavefunctions $\phi_n(k) = \langle n|\omega_k\rangle$ are expressed in terms of orthogonal polynomials of a discrete variable. Under the canonical transformation $\tilde{c}_k = \sum_{n=0}^{N} \phi_n(k) c_n$, the Hamiltonian is brought in the form $H = \sum_{k=0}^{N} \omega_k \tilde{c}_k^\dagger \tilde{c}_k$. With $|0\rangle\rangle$ the vacuum state defined by the property $c_n|0\rangle\rangle = 0$ for $n = 0, \ldots, N$, we readily have

$$H\tilde{c}_{k_1}^\dagger \ldots \tilde{c}_{k_m}^\dagger |0\rangle\rangle = \Big(\sum_{i=1}^{m} \omega_{k_i}\Big)\tilde{c}_{k_1}^\dagger \ldots \tilde{c}_{k_m}^\dagger |0\rangle\rangle. \tag{5}$$

2.2 Correlations and Entanglement

The ground state $|\Psi_0\rangle\rangle$ is obtained by filling the Fermi sea, that is by populating the vacuum with the excitations of energies up to $\omega_K < \omega_N : |\Psi_0\rangle\rangle = \tilde{c}_0^\dagger \ldots \tilde{c}_K^\dagger |0\rangle\rangle$.

The correlation matrix has elements

$$\bar{C}_{mn} = \langle\langle \Psi_0 | c_m^\dagger c_n | \Psi_0 \rangle\rangle = \sum_{k=0}^{K} \phi_m(\omega_k)\phi_n(\omega_k), \quad m, n = 0, \ldots, N, \tag{6}$$

thus showing that $\bar{C} = \sum_{k=0}^{K} |\omega_k\rangle\langle\omega_k|$, in other words that the correlation matrix $\bar{C}$ is the projector Π_E on the subspace of energy states in the Fermi sea.

In the following, we discuss entanglement between two complementary parts of the chain. Part 1 consists of the sites $\{0, 1, \ldots, \ell\}$, and part 2, the complement of part 1, is formed of the sites $\{\ell + 1, \ldots, N\}$. With the full system in the ground state and thus described by the density matrix $\rho = |\Psi_0\rangle\rangle\langle\langle\Psi_0|$, the entanglement of part 1 with part 2 is completely accounted for by the reduced density matrix $\rho_1 = \mathrm{Tr}_2|\Psi_0\rangle\rangle\langle\langle\Psi_0|$, where the trace is over the subspace of the Fock space generated by the creation operators $c_i^\dagger$ associated to part 2, i.e., with $i \in \{\ell + 1, \ldots, N\}$.

In the case of free-fermion chains, a considerable simplification [9] known as the Peschel trick occurs; namely the reduced density matrix, here a $2^{\ell+1} \times 2^{\ell+1}$ matrix, can be obtained from the chopped correlation matrix C. This matrix is the restriction of the full correlation matrix $\bar{C}$ to the subspace $S \subset \mathbb{C}^{N+1}$ spanned by the vectors $\{|0\rangle, \ldots, |\ell\rangle\}$ and thus, a $(\ell + 1) \times (\ell + 1)$ matrix. If we denote by Π_S the space projector on S, in view of the observation made before about $\bar{C}$, we arrive at the conclusion that the chopped correlation matrix C is the key entity and that it is given by the product of three projectors: $C = \Pi_S \Pi_E \Pi_S$. Indeed, the (von Neuman) entanglement entropy $\mathfrak{S} = -\mathrm{Tr}\, \rho_1 \log \rho_1$ is given by [9]

$$\mathfrak{S} = -\mathrm{Tr}\big[C \log C + (1 - C) \log (1 - C)\big]. \tag{7}$$

2.3 A Commuting (Heun) Operator

To compute the entanglement entropy $\mathfrak{S}$ we hence need to diagonalize the chopped correlation matrix $C = \Pi_S \Pi_E \Pi_S$. As a rule, this is a full matrix with eigenvalues near 0, a feature that does not facilitate a numerical treatment. We here wish to stress a useful observation, namely that for a class of fermionic chains, there exists a tridiagonal matrix T with a well-behaved spectrum and such that $[T, C] = 0$, see [1, 2, 10].

Such occurrences are particularly opportune because T shares its eigenvectors with C and is easier to diagonalize (numerically). We shall now discuss situations when such a commuting operator is present and indicate how it can be obtained.

The key is bispectrality. If the constants J_n and B_n in the Hamiltonian H are such that the orthogonal polynomials arising in the wavefunctions $\phi_n(k)$ belong to the Askey scheme [11], these wavefunctions are bispectral. (Note that there are many such choices.) In these cases, there is an operator X on $\mathbb{C}^{N+1}$ that is diagonal in the position basis and tridiagonal in the energy basis; that is, there is an X such that

$$X|n\rangle = \lambda_n|n\rangle, \qquad X|\omega_k\rangle = \bar{J}_{k-1}|\omega_{k-1}\rangle - \bar{B}_k|\omega_k\rangle + \bar{J}_k|\omega_{k+1}\rangle. \tag{8}$$

This follows from the fact that $\phi_n(k) = \langle n|\omega_k\rangle$, being bispectral, obeys a difference equation of the form

$$\lambda_n\phi_n(k) = \bar{J}_{n-1}\phi_n(k-1) - \bar{B}_k\phi_n(k) + \bar{J}_k\phi_n(k+1), \tag{9}$$

in addition to the three term recurrence relation that is implied by the reciprocal action of the operator Λ in the two bases:

$$\Lambda|\omega_k\rangle = \omega_k|\omega_k\rangle, \qquad \Lambda|n\rangle = J_{n-1}|n-1\rangle - B_n|n\rangle + J_n|n+1\rangle. \tag{10}$$

Under such circumstances, the commuting operator can be obtained as follows. To all bispectral problems, one may associate a so-called algebraic Heun operator defined as the most general bilinear expression in the two bispectral operators [12]. For simplicity, we shall here consider a special case and the operator

$$\bar{T} = \{X, \Lambda\} + \mu X + \nu\Lambda, \tag{11}$$

where the scalars μ and ν are for the moment unspecified. Clearly, $\bar{T}$ is tridiagonal in both the position and the energy bases. From the specific action of $\bar{T}$ in these bases, it is not difficult to see that by choosing for μ and ν the values

$$\mu = -(\omega_K + \omega_{K+1}), \qquad \nu = -(\lambda_\ell + \lambda_{\ell+1}), \tag{12}$$

$\bar{T}$ preserves the subspace S and the one spanned by the energy eigenvectors belonging to the Fermi sea. With T the restriction to S of $\bar{T}$, it follows that $[T, \Pi_S] = [T, \Pi_E] = 0$ and hence that $[T, C] = 0$ since $C = \Pi_S\Pi_E\Pi_S$.

This result stems from a striking parallel [13, 14] with the celebrated treatment [15] by Slepian et al. of the time and band limiting problem in signal processing. This deserves a short digression. The central and prototypical question is: how to best concentrate in a time interval $-T \leq t \leq T$ a signal $f(t)$ which is limited to a frequency band $[-W, W]$? This is in principle answered by looking for the eigenfunctions of the integral operator G with the sinc kernel, namely by solving

$$GF(p) = \int_{-W}^{W} dp' \frac{\sin(p-p')T}{\pi(p-p')} F(p') = \lambda F(p). \tag{13}$$

Note that G can also be written as the product of three projector: $G = \Pi_W^p \widetilde{\Pi}_T^p \Pi_W^p$ with $\Pi_L^x g(x) = [\Theta(x+L) - \Theta(x-L)]g(x)$, $\widetilde{\Pi}_T^p$ the Fourier transform of Π_T^t and $\Theta(x)$ the Heavyside function.

This should be the end of the story but G, a non-local operator, proves also intractable numerically. The way out came from the remarkable discovery [15] that the spheroidal wave operator

$$D = \frac{1}{2}\left\{\frac{d^2}{dp^2}, p^2\right\} - W^2 \frac{d^2}{dp^2} + T^2 p^2 \tag{14}$$

commutes with G. This result can be obtained [12] from the rather obvious observation that the Fourier function e^{ipx} are solutions to a simple bispectral problem with $\frac{d^2}{dp^2}$ and p^2 playing the roles of Λ and X in the scenario described before. The parameters of the Heun operator are here also fixed by demanding that the commutators with the projectors Π_W^p and $\widetilde{\Pi}_T^p$ be equal to 0.

The parallel between the entanglement analysis of free-fermion chains and the band and time limiting problem is thus quite clear. The filling of the Fermi sea in the construction of the ground state corresponds to the band limiting. Splitting space in two parts and restricting to one is akin to time limiting. The main problem regarding the fermionic chain is to diagonalize the chopped correlation matrix $C = \Pi_S \Pi_E \Pi_S$ while in the band and time limiting case, it is to diagonalize the integral operator $G = \Pi_W^p \widetilde{\Pi}_T^p \Pi_W^p$. The understanding that the existence of a commuting operator is rooted in underlying bispectral problems leads then in both contexts, to a simple identification of this operator from the respective algebraic Heun operators.

2.4 A Non-homogeneous Example: The Krawtchouk Chain

A nice example of a chain with bispectral features arises for the following choice of parameters:

$$J_n = \sqrt{(N-n)(n+1)p(1-p)}, \quad B_n = -(Np - n(1-2p)), \quad p \in [0, 1]. \tag{15}$$

In this case $\omega_k = \lambda_k = k$ and the wavefunctions are given as follows in terms of Krawtchouk polynomials (the ${}_2F_1$ part in the formula below) [11]:

$$\phi_n(k) = (-1)^n \sqrt{p^{n+k}(1-p)^{N-n-k}\binom{N}{n}\binom{N}{k}}\, {}_2F_1\left(\begin{matrix}-n, -k\\ -N\end{matrix}; \frac{1}{p}\right). \tag{16}$$

Note that B_n is a constant for $p = 1/2$. In this case, the non-zero matrix elements $T_{mn} = T_{nm}$ of the symmetric commuting operator are given by [1]:

$$T_{n,n} = \frac{N}{2}(2n - 2\ell - 1) - n(2K + 1), \quad T_{n-1,n} = (n - \ell - 1)\sqrt{n(N - n + 1)}. \tag{17}$$

As for the entanglement entropy for the half chain and at half filling it is well approximated by[1] [6]:

$$\mathfrak{S} = \frac{1}{6}\log\frac{N+1}{2} + a(p) - \frac{1}{2(N+1)}\frac{\cos(\frac{\pi}{2}\frac{N+1}{m(p)})}{\sin(\frac{\pi}{2m(p)})} + \dots \tag{18}$$

where

$$m(p) = \frac{1}{2}\left(1 - \log p + \frac{1 - \log 2}{2p}\right), \tag{19}$$

and $a(p)$ is a non-universal constant with respect to N.

3 Free Fermions on Graphs

We now consider free fermions on higher dimensional non-oriented graphs. Let $V = \{v_0, \dots, v_D\}$ be the set of vertices and $E \subset V \times V$, the set of edges. The orthonormal canonical basis $\{|v_0\rangle, \dots, |v_D\rangle\}$ will be referred to as the position basis with the vector $|v_i\rangle$ associated to the vertex v_i. The $(D + 1) \times (D + 1)$ symmetric adjacency matrix A has entries given by

$$A_{ij} = \langle v_i|A|v_j\rangle = \begin{cases} 1 & \text{if} \quad (v_i, v_j) \in E \\ 0 & \text{otherwise.} \end{cases} \tag{20}$$

The Hamiltonian $\mathcal{H}$ is taken to be $\mathcal{H} = \sum_{m,n=0}^{D} A_{m,n} c_m^\dagger c_n$ where c_n and $c_n^\dagger$ are the usual fermionic operators at v_n. Thus the adjacency matrix (made out of only 0s and 1s) plays a role similar to the matrix Λ introduced in the description of fermionic chains. We shall focus below on a natural family of graphs.

[1] Note that in (18) we fixed a small sign typo from the original publication [6].

3.1 The Hypercube and the Krawtchouk Chain

In the case of the hypercube Q_N in N dimensions, $V = \{0, 1\}^{\otimes N}$; that is, the vertices are strings of N bits. Two vertices $v_i, v_j \in V$ are linked if they differ by only one entry, i.e., if they are at Hamming distance $d(v_i, v_j) = 1$. The case $N = 1$ corresponds to the complete graph (where all vertices are connected to one another) with two vertices K_2. In general, Q_N is the N-fold Cartesian product of K_2, i.e. $Q_N = (K_2)^{\square N}$.

Let us now establish a connection between the system consisting of fermions on the hypercube and the Krawtchouk chain. Pick $0 = (0, \dots, 0)$ as a reference point on Q_N. Organize V in columns $V_n = \{x \in V \mid d(0, x) = n\}$ made out of all vertices at distance $n = 0, \dots, N$ of 0. It is easy to see that $k_n = \text{Card}(V_n) = \binom{N}{n}$. Let us label the vertices in the column V_n by V_{nm}, $m = 1, \dots, k_n$ and form the n-qbit Dicke or column vector states:

$$|col\ n\rangle = \frac{1}{\sqrt{k_n}} \sum_{m=1}^{k_n} |V_{nm}\rangle. \tag{21}$$

It is easy to see [16, 17] that $\langle col\ n+1|\ A\ |col\ n\rangle = \sqrt{(n+1)(N-n)}$ which is equal to twice the expression of J_n for the Krawtchouk chain when $p = \frac{1}{2}$ as per (15). In other words, for Λ corresponding to the Krawtchouk chain with $p = \frac{1}{2}$, we have $\langle col\ n+1|\ A\ |col\ n\rangle = 2\ \langle n+1|\ \Lambda\ |n\rangle$. When $p = \frac{1}{2}$, the diagonal term in the Hamiltonian of the Krawtchouk chain does not depend on n and thus yields a global constant that can be subtracted; we hence find that up to an overall multiplicative factor the hypercube system projects to the $p = \frac{1}{2}$ Krawtchouk chain. The reason for this is that Q_N is a distance-regular graph, implying that each vertex in column V_n is connected to the same number of vertices in the column V_{n+1} and vice versa. This observation suggests that the entanglement properties of free fermions on the hypercube bear a relation with those of the Krawtchouk chain. This connection can further be understood in terms of association schemes. The hypercube Q_N is part of the family of graphs on V consisting of those where it is the vertices at Hamming distance 0, 1, or 2, up to N that are connected by an edge. These are the graphs of the (binary) Hamming association scheme that have adjacency matrices whose action is closed on the space spanned by the column vector states. We discuss these constructs more generally next.

3.2 Association Schemes

An important concept in algebraic combinatorics is that of (symmetric) d-class association schemes [18] which can be considered as ensembles of $d + 1$ undirected graphs on a set of vertices V with cardinality $|V|$ satisfying certain axioms. Such an ensemble of graphs may be seen as colorings of the edges of the complete graph with d colors. In terms of the corresponding adjacency matrices A_i, $i = 0, \dots, d$, the axioms are equivalent to

$$\sum_{i=0}^{d} A_i = J, \quad A_0 = I, \quad A_i = A_i^T, \quad A_i A_j = \sum_{k=0}^{d} p_{ij}^k A_k, \tag{22}$$

where J is the all 1 matrix, I the identity and p_{ij}^k integers referred to as the intersection numbers. The commutative $d+1$ algebra thus generated by the adjacency matrices is called the Bose-Mesner algebra. Since the symmetric adjacency matrices all commute, they can be diagonalized simultaneously and admit the spectral decomposition

$$A_i = \sum_{j=0}^{d} \theta_i(j) E_j, \qquad E_i = \frac{1}{|V|} \sum_{j=0}^{d} \theta_i^*(j) A_j, \tag{23}$$

with E_j the idempotents projecting on the eigenspaces ($E_i E_j = \delta_{ij} E_i$ and $\sum_{i=0}^{d} E_i = I$). Finally, it is known that the distance matrices of distance regular graphs lead in a one-to-one way to association schemes that are P-polynomial in that $A_i = p_i(A_1)$ for p_i a polynomial of degree i. We shall often assume this situation to hold in the following. Dually, an association scheme is called Q-polynomial if there is an ordering such that its primitive idempotents E_i are given as polynomials of degree i of E_1 (under the entry-wise product).

3.3 *The Terwilliger Algebra and the Correlation Matrices*

An algebra introduced by Terwilliger [19] and extending the Bose-Mesner one can be attached to an association scheme and is relevant to our entanglement studies. Its definition requires picking a reference vertex v_0 and introducing the dual matrices A_i^* and E_i^* as the diagonal matrices with entries:

$$[A_i^*(v_0)]_{vv} = |V|[E_i]_{v_0 v}, \qquad [E_i^*(v_0)] = [A_i]_{v_0 v}. \tag{24}$$

Note that $E_i^* E_j^* = \delta_{ij} E_i^*$ and that E_i^* projects on the position subspace spanned by the vectors $|v\rangle$ corresponding to vertices connected to v_0 in the graph with adjacency matrix A_i, i.e., the column space at distance i from v_0. Recall that E_i projects on the energy eigenspaces of the adjacency matrices that intervene in the Hamiltonians. The Terwilliger algebra $\mathfrak{T}$ is generated by the adjacency matrices $\{A_0, \ldots, A_d\}$ and their duals $\{A_0^*, \ldots, A_d^*\}$ or equivalently by the sets of projectors $\{E_0, \ldots, E_d\}$ and $\{E_0^*, \ldots .E_d^*\}$.

It is appropriate to remark at this point that the entanglement analysis of fermions on graphs proceeds much as for fermion chains. The ground state is defined by filling the vacuum state $|0\rangle\rangle$ with the excitations corresponding to a subset SE of the eigenvalues $\theta(j)$ of the adjacency matrix A (or of combination) chosen for Hamiltonian. The correlation matrix $\bar{C}$ is given by

$$\bar{C} = \sum_j E_j = \Pi_{SE} \quad \text{for} \quad \theta(j) \in SE. \tag{25}$$

The bipartition of the vertices (or positions) V into SV (part 1) and its complement (part 2) is typically be done by picking columns at the successive distances from 0 to $\ell < d$ with respect to a vertex v_0. Since $E_i^* = \sum_{v|d(v_0,v)=i} |v\rangle\langle v|$, the projector on the position vectors of part 1 is $\Pi_{SV} = \sum_{i=0}^{\ell} E_i^*$ and the chopped correlation matrix thus reads $C = \Pi_{SV}\Pi_{SE}\Pi_{SV}$. It follows that this matrix C that needs to be diagonalized actually represents an element of the Terwilliger algebra $\mathfrak{T}$, a point worth underscoring.

A natural strategy to carry out the entanglement analysis of fermions on graphs of (P- and Q- polynomial) association schemes thus presents itself: (i) Identify the Terwilliger algebra $\mathfrak{T}$ for the scheme; (ii) Decompose the regular representation of $\mathfrak{T}$ on $\mathbb{C}^{|V|}$ into its irreducible components and as a result; (iii) Simplify the diagonalization of C by working on irreducible subspaces. This last step can further be aided by the presence of a commuting operator belonging also to $\mathfrak{T}$. For P- and Q- polynomial schemes, $\mathfrak{T}$ is generated by $A_1 = A$ and $A_1^* = A^*$ solely as all the others matrices are polynomials of one or the other. The energy and position bases are respectively the eigenbases of A and A^*. The action of each of these elements in the eigenbasis of the other is block tridiagonal. This implies that the overlaps between the two bases (the wavefunctions) are solutions of bispectral problems. Considering therefore the (generalized) algebraic Heun operator $\bar{T} = \{A, A^*\} + \mu A^* + \nu A$, it is possible to find μ and ν so that $[\bar{T}, \Pi_{SV}] = [\bar{T}, \Pi_{SE}] = 0$ and hence $[\bar{T}, C] = 0$.

3.4 Entanglement on Graphs of the (Binary) Hamming Scheme

Let us bring as an example the celebrated Hamming scheme which was referred to at the end of the subsection on the hypercube. It is known to be P- and Q- polynomial with the self-dual Krawtchouk polynomials arising in the expression of the (dual) adjacency matrices in terms of A and A^*. These two matrices hence generate $\mathfrak{T}$ which we easily recognize to be the Lie algebra $\mathfrak{su}(2)$. By applying the definitions of the adjacency matrix and its dual for K_2, it is seen that:

$$A = \begin{pmatrix} 0 & 1 \\ 1 & 0 \end{pmatrix} = \sigma_x, \qquad A^* = \begin{pmatrix} 1 & 0 \\ 0 & -1 \end{pmatrix} = \sigma_z, \tag{26}$$

where σ_x and σ_z are the usual notation for the Pauli matrices. Since as already noted, the hypercube $Q_d = (K_2)^{\square d}$ it follows that in this case:

$$A = \sum_{i=0}^{d-1} \underbrace{I \otimes \cdots \otimes I}_{i} \otimes \sigma_x \otimes I \cdots \otimes I, \quad A^* = \sum_{i=0}^{d-1} \underbrace{I \otimes \cdots \otimes I}_{i} \otimes \sigma_z \otimes I \cdots \otimes I. \tag{27}$$

This corresponds to the diagonal embedding of $\mathfrak{su}(2)$ into $\mathfrak{su}(2)^{\otimes d}$ with A and A^* resulting from the repeated application of the coproduct. The underlying representation is of course very well known. As anticipated, when part 1 is made out of columms the irreducible decomposition reduces the entanglement characterization problem on the graph to a combination of Krawtchouk chains, one of which being the chain discussed in Sect. 3.1.

4 Conclusion

The connections that have been highlighted between the entanglement analysis of fermionic systems, signal processing and algebraic combinatorics have been put to use in a variety of contexts but many avenues still remain to be explored. The entanglement of free fermions on the Hamming, Johnson, Hadamard and folded cube graphs have been examined [3–5, 8]. It would certainly be worth determining what other schemes [20] such as the dual polar [21] or Grassmann ones entail. In some cases the irreducible decompositions of the regular representation of the Terwilliger algebras have not yet been spelled out. P-multivariate association schemes and their graph descriptions are currently generating much interest [22–25]. Examining beyond the initial studies [6] the entanglement of fermionic systems built upon those structures should warrant attention. Finally, the bearing of graph symmetries on the entanglement properties of free fermions on these graphs is certainly a question that we plan to study in the future.

Acknowledgements LV is grateful to the organizers of the conference at the CRM in honour of Gordon Semenoff for having invited him to be part of this celebration. PAB holds an Alexander-Graham-Bell scholarship from the Natural Sciences and Engineering Research Council of Canada (NSERC). NC is partially supported by the IRP AAPT of CNRS. RN enjoys funding from the National Science Foundation under Grant No. PHY 2310594. GP holds a FRQNT and a CRM-ISM postdoctoral fellowship, and acknowledges support from the Mathematical Physics Laboratory of the CRM. The research of LV is funded in part by a Discovery Grant from NSERC.

References

1. N. Crampé, R. Nepomechie, L. Vinet, Free-Fermion entanglement and orthogonal polynomials. J. Stat. Mech., 093101 (2019)
2. N. Crampé, R. Nepomechie, L. Vinet, Entanglement in fermionic chains and bispectrality. Rev. Math. Phys. **33**(7) (2021)
3. P.-A. Bernard, N. Crampé, L. Vinet, Entanglement of free fermions on Hamming graphs. Nuclear Phys. B **986**, 116061 (2023)
4. P.-A. Bernard, N. Crampé, L. Vinet, Entanglement of free fermions on Johnson graphs. J. Mathemat. Phys. **64**, 061903 (2023)
5. N. Crampé, K. Guo, L. Vinet, Entanglement of Free Fermions on Hadamard Graphs. Nuclear Phys. B **360**, 115176 (2020)

6. P.-A. Bernard, N. Crampé, R. Nepomechie, G. Parez, L. Poulain d'Andecy, L. Vinet, Entanglement of inhomogeneous free fermions on hyperplane lattices. Nuclear Phys. B **984**, 115975 (2022)
7. G. Parez, P.-A. Bernard, N. Crampé, L. Vinet, Multipartite information of free fermions on Hamming graphs. Nuclear Phys. B **990**, 116157 (2023)
8. P.-A. Bernard, Z. Mann, G. Parez, L. Vinet, Absence of logarithmic enhancement in the entanglement scaling of free fermions on folded cubes. J. Phys. A: Math. Theor. **57**, 015002 (2024)
9. I. Peschel, Calculation of reduced density matrices from correlation functions. J. Phys. A: Math. Gen. **36**, L205 (2003)
10. V. Eisler, I. Peschel, Properties of the entanglement Hamiltonian for free-fermion chains. J. Stat. Mech., 104001 (2018)
11. R. Koekoek, P.A. Lesky, R.F. Swarttouw, *Hypergeometric Orthogonal Polynomials and Their q-Analogues*. Springer (2010)
12. F.A. Grünbaum, L. Vinet, A. Zhedanov, Algebraic Heun operator and band-time limiting. Commun. Mathem. Phys. **364**, 1041 (2018)
13. D. Gioev, I. Klich, Entanglement entropy of fermions in any dimension and the Widom conjecture. Phys. Rev. Lett. **96**, 100503 (2006)
14. V. Eisler, I. Peschel, Free-fermion entanglement and spheroidal functions. J. Stat. Mech., 04028 (2013)
15. D. Slepian, Some comments on Fourier analysis, uncertainty and modeling. SIAM Rev. **25**, 379 (1983)
16. M. Christandl, N. Datta, A. Ekert, A.J. Landahl, Perfect state transfer in quantum spin networks. Phys. Rev. Lett. **92**, 187902 (2004)
17. P.-A. Bernard, A. Chan, E. Loranger, C. Tamon, L. Vinet, A graph with fractional revival. Phys. Lett. A **382**, 259 (2018)
18. E. Bannai, E. Bannai, T. Ito, R. Tanaka, *Algebraic Combinatorics*. De Gruyter (2021)
19. P. Terwilliger, The subconstituent algebra of an association scheme. J. Algebraic Combinatorics (Part I) **1**(363) (1992), (Part II) **2**(73) (1993), (Part III) **2**(177) (1993)
20. A.E. Brouwer, A.M. Cohen, A. Neumaier, *Distance-Regular Graphs*. Springer (1989)
21. P.A. Bernard, N. Crampé, L. Vinet, The Terwilliger of symplectic dual polar graphs, the subspace lattices and $\mathcal{U}_q(sl_2)$. Discrete Mathem. **12**, 113169 (2023)
22. P.A. Bernard, N. Crampé, L. Poulain d'Andecy, L. Vinet, M. Zaimi, Bivariate P-polynomial association schemes. Discr. Math. **348**, 114332 (2025)
23. E. Bannai, H. Kurihara, D. Zhao, Y. Zhu, Multivariate P and/or Q-polynomial association schemes. J. Comb. Theory Ser. A **213**, 106025 (2023)
24. N. Crampé, L. Vinet, M. Zaimi, X. Zhang, A bivariate Q-polynomial structure for the non-binary Johnson scheme. J. Comb. Theory Ser. A **202**, 105829 (2024)
25. P.-A. Bernard, N. Crampé, L. Vinet, M. Zaimi, X. Zhang, m-distance regular graphs and their relation to multivariate P-polynomial association schemes. Discr. Math. **347**, 114179 (2024)

Ultralight Dark Matter Stars

Peter Suranyi and L. C. R. Wijewardhana

Abstract It has been proposed that ultra light bosons could provide a solution to the dark matter mystery. Most likely candidates are particles, similar to axions, postulated to solve the strong CP problem in QCD, and to axion-like particles present in string compactification models. They can be created via non-thermal processes in the early cosmos, and they can form gravitationally bound condensates during the cosmic evolution. In a contribution to this volume we summarize our recent research on ultra-light dark matter condensate stars.

Keywords Quantum field theory · Dark matter

1 Introduction

We consider being asked to contribute to Gordon's Festschrift as a great honor and an enormous privilege. Gordon has made substantial contributions to quantum field theory and condensed matter physics. We have previously had the pleasure of working with Gordon on several research projects, where we have profited from his keen scientific insight and mastery of quantum field theory. In this article we present an overview of our latest research on Bose condensate stars, a topic which we enjoyed discussing with Gordon on several occasions.

The microscopic quantum nature of dark matter (DM) is an open question. The most popular candidates, weakly interacting massive particles (WIMP), have not been detected in direct detection experiments or at particle accelerators. An alternative hypothesis, being widely studied, is that the dark sector is a macroscopic manifestation of quantum condensates of fundamental bosons. Boson particles which carry integer spin are prevalent in current models of particle physics. Spin zero bosons arise in field theories of phenomenological importance. The standard model Higgs is the well known example. An axion, which is postulated to solve the strong CP

P. Suranyi · L. C. R. Wijewardhana (✉)
University of Cincinnati, Cincinnati, OH, USA
e-mail: rohana.wijewardhana@gmail.com

M. Paranjape et al. (eds.), *Proceedings of Gravity, Strings and Fields*,
CRM Series in Mathematical Physics,
https://doi.org/10.1007/978-3-031-91266-5_24

problem, endemic to QCD, is another well motivated candidate. Axion like degrees of freedom generically arise in string theories, as Kaluza-Klein zero modes, from compactification of anti symmetric tensor fields defined in 10 dimensions. If long lived scalar degrees of freedom are abundant in the universe, they could form stable structures, due to their self gravitation. Below a critical mass, quantum pressure arising due to the uncertainty relation can balance all the attractive interactions. Due to Bose statistics, cold bosons tend to condense forming Bose-Einstein condensates (BECs). This leads to the natural hypothesis that free bosons, as well as BECs and boson stars, may be major components of dark matter. By studying the properties of these condensates and structures one can test whether they are viable dark matter candidates.

Over sixty years ago J.A.Wheeler surmised that electro-magnetic or gravitational waves could form bound states due to their gravitational self interactions [1], which he termed Geons. Further work on this was done by other investigators [2, 3]. The stability of such configuration is yet to be established [4]. Their analysis was purely classical and the role of photons and gravitons, as quanta of electromagnetic and gravitational fields, was not taken into account.

The investigation of gravitationally bound scalar boson states started with the work of Kaup [5] and Ruffini and Bonazzola [6]. Kaup solved the Klein-Gordon equation for a free complex scalar field in an asymptotically flat spherically symmetric background. He termed the resulting solutions Klein-Gordon Geons. He found that stable bound states exist if the mass of stars is less than $0.633{M_p}^2/m$ where m is the mass of the boson and M_p is Planck mass, which is a much smaller scale than the Chandrasekhar bound for fermion stars, ${M_p}^3/m^2$.

Ruffini and Bonazzola used a different approach to solve the same equations of motion as Kaup, assuming spherically symmetric metric and a field profile ansatz [6, 7]. They did this by assuming that the source of gravitational interactions was the expectation value of the normal ordered energy-momentum tensor operator of a second quantized free Hermitian scalar field theory, evaluated in the N-particle ground state. Their basis states for second quantization were the wave solutions of the scalar field equation. That approach yielded an equation of motion for the spherically symmetric wave function of the N-particle state, representing a condensate of N bosons in a single state. They found stable localized wave functions for the condensate, indicating the interesting possibility that lumps of self-gravitating scalar field configurations could exist in nature.

It is important to note that the particle number is not conserved for real fields. Then, provided interactions are included, condensates composed of real particles are unstable to decay. The Ruffiini-Banazzola approach has the significant benefit of providing a straightforward method for calculating the decay rate of boson stars composed of real fields. It also makes possible the incorporation of higher harmonic modes or ringing modes in the analysis.

Boson star solutons evade Derrick's theorem [8], which states that "no stable localized, static, regular scalar field solutions exist", by the fact that the solutions have explicit time dependence. For complex scalar fields the ansatz of the form $\Phi(x, t) = \phi(x)exp(i\omega t)$ makes the solution time dependent. However, with that ansatz the energy momentum tensor is time independent and the background metric is static. For real fields of the form $\Phi(x,t) = \phi(x)exp(i\omega t) + h.c.$ the energy momentum tensor is time dependent and the background metric also oscillates in time.

Bose stars made of interacting complex scalar fields with repulsive interactions were discussed by the authors of [9]. They found that they become unstable if their mass exceeds $0.232\sqrt{\lambda/4\pi}\,{M_p}^3/m^2$, where λ is the repulsive 4-scalar coupling, similar to the limiting mass found by Chandrasekhar for stars held by electron degeneracy pressure.

2 Ruffini Bonazzola Approch to Axion Stars

The Ruffini-Bonazzola approach can be readily generalized to study axion condensate stars [10]. The axion quantum field satisfies a nonlinear Klein-Gordon equation in gravitational background. The potential energy of the Hermitian axion field Φ is ${m_a}^2 {f_a}^2\,(1 - \cos(\Phi/f_a))$. Here m_a is the axion mass and f_a is its decay constant. We assume, that the critical temperature in the axion stars under consideration is so high that every axion is in the ground state. Therefore, we use the mode expansion $\Phi = R(r)exp(iE_0t)a_0 + h.c.$ where a_0 is the annihilation operator of an axion, and E_0 is the ground state energy. In the large N limit $\langle N|\cos(\Phi/f_a)|N\rangle \simeq J_0(2\sqrt{N}\,R\,/\,f_a)$, an effective potential that is different from the classical potential already in fourth order of wave function, R. We then obtain the equations of motion by taking the expectation value of the Einstein equations and the scalar field equation.

Assuming that the metric is of the form

$$ds^2 = -[1 + \delta\, b(r)]\, dt^2 + [1 + \delta\, a(r)]\, dr^2 + r^2\, d\Omega^2,$$

we calculated the expectation values of the rr and tt components of the Einstein equations in N-axion states $|N\rangle$, and of the nonlinear scalar field equations between N-axion and $N-1$ axion states $\langle N|$ and $|N-1\rangle$ to obtain a closed system of equations for the axion field and for the components of the metric tensor. These expectation values are calculated in tree approximation, which is justified because we consider weakly bound systems, in which particle-antiparticle production and loop contributions are suppressed.

We also assumed that the binding energy of axions in an axion star was small and the gravitational field was weak. That allowed us to perform a double expansion in the ratio $\delta = 8\pi\,{f_a}^2/{M_P}^2$, where M_P was the Planck mass, and in the binding energy $\Delta = \sqrt{{m_a}^2 - {E_0}^2}\,/\,m_a$, where E_0 was the ground state energy.

Our expansion plus appropriate scaling of variables allowed us to probe dilute axion stars. Other authors started with the RB method but only found denser solutions [11, 12], later found to be structurally unstable [13]. Our method reduces to the Gross-Pitaevski-Poisson equations used by other authors [14–16], to leading order, but we can go beyond that in a systematic way.

Though a dilute star sits at a metastable local minimum of the energy functional, we have shown in an appendix of Ref. [17] that the probability of tunneling out of the potential well is extremely small, until the mass becomes infinitesimally close to its maximum allowed value. For axions of mass m_a and decay constant f_a the maximum mass of the axion star was found to be of the order $M \simeq 10 M_P f_a/m_a$ and size $R \simeq 0.1 M_P/(f_a m_a)$. Axion star size scales inversely with its mass. For QCD axions with $f_a = 6 \times 10^{11}$ GeV and mass $m_a = 10^{-5}$ eV the maximum mass is $10^{-11} M_\odot$ with size $R \sim 200$ km. For axion stars made of Fuzzy Dark Matter axions with $f_a = 10^{17}$ GeV and $m_a = 10^{-22}$ eV the maximum mass is $10^{10} M_\odot$ with corresponding radius of $R \sim 100$ ly. Consequently, the mass and radius of a somewhat subcritical fuzzy axion star would be within the mass and size range of some ultra compact dwarf galaxies. They could also occupy the core of larger galaxies.

Once we incorporated higher harmonic modes in the field expansion [18] we were able to systematically probe the dense axion star region. In [19] structurally stable dense solutions were found, corresponding to the same QCD axion parameters, belong to an unstable and denser branch, with masses in the range of $10^{-16} M_\odot$ and sizes of a few meters.

We have generalized the Ruffini-Bonazzola [6] formalism for finding solutions for axion stars by including higher harmonics in the expansion of the field operator [18]. In the original Ruffini-Bonazzola approach the field operator was expanded in terms of the lowest bound state harmonic mode. The N particle condensate expectation value $\langle N|T_{\mu,\nu}|N\rangle$ of the energy momentum tensor operator entered into the right hand side of the Einstein field equations. The matrix element relation $\langle N|(\Box + m^2)\phi|N-1\rangle = 0$ was also imposed. All other matrix elements of the Klein Gordon operator equation, between different particle number states automatically vanish since the field expansion is linear in creation and annihilation operators. For spherically symmetric metrics, there are two independent Einstein equations in the tt and rr directions. Therefore, for non rotating solutions we have three equations for the three unknowns, tt and rr metric coefficients and the radial wave function R(r). Now, if we want to add interactions we need to modify this procedure, to match the number of independent equations with the number of variables. Say, in the presence of ϕ^4 interaction, to impose the condition $\langle N|(\Box + m^2)\phi + \lambda\phi^3|N-3\rangle = 0$, we need to introduce another function in the field expansion to get a consistent set of equations. That is taken care of by adding the third harmonic term in the field expansion. If we want to introduce all possible interactions with even powers, like for the cosine axion potential, an infinite number of functions need to be introduced. In our generalization an expansion was carried out including all the higher harmonic modes in terms of the binding energy parameter Δ. The coupled equations can be solved order by order. The system of equations were analyzed up to fifth order in Δ where interaction terms up to ϕ^6 were included. Recursion relations among the three coefficient functions were utilized to derive an effective equation for the lowest mode wave function. We have used this formalism to discuss axion stars. When $\Delta \leq f/M_p$ axion stars are in the dilute branch where gravity as well as the leading attractive ϕ^4 interaction are important. When $f/M_p \leq \Delta << 1$ gravity is decoupled but the leading ϕ^4 interaction plays the dominant role. This is the transition branch. When Δ approaches unity the dense branch is reached where the next to leading order repulsive interaction becomes comparable to the leading interaction. A recent paper also suggested the importance of including higher harmonics [20] when discussing the decay of dense axion stars. Our approach is a systematic way of including them within an expansion scheme involving a small parameter. Higher harmonics were also analyzed in [21] starting from the Gross-Pitaevski equation. Authors of references [21–23] developed approaches different from ours to deriving corrections to the axion star effective field theory. A comparison of these methods can be found in [24].

We have shown that axion self interactions, violating particle number conservation, cause axion condensates to decay, resulting in the emission of relativistic axions [25]. The rate of this process increases sharply with the binding energy of axions in the condensate. The decay mechanism also operates when an axion star is in the process of collapse, from a dilute state to a more dense state, when its mass becomes supercritical. The leading decay mechanism for dense axion stars, first pointed out in [25], involves three bound state axions transitioning to a single relativistic scattering state axion. If interactions permit it, the same process of multiple bound bosons turning to a single scattering state boson, would be relevant for the decay of structures termed oscillons, localized bound states of real scalar particles which are time dependent and long lived. Such a reaction cannot take place if the initial state axions are in momentum eigenstates, since, given that the bound axions have a very small momentum while the outgoing one is relativistic, both momentum and energy cannot simultaneously be conserved. The solution is that axion star itself and all its constituents are not in momentum eigenstates. The initial bound axions have a momentum distribution centered around zero momentum, but there is a small nonzero probability of three high momentum axions from the tail of the distribution joining together to cause this decay. The decay matrix element, which is the Fourier transform of the cube of the bound state axion wave function, has a leading singular behavior in the weak binding limit of the form $exp(-c/\Delta)$,where $\Delta = \sqrt{1 - e^2/m^2}$ is a measure of the binding energy and c is a constant. Here m is the axion mass and e is the energy eigenvalue of a bound axion. Subsequent to our publication other authors also started to make the same point, that it is possible for multiple bound hermitian bosons to turn into a single scattering state boson within oscillons [23] or axion stars [20]. The authors of [22] were critical of the 3 to 1 decay since it did not arise in their low energy effective field theory formalism. Their viewpoint has evolved since then, after the realization that the rate of the 3 to 1 decay computed in [25] is a non analytic function of axion binding energy. They have pointed out in the publication [24], the relevance of the 3 to 1 process to axion star or oscillon decays, by the emitting relativistic particles. For dilute QCD axion stars the 3 to 1 process is suppressed and they have lifetimes many orders of magnitude longer than the age of the universe. The 3 to 1 process becomes relevent for dense axion stars. Lifetime estimates of oscillons have been performed by computing the rate of emission of classical radiation from them and in references [26, 27] the decay rate is estimated to have a formula similar to ours with the same singular exponential dependence on binding energy. We have estimated that a QCD axion star with binding energy Δ less than 0.05 would have a lifetime greater than the age of the universe.

To summarize earlier work on QCD axion stars (single flavor, single level) predicted sizes on the order of meters and densities approaching that of nuclear matter. Such structures are structurally unstable towards collapse. By scaling the equations, and expanding them carefully, we discovered dilute solutions that are much larger, about asteroid size and mass. These solutions are structurally stable against collapse, and they are stable against decay (their lifetimes are much greater than the lifetime of the universe). Using our expansion procedure, we also studied dense states which are structurally stable but decay rapidly and intermediate density states that are structurally unstable.

The rotation curves of galaxies cannot be explained by observed luminous matter. Can boson stars explain them? Using the simplest types they cannot. Single flavor, single level bosonic dark matter (DM) models are severely constrained. For example, in a single flavor model the mass of the condensate is inversely proportional to its size, so the central density falls as the fourth power of its size. Such behavior is inconsistent with observations of large classes of galaxies. Consequently, the matter distribution of a massive galaxy like the Milky Way, where the addition of material tends to enlarge the galaxy, cannot be described by such condensates.

We have done a preliminary analysis of a scenario in which galactic dark matter consists of multiple flavors of bosons, with different masses and self couplings. Reference [28]. That is a reasonable scenario, as any theory, which gives rise to ultra light axion-like particles, also produces hundreds of them, examples being string compactifications [29] or clockwork theories [30]. Thus, theories in which multiple flavors of axion dark matter are produced in the early universe should be investigated seriously. A recent publication [31] investigated the possibility that the dark matter of dwarf galaxies consist of combinations of two or three flavors of axions with masses ranging from 10^{-22} eV to 10^{-18} eV. Gravitational collapse of multi flavor DM clouds due to density fluctuations would lead to bound condensates, or boson/axion stars consisting of multi boson flavors. We have explored a wide range of masses and self couplings with the constraint that the dimensions of the condensates make them structurally stable. Our analysis based on the variational method was restricted to two flavors and our expectation was that such condensates mainly constitute galactic cores. We also included an analysis of how the properties of a single flavor condensate gets modified due to having a large inner body made up of baryonic matter [28]. We were motivated by the work of Ref. [32] where the authors analyzed how the central densities ρ_c of a collection of galaxies scale with their core radii R_c using actual experimental data [33] and found the relation $\rho_c {R_c}^{1.2} = \text{constant}$. If you take a single component weakly bound axion BEC, the density would vary like M/R^3 and the mass scales like 1/R. Therefore the scaling of central density with the core size should be like $\rho_c {R_c}^4 = \text{constant}$. In [28] we found that structures formed from multiple flavors of bosons are qualitatively different in the way their sizes and masses scale compared to single flavor ones. We found ranges of parameters, in which the observed scaling is present. That offers opportunities for explaining the properties of multitudes of structures at different scales present in the universe, Observational biases would also limit the way scaling relations could be determined.

Another way to address these observational discrepancies is to assume that dark matter is made up from one axion flavor and the condensate is only expected to dominate the central region of a galaxy while particles in excited states make up the halo, forming the galactic dark matter as a bosonic atom, an idea being investigated by many researchers.

Acknowledgements The authors thank their collaborators J.Eby, M.Leembruggen, L.Street C.Vaz who participated in the work reported here. Work of L.C.R.W. is partially supported by the US. Department of Energy grant DE-SC1019775.

References

1. J.A. Wheeler, Geons. Phys. Rev. **97**, 511–536 (1955). https://doi.org/10.1103/PhysRev.97.511
2. D.R. Brill, J.B. Hartle, Method of the self-consistent field in general relativity and its application to the gravitational Geon. Phys. Rev. **135**, B271–B278 (1964). https://doi.org/10.1103/PhysRev.135.B271
3. P.R. Anderson, D.R. Brill, Gravitational geons revisited. Phys. Rev. D **56**, 4824–4833 (1997). https://doi.org/10.1103/PhysRevD.56.4824
4. F.I. Cooperstock, V. Faraoni, G.P. Perry, Probing the gravitational geon. Int. J. Mod. Phys. D **5**, 375-406 (1996). https://doi.org/10.1142/S0218271896000242, arXiv:gr-qc/9512025 [gr-qc]
5. D.J. Kaup, Klein-Gordon Geon. Phys. Rev. **172**, 1331–1342 (1968). https://doi.org/10.1103/PhysRev.172.1331
6. R. Ruffini, S. Bonazzola, Systems of selfgravitating particles in general relativity and the concept of an equation of state. Phys. Rev. **187**, 1767–1783 (1969). https://doi.org/10.1103/PhysRev.187.1767
7. J.D. Breit, S. Gupta, A. Zaks, Cold Bose stars. Phys. Lett. B **140**, 329–332 (1984). https://doi.org/10.1016/0370-2693(84)90764-0
8. G.H. Derrick, Comments on nonlinear wave equations as models for elementary particles. J. Math. Phys. **5**, 1252–1254 (1964). https://doi.org/10.1063/1.1704233
9. M. Colpi, S.L. Shapiro, I. Wasserman, Boson stars: gravitational equilibria of selfinteracting scalar fields. Phys. Rev. Lett. **57**, 2485–2488 (1986). https://doi.org/10.1103/PhysRevLett.57.2485
10. J. Eby, P. Suranyi, C. Vaz, L.C.R. Wijewardhana, Axion stars in the infrared limit. JHEP **03**, 080 (2015) [erratum: JHEP **11**, 134 (2016)]. https://doi.org/10.1007/JHEP11(2016)134 arXiv:1412.3430 [hep-th]
11. J. Barranco, A. Bernal, Self-gravitating system made of axions. Phys. Rev. D **83**, 043525 (2011). https://doi.org/10.1103/PhysRevD.83.043525 arXiv:1001.1769 [astro-ph.CO]
12. J. Barranco, A.C. Monteverde, D. Delepine, Can the dark matter halo be a collisionless ensemble of axion stars?. Phys. Rev. D **87**, 103011 (2013). https://doi.org/10.1103/PhysRevD.87.103011, arXiv:1212.2254 [astro-ph.CO]
13. J. Eby, M. Leembruggen, L. Street, P. Suranyi, L.C.R. Wijewardhana, Global view of QCD axion stars. Phys. Rev. D **100**(6), 063002 (2019). https://doi.org/10.1103/PhysRevD.100.063002 arXiv:1905.00981 [hep-ph]
14. P.H. Chavanis, Mass-radius relation of Newtonian self-gravitating Bose-Einstein condensates with short-range interactions: I. Analytical results. Phys. Rev. D **84**, 043531 (2011). https://doi.org/10.1103/PhysRevD.84.043531, arXiv:1103.2050 [astro-ph.CO]
15. P.H. Chavanis, L. Delfini, Mass-radius relation of Newtonian self-gravitating Bose-Einstein condensates with short-range interactions: II. Numerical results. Phys. Rev. D **84**, 043532 (2011). https://doi.org/10.1103/PhysRevD.84.043532, arXiv:1103.2054 [astro-ph.CO]

16. A.H. Guth, M.P. Hertzberg, C. Prescod-Weinstein, Do dark matter axions form a condensate with long-range correlation?. Phys. Rev. D **92**(10), 103513 (2015). https://doi.org/10.1103/PhysRevD.92.103513, arXiv:1412.5930 [astro-ph.CO]
17. J. Eby, M. Leembruggen, P. Suranyi, L.C.R. Wijewardhana, Stability of condensed fuzzy dark matter halos. JCAP **10**, 058 (2018). https://doi.org/10.1088/1475-7516/2018/10/058, arXiv:1805.12147 [astro-ph.CO]
18. J. Eby , P. Suranyi, L.C.R. Wijewardhana (eds)., Expansion in Higher Harmonics of Boson Stars using a Generalized Ruffini-Bonazzola Approach, Part 1: Bound States. JCAP 1804 (04), (2018) e-Print: arXiv:1712.04941 1805.12147 [astro-ph.CO]
19. E. Braaten, A. Mohapatra, H. Zhang, Dense axion stars. Phys. Rev. Lett. **117**(12), 121801 (2016). https://doi.org/10.1103/PhysRevLett.117.121801, arXiv:1512.00108 [hep-ph]
20. L. Visinelli, S. Baum, J. Redondo, K. Freese, F. Wilczek, Dilute and dense axion stars. Phys. Lett. B **777**, 64–72 (2018). https://doi.org/10.1016/j.physletb.2017.12.010, arXiv:1710.08910 [astro-ph.CO]
21. M.H. Namjoo, A.H. Guth, D.I. Kaiser, Relativistic corrections to nonrelativistic effective field theories. Phys. Rev. D **98**(1), 016011 (2018). https://doi.org/10.1103/PhysRevD.98.016011, arXiv:1712.00445 [hep-ph]
22. E. Braaten, A. Mohapatra, H. Zhang, Nonrelativistic effective field theory for axions. Phys. Rev. D **94**(7), 076004 (2016). https://doi.org/10.1103/PhysRevD.94.076004, arXiv:1604.00669 [hep-ph]
23. K. Mukaida, M. Takimoto, M. Yamada, On longevity of I-ball/Oscillon. JHEP **03**, 122 (2017). https://doi.org/10.1007/JHEP03(2017)122, arXiv:1612.07750 [hep-ph]
24. E. Braaten, A. Mohapatra, H. Zhang, Classical nonrelativistic effective field theories for a real scalar field. Phys. Rev. D **98**(9), 096012 (2018). https://doi.org/10.1103/PhysRevD.98.096012, arXiv:1806.01898 [hep-ph]
25. J. Eby, P. Suranyi, L.C.R. Wijewardhana, The lifetime of axion stars. Mod. Phys. Lett. A **31**(15,) 1650090 (2016). https://doi.org/10.1142/S02177323165009
26. G. Fodor, P. Forgacs, Z. Horvath, M. Mezei, Computation of the radiation amplitude of oscillons. Phys. Rev. D **79**, 065002 (2009). https://doi.org/10.1103/PhysRevD.79.065002, arXiv:0812.1919 [hep-th]
27. M.P. Hertzberg, Quantum radiation of oscillons. Phys. Rev. D **82**, 045022 (2010). https://doi.org/10.1103/PhysRevD.82.045022, arXiv:1003.3459 [hep-th]
28. J. Eby, M. Leembruggen, L. Street, P. Suranyi, L.C.R. Wijewardhana, Galactic condensates composed of multiple axion species. JCAP **10**, 020 (2020). https://doi.org/10.1088/1475-7516/2020/10/020, arXiv:2002.03022 [hep-ph]
29. A. Arvanitaki, S. Dimopoulos, S. Dubovsky, N. Kaloper, J. March-Russell, String Axiverse. Phys. Rev. D **81**, 123530 (2010). https://doi.org/10.1103/PhysRevD.81.123530, arXiv:0905.4720 [hep-th]
30. D.E. Kaplan, R. Rattazzi, Large field excursions and approximate discrete symmetries from a clockwork axion. Phys. Rev. D **93**(8), 085007 (2016). https://doi.org/10.1103/PhysRevD.93.085007, arXiv:1511.01827 [hep-ph]
31. H.N. Luu, S.H.H. Tye, T. Broadhurst, Multiple ultralight Axionic wave dark matter and astronomical structures. Phys. Dark Univ. **30**, 100636 (2020). https://doi.org/10.1016/j.dark.2020.100636, arXiv:1811.03771 [astro-ph.GA]
32. H. Deng, M.P. Hertzberg, M.H. Namjoo, A. Masoumi, Can light dark matter solve the core-cusp problem?. Phys. Rev. D **98**(2), 023513 (2018). https://doi.org/10.1103/PhysRevD.98.023513, arXiv:1804.05921 [astro-ph.CO]
33. D.C. Rodrigues, A. del Popolo, V. Marra, P.L.C. de Oliveira, Mon. Not. Roy. Astron. Soc. **470**, no.2, 2410-2426 (2017).https://doi.org/10.1093/mnras/stx1384, arXiv:1701.02698 [astro-ph.GA]

Surprises in Strongly Interacting Many-Body Conformal Dynamics

Fei Zhou

Abstract In this review article, we illustrate the main idea of many-body conformal dynamics and two surprises. Conformal dynamics are non-equilibrium phenomena occurring when inter particle interactions are scale invariant and the particular many-body systems we will focus on further exhibit a non-relativistic Galilean invariance. The first surprise is that in conformal expansion dynamics driven by unitary evolutions of an isolated quantum system, both the energy and entropy can be conserved as opposed to what happen in more familiar free expansions and adiabatic expansions in thermodynamics. Such dynamics in strongly interacting quantum systems where interactions are scale symmetric appear as a very rare example of reversible far-away-from equilibrium many-body dynamics. The second surprise is that conformal dynamics can also emerge in many strongly interacting systems in the large time limit as an asymptotic dynamic symmetry, even though scale symmetries are explicitly broken by strong interactions. The emergent infrared conformal dynamics, purely induced by real-time dynamic renormalization effects, can occur in both low and high dimensions but rely crucially on the scaling dimensions of interaction operators and infrared stability of the scale invariant fixed points.

1 Introduction

Quantum scale and conformal symmetries play a paramount role in many fascinating phenomena in strongly interacting quantum systems. These symmetries are ubiquitous in the studies of AdS/CFT, quantum phase transitions in condensed matter systems or lattices, bounded quantum hydrodynamics, Universal Planckian dynamics [1–12] as well as in the modern studies of interacting topological quantum states [13–16]. Very recently, the role of scale and conformal symmetries has also been investigated in far-away-from equilibrium quantum dynamics. In this article, I will review the main idea behind the so-called many-body conformal dynamics. I will emphasize on the two pleasant surprises that have been found in our research in recent years.

F. Zhou (✉)
Department of Physics and Astronomy, University of British Columbia, Vancouver, BC, Canada
e-mail: feizhou@phas.ubc.ca

M. Paranjape et al. (eds.), *Proceedings of Gravity, Strings and Fields*,
CRM Series in Mathematical Physics,
https://doi.org/10.1007/978-3-031-91266-5_25

2 Conformal Dynamics: The Main Idea

The conformal symmetry discussed in many-body quantum dynamics including the phenomena studied below is closely related to the covariance discussed in the context of non-relativistic quantum field theory decades ago [17, 18]. It was pointed out that a non-relativistic field theory, if *scale invariant*, exhibits a conformal symmetry or invariance under the following space-time transformation.

$$\mathbf{R} \to \tilde{\mathbf{R}} = \frac{\mathbf{R}}{\lambda(t)}, dt \to d\tau = \frac{dt}{\lambda^2(t)}$$
$$\psi(\mathbf{R}, t) \to \tilde{\psi}(\tilde{R}, \tau) = \frac{1}{\lambda^{\frac{d}{2}}(t)}\psi(\frac{\mathbf{R}}{\lambda(t)}, t)\exp[i\frac{\mathbf{R}^2}{2}\frac{\dot{\lambda}(t)}{\lambda(t)}] \tag{1}$$

if the dynamic re-parametrization factor $\lambda(t)$ is a solution in the following differential equation.

$$\frac{d}{dt}[\frac{\dot{\lambda}(t)}{\lambda(t)}] + [\frac{\dot{\lambda}(t)}{\lambda(t)}]^2 = 0. \tag{2}$$

Equation 2 indicates the following solutions with an arbitrary $c \in \mathcal{R}$;

$$\lambda(t) = 1 + ct, -\infty > t > +\infty. \tag{3}$$

The conformal time $\tau(t)$ is given as

$$\tau(t) = \frac{t}{1 + ct}. \tag{4}$$

The space-time mapping of the above transformation is illustrated in Fig. 1. In this transformation, $t = \pm\infty$ is always mapped into a finite $\tau(= 1/c)$ point in the conformal coordinate implying an effective freezing of any non-trivial quantum dynamics in the long time limit. This observation had been taken into account in a previous study of long time dynamics of N-body fermions. Based on this general analysis, it was proposed that in the long time limit, conformal symmetries always naturally lead to a stationary solution in a proper conformal reference frame even though its short time dynamics can be rather complex and further depend on initial preparations [19].

Two types of quench dynamics where initial states are spatially localized have been the focuses of quantum many-body studies in connections to conformal symmetries. They are listed below.

I: Free expansion: For the studies of quantum dynamics in free space *with translation symmetry*, starting with an initial state typically associated with an equilibrium state in an external harmonic trapping potential $V(r) = \frac{1}{2}\omega_i^2 r^2$ with a frequency ω_i, it is much more convenient to force the re-parametrization factor to be the solution of the follow differential equation,

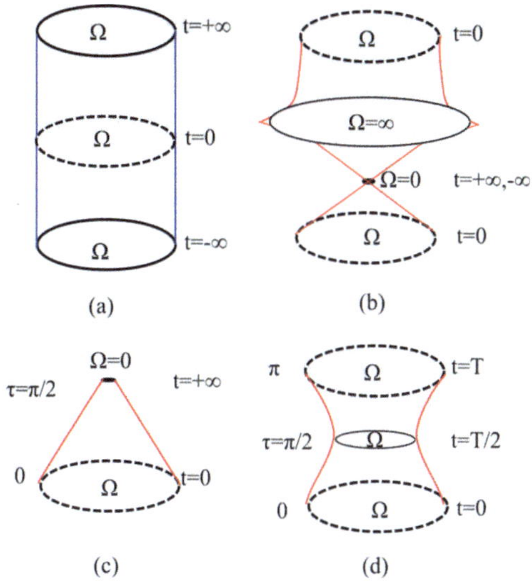

Fig. 1 Schematics of time evolution of a constant volume in a Euclidean space (**a**), in conformal coordinates introduced in non-relativistic quantum field theory (**b**) and in conformal coordinates suggested by many-body quantum dynamics [(c), (d)]. In (b), (c) and (d), we show the transformations of a finite volume indicated by Ω along the time domain (defined as the vertical direction); $\tau(t)$ is the time defined in a conformal co-moving frame (see the main text). (b) is the transformation introduced for the studies of invariance of actions in $(d+1)$ dimension. **c**, **d** are conformal co-moving coordinates introduced for the studies of N-body fermion dynamics in a free space and in a soft harmonic trap respectively

$$\frac{d}{dt}[\frac{\dot{\lambda}(t)}{\lambda(t)}] + [\frac{\dot{\lambda}(t)}{\lambda(t)}]^2 = \frac{\omega_i^2}{\lambda^4(t)} \tag{5}$$

where its solution $\lambda(t)$ can be shown to be

$$[\lambda(t)]^2 = \frac{1}{1+\omega_i^2 t^2}. \tag{6}$$

The conformal time coordinate τ defined in Eq. 1 for this choice of $\lambda(t)$ is

$$\tau(t) = \frac{1}{\omega_i}\arctan(\omega_i t), \tau \in [0, \frac{\pi}{2}], \tag{7}$$

which approaches $\pi/2$ when $t \to \infty$. Conformal coordinates effectively compactify quantum dynamics in its time domain (see Fig. 1b) if the interactions are scale invariant

II: Expansion in a shallow external harmonic potential The quantum many-body dynamics in the presence of a soft external harmonic trap with frequency ω_f that explicitly breaks the translation symmetry can be studied via the conformal coordinates defined in Eq. 1 by further taking into the external trapping potential in the dynamics of $\lambda(t)$,

$$\frac{d}{dt}[\frac{\dot{\lambda}(t)}{\lambda(t)}]+[\frac{\dot{\lambda}(t)}{\lambda(t)}]^2+\omega_f^2=\frac{\omega_i^2}{\lambda^4(t)}. \tag{8}$$

Equation 8 leads to the following factor of $\lambda(t)$ with a frequency of $2\omega_f$ and the conformal time domain defined by $\tau(t)$.

$$\begin{aligned}[\lambda(t)]^2 &= \cos^2(\omega_f t)+\frac{\omega_i^2}{\omega_f^2}\sin^2(\omega_f t)\\ \tau(t) &= \frac{1}{\omega_i}\arctan\big(\frac{\omega_i}{\omega_f}\tan\omega_f t\big),\ \tau(t)\in[0,\pi]\end{aligned} \tag{9}$$

where $\tau(t)$ effectively compactifies the quantum dynamics into a domain $[0, \pi]$ with $t=+\pi/2\omega_f$ being mapped into $\tau=\pi/2$.

The dynamics of strongly interacting scale invariant non-relativistic quantum gases is constrained by an emergent SO(2,1) [17–25] or conformal symmetry thanks to a nearby scale symmetric fixed point. As a result, a large class of dynamic phenomena only depend on the conformal symmetry where the initial conditions play very little role. Conformal symmetry has been exploited numerous times to examine the dynamics of both strongly interacting Bose and Fermi gases in various dimensions [19, 26–42]. These dynamics shall be generic in a broad class quantum critical phenomena with dynamic critical exponent $z=2$ [43] where Galilean symmetry can emerge in lattices. The emergent Galilean symmetry along with the scale symmetry are sufficient to result in conformal symmetries studied here.

One important consequence of conformal symmetry is the possibility of having quantum dynamics where the energy and Gibbs entropy are both preserved (see below) [19, 37]. Such energy-entropy conserving dynamics are so far known to only occur in conformally invariant systems, but not in generic strongly interacting systems [19, 33, 37, 40]. This feature was illustrated previously in the context of cyclic quantum states of three-dimensional (3D) Fermi gases with resonantly interactions in quenched harmonic potentials [37]. Provided that the initial state is a thermal mixture in a harmonic trap with frequency ω_i, the resulting far-away from equilibrium dynamics in a quenched soft harmonic trap with frequency $\omega_f(\ll \omega_i)$ conserve both energy and entropy. In this situation, the gas will expand and contract in the soft harmonic trap at a frequency of $2\omega_f$ perpetually.

These large-amplitude oscillations are one signature of a quantum system undergoing a full quantum revival. In these quantum revivals, one can show [37] that the N-body density matrix, where N is the number of particles, is completely periodic with frequency $2\omega_f$, a consequence of the non-relativistic conformal symmetry and dynamics of the resulting conformal tower states [21–23]. In fact, the time-dependence of the density matrix in position space is simply a time-dependent rescaling, up to a time-dependent coherent flow factor that represents onset of coherent flow induced during the expansion. Below we will focus on a few surprising aspects that had been found in the previous detailed studies of conformal dynamics [19, 33, 37, 40].

3 Surprise One: Reversibility and Energy-Entropy Conservation in Conformal Dynamics

Quantum conformal dynamics are a phenomenon that occurs when a quantum many-body state is under unitary evolution set by scale symmetric interactions. A convenient context to study such dynamics is expansion or compression of a 3D quantum gas when strongly interacting via resonant interactions [44, 45]. A very puzzling aspect of such dynamics is that both energy and entropy are conserved, perfectly and simultaneously. This phenomenon had been named as a *quantum Boltzmann Breather in* Ref. [37],*an analogue of the classical Boltzmann breather of weakly interacting hot gases in quantum and strongly interacting regimes.* Let us take a close look at this aspect of the phenomenon from the view of the second law of thermodynamics.

Let us consider possible ways of expansions of a generic thermal gas (see Fig. 2). The only energy conserved expansion processes are free expansion, where the gas expands freely as an isolated system and hence energy is obviously conserved. However, this is also the example we usually apply to teach physics students about the second law of thermodynamics. When the gas expanding into a larger volume, there are many more, exponentially more micro-states become available at a given energy E. Once the vast amount of new micro-states all become available to the expanding gas, which always occur because of scattering between particles in the gas, the gas has an increased amount entropy. In the simplest approximation, we say the entropy increases logarithmically as a function of the volume of the expanded gas.

The isothermal process maintains a constant temperature by heat exchange. Although for idea gases, this also indicates a constant energy as an open system, in general the energy of the gas is not conserved once interactions are taken into account. Of course, entropy also varies in such a process because the micro-states available at a given temperature change dramatically as the volume changes.

The only thermodynamic process that fully conserves the thermal entropy is an adiabatic process. For instance, in an adiabatic expansion, although the number of micro-states at a given energy E increases exponentially, because of the work done by the gas, the energy E also drops quickly enough so that the number of micro-states available to the gas remains to be a constant hence conserves the entropy S.

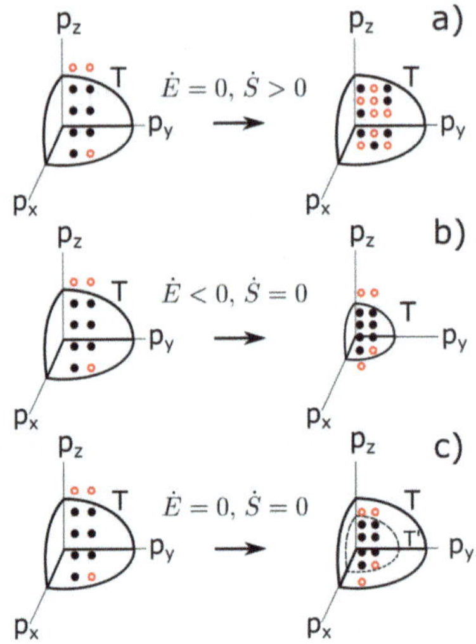

Fig. 2 Schematics of a free expansion (**a**) and adiabatic expansion (**b**). Here solid (black) dots are occupied states and empty (red) circles are unoccupied ones. Solid curves indicate states of thermal energy κT. **c** is for conformal dynamics where T' implies the temperature sets the entropy S while the total energy E is conserved. See the main text for discussions. (*Adapted from* Ref. [37]))

A fundamental assumption in an adiabatic process is that the gas can not be a fully isolated quantum system and it has to interact with another party or an environment so that it can perform work on or be performed work on. Adiabatic expansion (compression) thus always leads to cooling (heating) of the gas, i.e. removing (adding) thermal energy to the gas by a third party.

So in the standard thermodynamic paradigm, only one of these two quantities, E and S, can be conserved but not both of them in one single process. This seems to be one of the most useful lessons in thermodynamics.

Now we can turn to the conformal dynamics. The main conclusion of our studies when it comes to conformal dynamics is that its density matrix can have a very simple universal structure, The unitary evolution of a state (specified by a density matrix ρ_0) at $t > 0$ has the following generic covariant structure

$$\rho(t \geq 0) = \frac{1}{\lambda^{Nd}} \rho_0 \left(\{\frac{\mathbf{r}_i}{\lambda(t)}\}, \{\frac{\mathbf{r}'_i}{\lambda(t)}\}, i, j = 1, ..., N; \frac{\omega_i}{T} \right) \\ \otimes \exp[\sum_i (\frac{|\mathbf{r}_i|^2}{2} - \frac{|\mathbf{r}'_i|^2}{2}) h(t)], \tag{10}$$

where $\lambda(t)$ is a dynamic scaling factor and $\lambda(t) = 1$ at $t = 0$. $h(t)$ is a measure of the rate of conformal expansion or contraction. $h(t)$ is defined as

$$h(t) = \frac{\dot{\lambda}}{\lambda}, \dot{\lambda}(t) = \frac{d\lambda(t)}{dt}. \tag{11}$$

d is the spatial dimension which usually takes values of $d = 1, 2, 3$. However, $d = 2$ has quantum anomalies and the concept of conformal dynamics can only be applied approximately when interactions are present. The parameter ω_i/T in ρ_0 illustrates an initialization of the quantum state.

At last, in Eq. 10 we also have introduced $\otimes$ to refer to a product of degrees of freedom in two subspaces: one for internal thermal degrees of freedom and one for the external motion due to coherent flow. The coherent flow at point $\mathbf{r}_i$ is given by the gradient of the phases in the exponent,

$$\mathbf{v}(\mathbf{r}_i) = \nabla[\frac{|\mathbf{r}_i|^2}{2}h(t)] = \mathbf{r}_i h(t), i = 1, 2, 3...N. \tag{12}$$

Equation 10 is valid in the following two limits.

(1) It is exact for a limited class of initial states that can be related to tower states of primary conformal operators via state-operator correspondence;

(2) Or it can represent the long time asymptotics of generic spatially localized initial states.

The structure in Eq. 10 contains two distinct terms. The first term on the right hand side of equation is of the simple scaling form. The function ρ_0 is set by the initial conditions. This term has a simple scaling structure and completely dictated by the dynamic scaling factor of $\lambda(t)$.

It has been proved that all the Gibbs entropy or the information associated with the N-particle density matrix in Eq. 10 is fully coded in this term. Furthermore, if we set $\rho_0 = \rho_{eq}$ so that at $t = 0$ the state is simply related to a local equilibrium state in a harmonic trap with frequency ω_i at temperature T, the structure of the first term in Eq. 10 turns out to be identical to the structure in an adiabatic expansion. That is

$$\begin{aligned}
\rho(t \geq 0) &= \rho^{(1)}(t \geq 0) \otimes \exp[\sum_i (\frac{r_i^2}{2} - \frac{r_i'^2}{2})h(t)], \\
\rho^{(1)}(t \geq 0) &= \rho_{ad}(t \geq 0) \\
&= \frac{1}{\lambda^{Nd}}\rho_{eq}\left(\{\frac{\mathbf{r}_i}{\lambda(t)}\}, \{\frac{\mathbf{r}_i'}{\lambda(t)}\}, i, j = 1, ..., N; \frac{\omega_i}{T}\right)
\end{aligned} \tag{13}$$

where the subscript *(1)* refers to the first term in Eq. 10 and the subscript *ad* is for an adiabatic process. Not surprisingly, the entropy of the system is conserved in conformal dynamics given that the part of density matrix that sets the entropy is identical to that in an adiabatic expansion.

However, if we just single out this part of the density matrix in Eq. 10 and compute the energy associated with it, we have

$$E^{(1)}(t) = Tr\{H\rho^{(1)}(t \geq 0)\} = \frac{E(0)}{\lambda^z(t)}, z = 2 \quad (14)$$

which is time dependent and decreases as $\lambda(t)$ increases as in adiabatic cooling. $z = 2$, sometimes called a dynamic critical exponent, reflects the scaling dimension of energy in our problem as all the spatial coordinates scale as $1/\lambda(t)$ in our dynamics.

So although all the entropy information is coded in the first term in Eq. 10, it is essential to include the second term involving $h(t)$ in the density matrix. This unusual but universal structure actually contains all the information of marcoscopic flow generated during the conformal expansion. Its simple structure which only depends on $h(t)$ is represented by a pure universal coherent flow factor, independent of details of quantum states except $h(t)$, the rate of expansion flow and is therefore fully coherent, in contrast to the first term which contains all the thermal information such as temperatures etc. Only after the energy associated with this coherent coherent flow factor $\sum_i \frac{\mathbf{r}_i \cdot \mathbf{v}(\mathbf{r}_i)}{2} h(t)$ due to a coherent flow $\mathbf{v}(\mathbf{r}_i) = \mathbf{r}_i h(t)$ is taken into account, the total energy is conserved, i.e.

$$E(t) = Tr\{H\rho(\geq 0)\} = E(0) = const. \quad (15)$$

The difference between $E(t)$ in Eq. 15, the conserved energy in the conformal dynamics, and $E^{(1)}(t)$ Eq. 14, the energy in an adiabatic cooling, is a measure of the amount of the energy appearing in the coherent expansion flow (CEF).

$$E^{CEF}(t \geq 0) = E(0)[1 - \frac{1}{\lambda^z(t)}], \lambda(t) \geq 1. \quad (16)$$

Equations 14, 15, 16 indicate a simple and transparent relation between the thermal energy and entropy coded in the first term in Eq. 10, and the energy pumped into the coherent expansion flow due to scale invariant interactions. When one starts with a local equilibrium state given by ρ_{eq}, all the energy and entropy are stored in the thermal form. In the subsequent conformal dynamics, the thermal energy is gradually transformed into the energy of coherent flow which is reflected in the second coherent flow term (see also Eq. 12) of the density matrix in Eq. 10. During this whole conversion process, there is no entropy production. In the dynamics where $\lambda(t)$ increases as a function of time, all the energy originally stored in the thermal form will eventually be transformed into the energy in a coherent flow as $\lambda(t) \to \infty$. Furthermore, in the dynamics where $\lambda(t)$ is oscillating as a function of time t, conversion can occur along the opposite direction: the coherent flow energy can be further converted back to the thermal energy and so forth leading to fully reversible and revivable dynamics.

This is quite remarkable as generically thermalization shall occur in an isolated interacting quantum system. Thermalization implies entropy production and irreversibility so zero entropy production in a state excludes the possibilities of being thermalized. Instead, in this case, scale symmetric strong interactions can fully convert the thermal energy released during the expansion into a coherent form as if the gas were doing work on the expanding boundary of the gas driving the flow outwards. The conversion between the thermal energy and energy of the coherent flow is perfect and full reversible. This is the first surprising aspect of conformal dynamics.

4 Surprise Two: Conformal Symmetry as an Emergent Infrared Quantum Phenomenon

So far we have discussed the conformal dynamics in an ideal situation when the interactions are fully scale symmetric. We have illustrated that conformal dynamics are fully reversible and the energy can be transformed between incoherent thermal ones and fully coherent-flow energies, back-and-forth, without any losses and with zero entropy production. As to have scale symmetric interactions one needs to fine-tine the interaction to some critical form, one might wonder whether physically this can only happen in a very restricted class of fine-tuned systems. Recently, we show very similar dynamics can also occur in much broader classes of quantum systems that do not have exact scale symmetric interactions.

Although it is true that the limit we have discussed above can be reached only if the Hamiltonian is a scale symmetric fixed-point, in this section we further show that there are much broader classes of Hamiltonians or physics systems where conformal symmetry can naturally appear in the infrared limit or long time limit of dynamic states. In these systems, the scale symmetry can be explicitly broken by interactions etc, but the strong renormalization effects of interactions during the course of quantum dynamics can drive the dynamic states into conformal symmetric ones. We have recently named such phenomena as emergent infrared conformal dynamics (EIRCD) [40], to contrast with exact conformal dynamics(ECD) discussed in the previous section.

A few important remarks follow.

A: High vs low dimensions One is that EIRCD is a general emergent dynamic phenomenon that can occur both in high dimensional quantum systems ($d = 3$) and low dimension ones ($d = 1, 2$).

B: Strong interactions vs free particles Furthermore, EIRCD can also occur near both strongly coupled fixed points and free particle fixed points, as far as the scaling dimensions of symmetry breaking interaction operators are high enough so that they become *dynamically* irrelevant in the long time limit of conformal dynamics.

However, if one is only interested in strong coupling physics of the same interaction Hamiltonian, strong coupling fixed points in high- and low-dimensions of the same interacting model usually have different stabilities. It is usually true that EIRCD only occurs in low-dimensions because strong coupling fixed points like the Wilson-Fisher one are usually more stable below an upper critical dimension.

C: Dynamics vs equilibrium stability Third point is that there is a subtle but very important difference between being dynamically infrared irrelevant and being thermal dynamically irrelevant. The Wilsonian approach to the effective field theories provides a powerful way to classify operators in terms of scaling dimensions and plays a critical role in our understanding of thermal dynamics near scale invariant fixed points. The Wilsonian classification also plays a very important role in the study of EIRCD although it is applied in a very different way. The relevance or irrelevance of a particular interaction operator in conformal dynamics further depends on the scaling dimension of the speed of expansion $h(t)$ introduced before, as we are interested in dynamic states with evolving densities rather a ground state with a constant density. This was emphasized in a few previous studies [19, 33, 37].

One important dynamic *order parameter* to characterize conformal dynamics and its breaking is the entropy production rate. Conformal symmetry implies that there exist a class of expansion or contraction dynamics with zero entropy production. This is in stark contrast to the expansion of generically interacting gases which usually results in a finite amount of entropy production and are irreversible [37]. The amount of entropy production for a nearly scale invariant system depends on whether the scale-breaking interactions are relevant or irrelevant in the long time limit.

The relevancy of scale breaking interactions on the low-energy physics depends on how these interactions rescale as one approaches long wavelengths or long time scales, i.e. the infrared (IR) limit. If the effects of breaking scale symmetry are IR irrelevant, entropy production is restricted to short time scales, while the long time dynamics can still be isentropic and conformal. Such *emergent infrared conformal dynamics* (*EIRCD*) are not associated with the exact symmetry of the microscopic interactions, as in the case of resonance, rather it is an IR symmetry which only appears in the long wavelength or low frequency dynamics. On the other hand, if the scale breaking interactions are IR relevant, one expects an appreciable entropy production rate in the low frequency dynamics.

Entropy production can also play an important role in the study of thermalization. Zero entropy production which only occurs in conformal dynamics is a sufficient condition for absence of thermalization, although not a necessary one. This implies the absence of thermalization in conformal dynamics. For instance, when a 1D integrable system is subject to an additional harmonic potential, an otherwise-integrable system can potentially thermalize towards standard Gibbs ensemble unless the integrable system (without trapping potential) are further protected by a scale symmetry and entropy production in a trapping potential is forbidden (See below). So in this particular case, it is the scale or conformal symmetry that turns out to prevent a system from producing entropy and being thermalized. On the other hand, in the presence of *EIRCD*, it has yet to be quantified how the vanishingly small entropy production rate appears whenever the scale symmetry is broken explicitly,

To characterize the presence of the emergent conformal dynamics, we examine the energy dissipation associated with macroscopic flow of the dynamics. This quantity effectively acts as an order parameter for the conformal dynamics. We show that the dissipation in 1D dynamics generically vanishes in the IR limit, regardless of the initial conditions, leading to asymptotic conformal dynamics.

EIRCD is purely driven by infrared renormalization dynamics. To discuss the possibility of *EIRCD* and visualize the renormalization effects, we employ a model Hamiltonian for a quantum gas (either spin 1/2 fermions or bosons) with contact interactions in d dimensions and in the presence of a quenched harmonic potential:

$$\begin{aligned}\mathcal{H}_{\omega(t)} &= \mathcal{H} + \omega^2(t)C, \\ \omega(t) &= \omega_i\theta(-t) + \omega_f\theta(t)\end{aligned} \tag{17}$$

where $\omega(t)$ is the time-dependent frequency of the harmonic trap, and

$$C = \int d^d r \frac{r^2}{2}\psi^\dagger(\mathbf{r})\psi(\mathbf{r}). \tag{18}$$

In Eq. 17, the frequency of the external confining potential switched from an initial value of ω_i to the final value $\omega_f(< \omega_i)$ at $t = 0^+$. We also define $\mathcal{H}$ as the Hamiltonian that is translationally invariant:

$$\begin{aligned}\mathcal{H} = &\int d^d\mathbf{r}\, \frac{1}{2}\psi^\dagger(\mathbf{r}) \left(-\frac{1}{2}\nabla^2\right)\psi(\mathbf{r}) \\ &+ \int d^d\mathbf{r}\, g(\Lambda)\psi^\dagger(\mathbf{r})\psi^\dagger(\mathbf{r})\psi(\mathbf{r})\psi(\mathbf{r}).\end{aligned} \tag{19}$$

In Eq. (19), $\psi^\dagger(\mathbf{r})$ is the creation operator, and $g(\Lambda)$ is the coupling constant which depends on the ultraviolet (UV) scale of the theory, Λ. (We have also muted the spin indices in the case of spin 1/2 fermions and have set $\hbar = m = 1$.)

Broadly speaking, Eq. 19 is also an effective field theory of quantum critical points with dynamic critical exponent $z = 2$ [49]. So conformal dynamics are also relevant to quantum critical dynamics of $z = 2$ although the main context we have in mind here is atomic gases with resonant interactions. Indeed, Eq. (19) can be applied to an atomic gas at low-energies, provided that the low-energy scattering amplitude calculated from the effective theory is matched with the physical one. Such a procedure relates the running coupling constant $g(\Lambda)$ to the d-dimensional scattering length, a. We discuss the possibility of *EIRCD* using this model Hamiltonian for a spin-$1/2$ Fermi gas with 4-fermion interactions in d dimensions and in the presence of a time-dependent harmonic trap: $\mathcal{H}_{\omega(t)} = \mathcal{H} + \omega^2(t)C$. $\mathcal{H}$ is the free-space Hamiltonian which contains the kinetic energy and interactions with strength $g(\Lambda)$, that depends on the ultraviolet (UV) scale of the theory, and $\omega^2(t)C$ is the harmonic trapping potentials.

Since the dynamics are intimately related to the behaviour of the scale symmetry breaking interactions in the IR limit, it is beneficial to briefly recall how the Hamiltonian in Eq. (19) (in units of Λ^2 *so it is dimensionless*) behaves under a scale transformation. We rewrite the interaction as $g(\Lambda)\Lambda^{d-2} = c_d\tilde{g}(\Lambda)$ where $\tilde{g} = \tilde{g}^* + \delta\tilde{g}$ and

c_d is a non-zero positive constant that only depends on the dimension d. The dimensionless interaction constant $\tilde{g}^*$ represents the scale invariant part of the interactions while $\delta\tilde{g}$ is the part that breaks scale symmetry.

The value of $\tilde{g}^*$ can be obtained via the standard renormalization group equations (RGEs) [19]. Apart from the trivial non-interacting fixed point, $\tilde{g}^* = 0$, for dynamic critical exponent $z = 2$ cases there are further strongly interacting fixed points

$$\tilde{g}^* = 2 - d, d = 1, 2, 3. \tag{20}$$

$d = 2$ is the upper critical dimension of the theory with dynamic critical exponents $z = 2$. In $d = 2$, two fixed points in 1D or 3D merge into each other and there appears to be only one single fixed point with $\tilde{g}^* = 0$. In 1D and 3D, there are strong coupling fixed points given by $\tilde{g}^* = +1, -1$ respectively apart from the trivial non-interacting fixed point $\tilde{g}^* = 0$.

We can expand the dimensionless Hamiltonian as $\mathcal{H} = \mathcal{H}^* + \delta\mathcal{H}_I$, where $\mathcal{H}^*$ represents the scale symmetric interacting Hamiltonian while $\delta\mathcal{H}_I$ represents the interactions that break scale symmetry and is proportional to $\delta\tilde{g}(\Lambda)$. Under a scale transformation the residual interaction term $\delta\mathcal{H}_I$ changes according to:

$$\Lambda \to \frac{\Lambda}{\lambda}, \qquad \circ\mathcal{H}_I \to \frac{1}{\lambda^\eta}\circ\mathcal{H}_I \tag{21}$$

where λ is a scaling factor set to be larger than unity for our discussions, i.e. $\lambda > 1$, as this generates a flow towards the IR limit.

In 3D or $d = 3$, there is a scale invariant attractive interaction, $\tilde{g}^* = -1$, which has been identified as a Feshbach resonance with an infinite scattering length $a_{3D} = +\infty$. The scaling dimension for $\delta\mathcal{H}_I$ is $\eta = -1 < 0$. The scale invariant fixed point Hamiltonian $\mathcal{H}^*$ is IR unstable under a scale transformation; any small deviations of $\delta\mathcal{H}_I$ from the scale invariant fixed point will be amplified during the course of rescaling towards the IR limit [19]. In 1D, the scale invariant interactions are repulsive, $\tilde{g}^* = +1$, and represent infinite strength contact interactions, i.e. the 1D scattering length $a_{1D} = 0$. The associated scaling dimension is $\eta = +1 > 0$, and the effect of $\circ\mathcal{H}_I$ diminishes as the running scale Λ is lowered (or equivalently the scaling factor λ increases); $\mathcal{H}^*$ is IR stable.

The scaling dimension of $\circ\mathcal{H}_I$, $\eta = \eta(d, \tilde{g}^*)$, is universal only depending on the spatial dimension d and $\tilde{g}^*$. It can again be directly obtained by solving the RGEs around $\tilde{g}^*$. It plays a critical role in the discussions of dynamics and *EIRCD*. For this particular Hamiltonian, we summarize the scaling dimensions in $d = 1, 2, 3$ for both free-particle fixed points $\tilde{g}^* = 0$ and strong coupling fixed points $\tilde{g}^* = 2 - d$ as

$$\eta(\tilde{g}^*, d) = d - 2 \ \ \text{if} \tilde{g}^* = 0;$$

$$\eta(\tilde{g}^*, d) = 2 - d \;\; \text{if} \tilde{g}^* = 2 - d. \tag{22}$$

Equation 22 implies a duality between 3D and 1D physics. As far as the scaling dimension $\eta(\tilde{g}^*, d)$ is concerned, 1D strong coupling fixed point at $\tilde{g}^* = +1$ is mapped into 3D free particle fixed point at $\tilde{g}^* = 0$; 1D free particle fixed point $\tilde{g}^* = 0$ is mapped into 3D strong coupling fixed point $\tilde{g}^* = -1$.

In 2D, there are only one non-interacting fixed point of $\tilde{g}^* = 0$. $\eta = 0$ implies the marginal relevance (either relevant or irrelevant) of interactions in 2D. This in fact is intimately related to the well known results of 2D quantum anomalies of interacting many-body systems. That is the scale symmetry that appears at the tree or classical level of the actions is violated once quantum loop corrections are included. Below we will focus on the strong coupling physics and contrast 1D against 3D.

We can illustrate how the dynamics induce such renormalization by considering the following transformation:

$$\psi(\mathbf{x}, t) = \frac{1}{\lambda^{d/2}(t)} e^{i \frac{|\mathbf{x}|^2}{2} \frac{\dot{\lambda}(t)}{\lambda(t)}} \psi\left(\frac{\mathbf{x}}{\lambda(t)}, \tau(t)\right). \tag{23}$$

where $\lambda(t)$ and $\tau(t)$ at $t > 0$ satisfy:

$$\ddot{\lambda}(t)\lambda^3(t) + \omega_f^2 \lambda^4(t) = \omega_i^2, \qquad \dot{\tau}(t) = \frac{1}{\lambda^2(t)} \tag{24}$$

and the boundary conditions $\lambda(0) = 1$ and $\dot{\lambda}(0) = 0$. Such a transformation has been used extensively in the literature. This transformation involves a time-dependent scale transformation, $\lambda(t)$, which removes the isentropic dynamics of the system.

Under Eq. (23) the Hamiltonian is mapped onto a similar Hamiltonian with a static harmonic trapping potential of frequency $\omega_i = \omega(t = 0^-)$, and a time-dependent interaction of the form: $g(\Lambda/\lambda(t))$. This is equivalent to looking at the value of the coupling constant at a time-dependent running scale $\Lambda/\lambda(t)$. Hence the expansion dynamics in a sense follows the renormalization group flow. Provided that the scale-breaking interaction $\delta\mathcal{H}$ is irrelevant in the renormalization group sense, i.e. $\eta > 0$, one can probe emergent conformal dynamics in the IR limit during an expansion when $\lambda(t) \gg 1$.

In the quantum dynamics of interest, one can further relate the dynamic scaling factor $\lambda(t)$ directly to the time dependence of the gas density or chemical potential which characterizes the dynamic states of the gas. The particle density $n(t)$ and the characteristic chemical potentials $\mu(t) \sim n^{2/d}(t)$ depend on time t and can dramatically differ from their initial values at $t = 0$. We quantify the effects of $\circ\mathcal{H}_I$ on the dynamics by evaluating it at the energy scale $\mu(t)$, or equivalently at a momentum scale set by the instantaneous Fermi momentum: $k_F(t) \propto n^{1/d}(t)$. It is then natural to define a time dependent running scale at which we evaluate $\circ\mathcal{H}_I$ that is proportional to the ratio of the density at time t to the initial density: $\Lambda(t) \approx (n(t)/n(0))^{1/d} \Lambda(0)$. In this way we can define a dynamic rescaling factor $\lambda(t) \sim (n(t)/n(0))^{1/d}$.

From this point of view, expansion or contraction dynamics can be studied by examining the renormalization flow of the symmetry breaking action $\delta\mathcal{H}_I$, which describes how relevant $\delta\mathcal{H}_I$ is with respect to the dynamic rescaling factor $\lambda(t)$. If the gas becomes more and more rarified, $\lambda(t)$ increases well beyond unity and the system flows into the IR limit, probing the long-wavelength physics while instances of compression allow the system to flow into the UV limit, probing the short-distance physics. If $\eta < 0$, the effects of the scale breaking action $\delta\mathcal{H}_I$ can be dominating in dynamics when $\lambda(t)$ becomes much larger than one. However, if $\eta > 0$, we anticipate that the effects of $\delta\mathcal{H}_I$ are only relevant when $\lambda(t)$ remains relatively small; and in expansion dynamics when $\lambda(t)$ increases, the dynamic states shall always be mainly determined by $\mathcal{H}^*$, signalling *emergent infrared conformal dynamics*.

In order to observe substantial renormalization effects during the course of the dynamics, it is thus imperative to consider the far-away-from-equilibrium expansion (contraction) dynamics, where the density varies over a broad range, so that the rescaling factor $\lambda(t)$ can be much larger (smaller) than unity, its initial value. This is contrast to the perturbative linear response regime where $\lambda(t)$ always remains close to its initial value of unity.

For this purpose, one can then directly study *ECD* and the possibility of *EIRCD* by considering a thermally equilibrated quantum gas that is released from a very tight isotropic d-dimensional harmonic trap of frequency ω_i into a much softer harmonic trap with frequency ω_f. Provided $\omega_f \ll \omega_i$, one can then generate an appreciable dynamic renormalization group flow towards the IR limit.

When the interactions are exactly scale symmetric, there are *ECD*. One can show that the scale parameter for this quench protocol is given by [22, 27, 37]:

$$\lambda^2(t) = \cos^2(\omega_f t) + \frac{1}{\tilde{\omega}_f^2}\sin^2(\omega_f t), \qquad \tilde{\omega}_f = \frac{\omega_f}{\omega_i}. \tag{25}$$

Equation (25) is a periodic function with period $T_f = \pi/\omega_f$. It takes on values: $\lambda(t) \in [1, \lambda_{max}]$, with a maximum value: $\lambda_{max} = 1/\tilde{\omega}_f$ and a minimum value: $\lambda_{min} = 1$.

We can examine the effects of $\delta\mathcal{H}_I$ on the conformal dynamics by treating $\delta\mathcal{H}_I$ perturbatively. At leading order, the scale breaking perturbation effectively becomes time dependent during the course of the dynamics. We can capture the time-dependence by rescaling $o\mathcal{H}_I$ by $\lambda(t)$ defined in Eq. (25). Thus the time-dependence of the scale breaking perturbation sensitively depends on the scaling dimension η introduced in Eq. (21). In 3D, $\eta = -1$ and $\delta\mathcal{H}_I(t)$ is appreciable over the entire period and becomes more relevant when $\lambda(t)$ approaches λ_{max}, i.e. when $t \simeq T_f/2$. In 1D, $\eta = +1$ and $o\mathcal{H}_I(t)$ is only appreciable for small time windows $\delta t \sim 1/\omega_i$ near $t = 0$ or $t = T_f$ when the scaling factor $\lambda(t) \sim 1$ is at a minimum and the gas is most dense; $o\mathcal{H}_I$ is strongly suppressed around $t = T_f/2$ when the gas is most dilute, $\lambda(t) \approx \lambda_{max}$. Accordingly, there are *EIRCD* in 1D, provided $\tilde{\omega}_f$ is sufficiently small, while in 3D *EIRCD* are absent.

Now let us turn to entropy production in EIRCD. When the interactions break scale symmetry, the oscillatory conformal dynamics in Eq. (25) will become damped, a signal of entropy production. More importantly the dissipation rate is directly proportional to the entropy production rate. We quantify the extent of the breaking of scale symmetry and the onset of the *EIRCD* using $\Gamma(\tilde{\omega}_f)$, the decay rate of the oscillatory dynamics averaged over the period. When $\tilde{\omega}_f \to 0$, we can expand the decay rate in terms of $\tilde{\omega}_f$,

$$\Gamma(\tilde{\omega}_f) = \Gamma_0 + \Gamma_1\tilde{\omega}_f + \Gamma_2\tilde{\omega}_f^2 + ...$$
$$\Gamma_{0,1,2} = \omega_i^2 \tau_F \gamma_{0,1,2}\left[\left(\frac{a_{sc}^2}{\tau_F}\right)^{2-d}, T\tau_F, \omega_i \tau_F\right] \tag{26}$$

where $\tau_F \propto n(0)^{-2/d}$ is a Fermi time defined in terms of the initial density at the center of the trap, and $\gamma_{0,1,2}$ are dimensionless functions. Note that $\Gamma_{0,1,2}$ are constants independent of the frequency of the dynamics ω_f but can depend on interactions and initial density etc. The general structure of these coefficients in Eq. (26) is highly complex and we will focus on the strong coupling limit where $(a_{sc}^2/\tau_F)^{2-d} \ll 1$, with a_{sc} being the d-dimensional scattering length: $a_{sc} = a_{3D}$ for $d = 3$ and $a_{sc} = a_{1D}$ for $d = 1$.

When $\mathcal{H}$ is fine tuned to have scale invariant interactions, i.e. $\mathcal{H} = \mathcal{H}^*$, it was shown before that for an arbitrary temperature T, high or low,

$$\Gamma(\tilde{\omega}_f) = 0 \tag{27}$$

for all values of $\tilde{\omega}_f$ and arbitrary temperatures T [19, 37]. It is the direct consequence of the conformal correlations indicated explicitly in the density matrix in Eq. 10 [19]. This is a general signature of *ECD*; and it is also fully consistent with the absence of bulk viscosity in the high temperature thermal phase specially studied in [25]. For generically interacting systems Γ will be finite, and will be proportional to $(a_{sc}^2/\tau_F)^{2-d}$ near the strongly interacting fixed point. However, the dissipation in the IR limit sensitively depends on whether Γ_0 or $\Gamma_{1,2,...}$ is the leading contribution to the decay rate.

Thus for scale invariant interactions, the decay rate of the oscillatory motion, or dissipation rate Γ, is identically zero and the gas does not thermalize. If the interactions break the symmetry , we shall expect a non-zero Γ, leading to thermalization where energy associated with the macroscopic flow can be converted into internal energy, leading to an increase in entropy and an eventual thermalization. Although there will undoubtedly be some entropy production, as discussed above, the $\tilde{\omega}_f$ dependence of dissipation rates depends on whether the scale invariant point is IR stable or unstable. This is where EIRCD can set in.

The leading behaviour of Γ for finite $\circ\mathcal{H}_I(t)$ in the IR limit is also directly related to the value and sign of the scaling dimension η defined in Eq. (21). When $\eta = -1$, as in 3D Fermi gases near resonance, the symmetry breaking action of $\delta\mathcal{H}_I(t)$ is relevant as the gas expands. The dissipation occurs approximately uniformly over the whole period T_f leading to an ω_f-independent time-averaged dissipation rate, or a finite Γ_0(see detailed discussions below). Thus in the low frequency limit,

$$\Gamma(\tilde{\omega}_f) = \Gamma_0 + ..., \tag{28}$$

is approximately a constant as $\tilde{\omega}_f \to 0$, as in generic strongly interacting systems. Furthermore in the strong coupling limit, $\Gamma_0 \sim \omega_i^2 \tau_F \gamma_0$, $\gamma_0 \sim (a_{sc}^2/\tau_F)^{2-d}$. This signifies finite dissipation in the IR limit as $\tilde{\omega}_f \to 0$ and the absence of *EIRCD*.

By contrast, when $\eta = 1$, as in 1D, $\delta\mathcal{H}_I(t)$ becomes strongly suppressed in the bulk of the period, when $\lambda(t) \gg 1$. The dissipation due to $\delta\mathcal{H}_I(t)$ is then confined to a small time window of $\delta t \sim \pi/\omega_i \ll T_f$ when $\lambda(t) \approx 1$, i.e. $t = 0, T_f, 2T_f$ etc. The averaged dissipation rate therefore is inversely proportional to T_f or proportional to ω_f. So we expect that $\Gamma_0 = 0$ while Γ_1 remains finite. This scaling implies that if one approaches the IR limit, $\tilde{\omega}_f \to 0$, Γ becomes vanishingly small indicating an unexpected dynamical phase with *EIRCD*, i.e.

$$\Gamma(\tilde{\omega}_f) = 0 + \Gamma_1 \tilde{\omega}_{f+...} \to 0 \tag{29}$$

The vanishingly small dissipation rate in low frequency dynamics is a unique characteristic of EIRCD. Such nearly dissipationless dynamics can be directly studied in experiments of strongly interacting 1D quantum gases. This limit recently has been studied more quantitatively in Ref. [40] along with a two-quench protocol proposed for experimental detections of $\Gamma(\tilde{\omega}_f)$.

Let us emphasize that so far we have restricted ourselves to strong coupling fixed points around g^* in both 1D and 3D with the intention to understand strongly interacting many-body systems. However, *EIRCD* as a general phenomenon can occur in any spatial dimensions although our discussions have been focused on 1D and 3D for the purpose of a comparison. For instance, EIRCD can also occur in 3D near a non-interacting fixed point, in 2D repulsive quantum gases. Furthermore, EIRCD also occurs in the 3D strong coupling limit of Feshbach resonance when one deviates from a scale symmetric fixed point by hypothetically varying the atom-dimer coupling. However, practically it is less feasible to experimentally probe EIRCD in this limit.

5 More Surprises to Come

So far I have talked about a few elementary results of quantum conformal dynamics that have emerged during the last few years in our research. Let me now mention a few very recent results we have obtained [46] and are still consolidating. These results are related to imaginary time evolution of primary operators and conformal Bootstrap to N-body autocorrelation functions in quantum dynamics. The conformal bootstrap had been applied to study the non-relativistic conformal field theory during the last decay [47, 48].

N-body autocorrelation function which is an overlap function between the initial and final N-body states in quantum dynamics is usually very challenging to obtain if not impossible. The conformal symmetry only the other makes it possible for us to suggest a general universal structure of the function. Bootstrap seems to be a very power tool to explore such a complex correlation in many-body dynamics.

Another very exciting progress is the relation between an initial state prepared in a physical trapping harmonic potential with frequency ω_i and imaginary time evolution of an N-body primary operator $O^\dagger$ in the presence of an external trapping potential with frequency ω_f. Let $|\psi_0\rangle$ be an initial state prepared as the ground state of the Hamiltonian H_{ω_i} in experiments.

We have shown that the an initial state as the ground state of the H_{ω_i} can always be related to imaginary time-evolution of $O^\dagger$, a primary conformal N-particle operator, (i.e. an operator that creates N-particle states with the lowest scaling dimension) in an external trap with frequency ω_f, or under the Hamiltonian H_{ω_f}:

$$|\psi_0\rangle = e^{-H_{\omega_f}\tau} O^\dagger |vac\rangle, \tanh(\omega_f \tau) = \frac{\omega_f}{\omega_i} \tag{30}$$

where τ is the amount imaginary time needed so that a desired initial state $|\psi_0\rangle$ is evolved into starting with a primary operator $O^\dagger$, and $|vac\rangle$ is the many-body vacuum. The scaling dimensions of a primary operator $O^\dagger$ were previously discussed [23]. Note τ is a function of both ω_f and ω_i given by the solution to the second equation above. [See the definition of $H_\omega(t)$ in Eq. 17; H_{ω_f} is also equal to the Hamiltonian defined there with $\omega(t)$ at $t > 0$ being taken to be equal to the constant frequency ω_f.]

Equation 30 has two implications. The first is that for a given initial state $|\psi_0\rangle$ or H_{ω_i}, one can always find a τ so that it can be a state evolved out of a primary conformal operator $O^\dagger$ in the presence of different soft trapping potentials or different H_{ω_f}. Second implication is that for any given H_{ω_f}, the evolution of $O^\dagger$ over different time duration τ leads to different initial states, physically prepared in different trapping potentials with frequency ω_i.

The conformal dynamics of an initial N-body state $|\psi_0\rangle$ hence can be directly related to $U(t)$, a unitary evolution of a primary operator $O^\dagger$ with a complex time $\tilde{t}$, i.e.

$$U(t) = \exp(-iH_{\omega_f}\tilde{t}), \tilde{t} = t - i\tau \tag{31}$$

where t is the amount of the time over which $|\psi_0\rangle$ has evolved. The complex time evolution can potentially lead to many other very exciting emergent geometries that will be further studied in the future.

6 Conclusion

In this article, we discuss the basic principle of conformal symmetry in the non-relativistic field theory and illustrate a few surprises that have appeared in recent studied of conformal dynamics in many-body physics. The strong coupling fixed point that lead to conformal dynamics can further result in the Planckian dynamics where microscopic dynamics and correlations can be shown to be universal and are uniquely set by $\frac{\hbar}{T}$ [49]. The microscopic dynamics discussed there reminisce dynamics near an AdS-CFT black hole [7–12, 50–52]. It is fair to say more efforts are needed to further understand these peculiar connections.

I am glad to have an opportunity to present these interesting ideas in the occasion of celebrating Gordon W. Semenoff's distinguished career in theoretical physics. Gordon and I had regular morning coffee breaks in the main office in our department. Among many exciting theoretical topics we have casually explored during those conversations ranging from condensed matter and statistical physics, quantum field and string theories to gravity, the Wilsonian approach in QFT and renormalization effects have been one of topics we have frequently touched upon. So I thought conformal dynamics which can be directly studied in atomic labs would be an excellent topic to share in this occasion. Happy birthday, Gordon.

I want to thank Prof. Manu Paranjape and other organizers for inviting me to the workshop on *Gravity, strings and fields* at CRM, Montreal in July, 2023. It had been my enormous pleasure to be there and share and discuss exciting ideas with my colleagues from different research fields. Some of the discussions during the conference have been leading to new efforts to apply the idea of emergent conformal dynamics beyond the current context of many-body physics.

Finally, I want to thank two former PhD students at UBC, Dr. Jeff Maki for a decade long collaboration on this extremely fascinating topic and for tirelessly pursuing, with me, these surprises and sometimes other more crazy ideas, and Fan Yang for being willing to tackle very challenging issues in theoretical physics and exploring unchartered territories. This research has been supported by a Discovery Grant from NSERC, Canada under RGPIN-2020-07070.

References

1. J.M. Maldacena, Adv. Theor. Math. Phys. **2**, 231 (1998)
2. E. Witten, Adv. Theor. Math. Phys. **2**, 231 (1998)
3. S. Gubser, I.R. Klebanov, A.M. Polyakov, Phys. Lett. B **428**, 105 (1998)
4. O. Aharony, S.S. Gubser, J.M. Maldacena, H. Ooguri, Y. Oz, Phys. Rept. **323**, 183 (2000)
5. S. Sachdev, *Quantum Phase Transitions* (Cambridge University Press, Cambridge, UK, 2011)
6. P.K. Kovtun, D.T. Son, A.O. Starinets, Phys. Rev. Lett. **94**, 111601 (2005)
7. A. Kitaev, A simple model of quantum holography, in *KITP Strings and Entanglement* (2015)
8. J. Maldacena, S.H. Shenker, D. Stanford, JHEP **2016**, 106 (2016); J. Maldacena, D. Stanford, Phys. Rev. D **94**, 106002 (2016)
9. J. Polchinski, V. Rosenhaus, JHEP **2016**, 1 (2016). ([arXiv: 1601.06768])
10. P. Coleman, A.J. Schofield, Nature **433**, 226 (2005)
11. S. Sachdev, Phys. Rev. Lett. **105**, 151602 (2010)
12. H. Liu, J. McGreevy, D. Vegh, Phys. Rev. D **83**, 065029 (2011)
13. W. Ji, X.-G. Wen, Phys. Rev. Res. **2**, 033417 (2020)
14. R. Verresen, R. Thorngren, N. Jones, F. Pollmann, Phys. Rev. X **11**, 041059 (2021)
15. R. Thorngren, A. Vishwanath, R. Verresen, Phys. Rev. B **104**, 075132 (2021)
16. A. Chatterjee, W. Ji, X.-G. Wen, ArXiv: **2212.14432**
17. C.R. Hagen, Phys. Rev. D **5**, 377 (1972)
18. U. Niederer, helv. Phys. Acta **45**, 802 (1972)
19. J. Maki, F. Zhou, Phys. Rev. A **100**, 023601 (2019)
20. L.P. Pitaevskii, A. Rosch, Phys. Rev. A **55**, R853(R) (1997)
21. Y. Castin, Comp. Rend. Phys. **5**, 407 (2004)
22. F. Werner, Y. Castin, Phys. Rev. A **74**, 053604 (2006)
23. Y. Nishida, D.T. Son, Phys. Rev. D **76**, 086004 (2007)
24. Y. Castin, F. Werner, Lect. Notes Phys. **836**, 127 (2012) in *BEC-BCS Crossover and the Unitary Fermi gas* (Springer, 2012, edited by W. Zwerger)
25. D.T. Son, Phys. Rev. Lett. **98**, 020604 (2007)
26. V. Gritsev, P. Barmettler, E. Demler, New J. Phys. **12**, 113005 (2010)
27. S. Moroz, Phys. Rev. A **86**, 011601(R) (2012)
28. E. Elliott, J.A. Joseph, J.E. Thomas, Phys. Rev. Lett. **112**, 040405 (2014)
29. W. Rohringer, D. Fischer, F. Steiner, I.E. Mazets, J. Schmiedmayer, M. Trupke, Sci. Rep. **5**, 9820 (2015)
30. S. Deng, Z.-Y. Shi, P. Diao, Q. Yu, H. Zhai, R. Qi, H. Wu, Science **353**, 371 (2016)
31. C.L. Qu, L.P. Pitaevskii, S. Stringari, Phys. Rev. A **94**, 063635 (2016)
32. S.E. Gharashi, D. Blume, Phys. Rev. A **94**, 063639 (2016)
33. J. Maki, L.M. Zhao, F. Zhou, Phys. Rev. A **98**, 013602 (2018)
34. S. Deng, P. Diao, Q. Yu, A. del Campo, H. Wu, Phys. Rev. A **97**, 013628 (2018)
35. P. Diao, S. Deng, F. Li, S. Yu, A. Chenu, A. del Campo, H. Wu, New J. Phys. **20**, 105004 (2018)
36. R. Saint-Jalm, P. Castilho, E. Le Cerf, B. Bakkali-Hassani, J.L. Ville, S. Nascimbene, J. Beugnon, J. Dalibard, Phys. Rev. X **9**, 021035 (2019)
37. J. Maki, F. Zhou, Phys. Rev. A **102**, 063319 (2020)
38. C. Lv, R. Zhang, Q. Zhou, Phys. Rev. Lett. **125**, 253002 (2020)
39. J. Maki, S. Zhang, F. Zhou, Phys. Rev. Lett. **128**, 040401 (2021)
40. J. Maki, F. Zhou, ArXiv: **2305.19061**
41. V. Bekassy, J. Hofmann, Phys. Rev. Lett. **128**, 193401 (2022)
42. T. Enss, N.C. Braatz, G. Gori, Phys. Rev. A **106**, 013308 (2022)
43. S. Sachdev, *Quantum Phase Transitions* (Cambridge University Press, Cambridge, UK, 2011)
44. I. Bloch, J. Dalibad, W. Zwerger, Rev. Mod. Phys. **80**, 885–964 (2008)
45. C. Chin, R. Grimm, P. Julienne, E. Tiesinga, Rev. Mod. Phys. **82**, 1225 (2010)
46. J. Maki, F. Zhou, to appear (2024)
47. S. Golkar, D.T. Son, JHEP **12**, 063 (2014)
48. W.D. Goldberger, Z.U. Khandker, S. Prabhu, JHEP **12**, 048 (2015)

49. S.J. Jiang, F. Zhou, ArXiv: **2301.12657**
50. S. Sachdev, J.-W. Ye, Phys. Rev. Lett. **70**, 3339 (1993)
51. L. Zhu, S. Kirchner, Q. Si, A. Georges, Phys. Rev. Lett. **93**, 267201 (2004); L. Zhu and Q. Si, Phys. Rev. B **66**, 024426 (2002)
52. J. Zaanen, Nature **430**, 512 (2004); Scipost Phys. **6**, 061 (2019)

MIX
Papier aus verantwortungsvollen Quellen
Paper from responsible sources
FSC® C105338

If you have any concerns about our products,
you can contact us on
ProductSafety@springernature.com

In case Publisher is established outside the EU,
the EU authorized representative is:
Springer Nature Customer Service Center GmbH
Europaplatz 3, 69115 Heidelberg, Germany

Printed by Libri Plureos GmbH
in Hamburg, Germany